# MEANS
# ELECTRICAL
# ESTIMATING

## Standards
## and
## Procedures

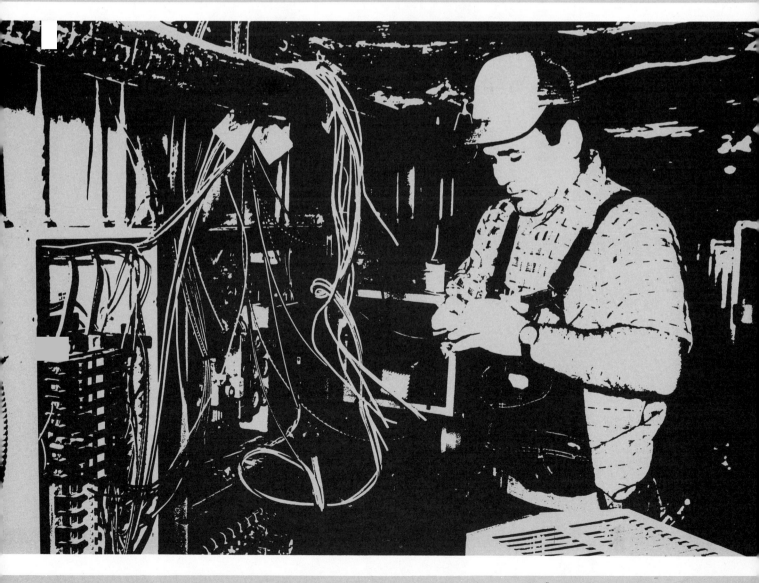

# MEANS
# ELECTRICAL
# ESTIMATING

## Standards
## and
## Procedures

**Publisher**
E. Norman Peterson, Jr.

**Editor-In-Chief**
William D. Mahoney

**Senior Editor**
Paul H. DeLong

**Contributing Editor**
Mary P. Greene

**Technical Services**
Joyce A. Baron
Sharon L. Monty

**Illustrator**
Carl W. Linde

Copyright 1993

**R.S. MEANS COMPANY, INC.**
CONSTRUCTION PUBLISHERS & CONSULTANTS

100 Construction Plaza
P.O. Box 800
Kingston, MA 02364-0800
(617) 585-7880

Southam
Construction
Information
Network

Library of Congress Catalog Number 86-216696

ISBN 0-911950-83-4

# TABLE OF CONTENTS

# FOREWORD

For over 45 years, R.S. Means Co., Inc. has researched and published cost data for the building construction industry. *Means Electrical Estimating: Standards and Procedures* applies that valuable experience to aid the construction professional in understanding the estimating process. This book is one of a series of estimating reference books published by R.S. Means and designed to benefit all who are involved with the construction industry: the contractor, engineer, owner, developer, architect, designer, or facilities manager.

The goal of this book is to provide sound, practical methods and standards for accurate electrical estimating. The entire process is demonstrated, step by step, through to the finished estimate and includes activities such as drawing takeoff, pricing, labor costing, figuring of overhead and profit, bidding, and scheduling.

The book is presented in three parts. The first part of the book, "The Estimating Process", provides guidelines for estimating. This section begins with the basics, such as the types and elements of estimates. The discussion proceeds sequentially from preliminary activities through the takeoff, pricing, and estimate summary. Suggested scheduling, bidding, and project management techniques wrap up the estimating process.

The second part, "Components of Electrical Systems", is devoted to a detailed description of the individual components of electrical systems. Sixty-nine electrical installation categories are covered here. A discussion of each system or component addresses the characteristics and the appropriate conditions for that component's use. A delineated takeoff procedure is provided, together with guidelines for listing and pricing materials and labor. Basic working conditions and factors are given for cost adjustments — both for increases — to allow for work over a certain height, for example — and decreases — based on economies of scale.

The third part of the book, the "Sample Estimate", uses a set of electrical building plans as the basis for a working model. Takeoff forms and procedures shown and explained in the earlier parts of the book are utilized here to complete a typical electrical estimate. The prices used in the sample estimate are from the 1986 edition of *Means Electrical Cost Data*. This annual Means publication — its format, use as an estimating tool, and the origin of its costs — is explained in Part Three.

An Appendix contains useful reference information. Included in this section are tables, charts, and graphs which may aid the estimator in projecting costs. Also included is an extensive list of commonly used graphic symbols for electrical drawings and a listing of common abbreviations.

# Part I
# THE ESTIMATING PROCESS

# Chapter 1
# COMPONENTS OF AN ESTIMATE

The term "estimating accuracy" is a relative concept. What is the correct or accurate cost of a given construction project? Is it the total price paid to the contractor? Might not another reputable contractor perform the same work for a different cost, whether higher or lower? There is no one correct estimated cost for a given project. There are too many variables in construction. At best, the estimator can determine a very close approximation of what the final costs will be. The resulting accuracy of this approximation is directly affected by the amount of detail provided and the amount of time spent on the estimate.

Every cost estimate requires three basic components. The first is the establishment of standard *units of measure*. The second component of an estimate is the determination of the *quantity* of units for each component, an actual counting process: how many linear feet of wire, how many outlet boxes, etc. The third component, and perhaps the most difficult to obtain, is the determination of a reasonable *cost* for each unit.

The first element, the designation of measurement units, is the step which determines and defines the level of detail, and thus the degree of accuracy of a cost estimate. In electrical construction, such units could be as all-encompassing as the number of watts per square foot of floor area, or as detailed as a linear foot of wire. Depending upon the estimator's intended use, the designation of the "unit" may describe a complete system, or it may imply only an isolated entity. The choice and detail of units also determines the time required to do an estimate.

The second component of every estimate, the determination of quantity, is more than the counting of units. In construction, this process is called the "quantity takeoff". In order to perform this function successfully, the estimator should have a working knowledge of the materials, methods, and codes used in electrical construction. An understanding of the design specifications is particularly important. This knowledge helps to assure that each quantity is correctly tabulated and that essential items are not forgotten or omitted. The estimator with a thorough knowledge of construction is also more likely to account for all requirements in the estimate. Many of the items to be quantified (counted) may not involve any material but, rather, entail labor costs only. Testing is an example of a labor-only item. Experience is, therefore, invaluable to ensure a complete estimate.

The third component is the determination of a reasonable cost for each unit. This aspect of the estimate is significantly responsible for variations in estimating. Rarely do two estimators arrive at exactly the same material cost for a project. Even if material costs for an installation are the same for competing contractors, the labor costs for installing that material can vary considerably, due to productivity and the pay scales in different areas. The use of specialized equipment can decrease installation time and, therefore, cost. Finally, material prices do fluctuate within the market. These cost differences occur from city to city and even from supplier to supplier in the same town. It is the experienced and well-prepared estimator who can keep track of these variations and fluctuations and use them to his best advantage when preparing accurate estimates.

This third component of the estimate, the determination of costs, can be defined in three different ways by the estimator. With one approach, the estimator uses a unit cost which includes all the elements (i.e., material, installation, overhead, and profit) into one number expressed in dollars per unit. A variation of this approach is to use a unit cost which includes total material and installation only, adding a percent mark-up for overhead and profit to the "bottom line".

A second method is to use unit costs — in dollars — for material and for installation. This is done separately for each item . . . without mark-ups. These are sometimes called "bare costs". Different profit and overhead mark-ups are applied to each before the material and installation prices are added; the result is the total "selling" cost.

A third method of unit pricing uses unit costs for materials, and man-hours as the units of labor. Again, these figures are totalled separately; one represents the value for materials (in dollars), and the other shows the total man-hours for installation. The average cost per hour of craft labor is determined by allowing for the expected ratios of foremen, journeymen, and apprentices. This is sometimes called the "composite labor rate". This rate is multiplied by the total man-hours to get the total bare cost of installation. Different overhead and profit mark-ups can then be applied to each, material and labor, and the results added to get the total selling cost.

The word "unit" is used in many ways, as can be seen in the above definitions. Keeping the concepts of units clearly defined is vital to achieving an accurate, professional estimate. For the purposes of this book, the following references to different types of units are used:

**Unit of Measure:** the standard by which the quantities are counted, such as *linear feet* of conduit, *number* of boxes.

**Cost Units:** the total dollar price per each installed unit of measure — including the costs of material and installation; this figure may or may not include overhead and profit.

**Material Unit Cost:** the cost to purchase each unit of measure: this cost represents material dollars only — with no overhead and profit.

**Installation Unit Cost:** the cost for installing each unit of measure: this cost includes labor dollars only — with no overhead and profit.

**Labor Unit:** the man-hours required to install a unit of measure. (Note: Labor units multiplied by the labor rate per hour equals the installation unit cost in dollars.)

Very often, two or three of these different numerical values may appear on the same estimate sheet under columns all titled simply "UNITS". The estimator must, therefore, exercise care to utilize these columns correctly and consistently for the format of each particular estimate.

# Chapter 2
# TYPES OF ESTIMATES

Estimators use four basic types of estimates. These types may be referred to by different names and may not be recognized by all as definitive. Most estimators, however, will agree that each type has its place in the construction estimating process. These types of estimates are as follows:

1. **Order of Magnitude Estimates:** The order of magnitude estimate could be loosely described as an educated guess. It can be completed in a matter of minutes. Accuracy may be plus or minus 20%.

2. **Square Foot and Cubic Foot Estimates:** This type is most often useful when only the proposed size and use of a planned building is known. This method can be completed within an hour or two. Accuracy may be plus or minus 15%.

3. **Systems (or Assemblies) Estimate:** A systems estimate is best used as a budgetary tool in the planning stages of a project when some parameters have been decided. This type of estimate could require as much as one day to complete. Accuracy is expected at plus or minus 10%.

4. **Unit Price Estimate:** Working drawings and full specifications are required to complete a unit price estimate. It is the most accurate of the four types but is also the most time consuming. Used primarily for bidding purposes, the accuracy of a unit price estimate can be plus or minus 5%.

Figure 2.1 graphically demonstrates the relationship of required time versus resulting accuracy of a complete building estimate for each of these four basic estimate types. It should be recognized that, as an estimator *and* his company gain repetitive experience on similar or identical projects, the accuracy of all four types of estimates should improve dramatically. In fact, given enough experience, Order of Magnitude and Square Foot estimates may closely approach the accuracy of Unit Price estimates.

# Order Of Magnitude Estimates

The Order of Magnitude Estimate can be completed with only a minimum of information. The "units", as described in the first chapter of this book, can be very general for this type and need not be well defined. For example: "The electrical work for the office building of a small service company in a suburban industrial park will cost about $20,000". This type of statement (or estimate) can be made after a few minutes of thought used to draw upon experience and to make comparisons with similar projects from the past. While this rough figure might be appropriate for a project in one region of the country, substantial adjustments may be required for a change of geographic location and for cost changes over time, due to, for example, changes of materials, inflation, or code changes.

Figure 2.2, from *Means Electrical Cost Data*, includes data for a refined approach to the Order of Magnitude estimate. This format is based on unit of use. Please note at the bottom of the category "Hospitals", for example, that costs are given "per bed" or "per person". The proposed use and magnitude of the planned structure — such as the number of beds for a hospital building or the number of apartments in a complex — may be the only parameters known at the time the Order of Magnitude estimate is done. The data given in Figure 2.2 does not require that details of the proposed project be known in order to determine rough costs; the only required information is the intended use and capacity of the building. What is lacking in accuracy (plus or minus 20%) is more than compensated by the minimal time required to complete the Order of Magnitude estimate — a matter of minutes.

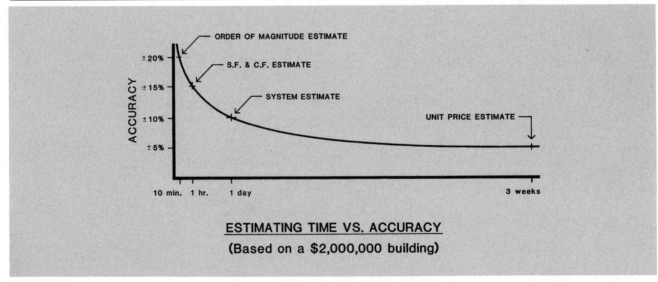

**ESTIMATING TIME VS. ACCURACY**
**(Based on a $2,000,000 building)**

Figure 2.1

8

| 17.1 S.F., C.F. and % of Total Costs | | UNIT | UNIT COSTS | | | % OF TOTAL | | |
|---|---|---|---|---|---|---|---|---|
| | | | 1/4 | MEDIAN | 3/4 | 1/4 | MEDIAN | 3/4 |
| 290 | Electrical | S.F. | 3.14 | 7.09 | 7.91 | 8% | 11.30% | 15.80% |
| 310 | Total: Mechanical & Electrical | ↓ | 8.45 | 12.95 | 19.40 | 16.50% | 25.90% | 33% |
| 36-001 | **FIRE STATIONS** | ↓ | 47.25 | 63.55 | 77.60 | | | |
| 002 | Total project costs | C.F. | 3.02 | 4.06 | 5.10 | | | |
| 272 | Plumbing | S.F. | 3.06 | 4.78 | 6.50 | 5.90% | 7.30% | 9.50% |
| 277 | Heating, ventilating, air conditioning | | 2.62 | 4.22 | 6.70 | 4.80% | 7.30% | 9.20% |
| 290 | Electrical | | 3.48 | 5.95 | 8.55 | 7.10% | 9.70% | 12.10% |
| 310 | Total: Mechanical & Electrical | | 8.95 | 14.10 | 19.40 | 17.50% | 22.60% | 27.60% |
| 37-001 | **FRATERNITY HOUSES** And Sorority Houses | ↓ | 47.20 | 56.35 | 62.85 | | | |
| 002 | Total project costs | C.F. | 4.51 | 5.50 | 6.10 | | | |
| 272 | Plumbing | S.F. | 3.56 | 4.16 | 5.55 | 5.90% | 8% | 10.80% |
| 290 | Electrical | | 3.11 | 4.10 | 7.45 | 6.50% | 8.80% | 10.40% |
| 310 | Total: Mechanical & Electrical | ↓ | 8.80 | 12.50 | 15.10 | 14.60% | 20.70% | 24.20% |
| 38-001 | **FUNERAL HOMES** | S.F. | 44.90 | 56.45 | 82.75 | | | |
| 002 | Total project costs | C.F. | 3.17 | 4.56 | 5.40 | | | |
| 272 | Plumbing | S.F. | 1.78 | 2.47 | 2.70 | 4.10% | 4.40% | 4.70% |
| 277 | Heating, ventilating, air conditioning | | 3.96 | 4 | 4.81 | 7% | 9.20% | 10.40% |
| 290 | Electrical | | 2.97 | 3.69 | 5.75 | 6.20% | 7.70% | 11% |
| 310 | Total: Mechanical & Electrical | | 8.10 | 10.70 | 12.40 | 18.80% | 20.80% | 27.20% |
| 39-001 | **GARAGES, COMMERCIAL** | ↓ | 26.90 | 42.60 | 56.55 | | | |
| 002 | Total project costs | C.F. | 1.71 | 2.50 | 3.57 | | | |
| 272 | Plumbing | S.F. | 1.74 | 2.74 | 5.50 | 4.90% | 7.30% | 11% |
| 273 | Heating & ventilating | | 2.81 | 3.91 | 4.40 | 7% | 11.20% | 11.30% |
| 290 | Electrical | | 2.41 | 3.97 | 5.55 | 7.10% | 9% | 11.40% |
| 310 | Total: Mechanical & Electrical | | 5.60 | 9.95 | 14.35 | 15.70% | 21.90% | 27.80% |
| 40-001 | **GARAGES, MUNICIPAL** | ↓ | 28.95 | 45.25 | 63.20 | | | |
| 002 | Total project costs | C.F. | 2.07 | 2.80 | 3.76 | | | |
| 272 | Plumbing | S.F. | 1.80 | 3.40 | 5.55 | 4.10% | 6.90% | 8.60% |
| 273 | Heating & ventilating | | 1.97 | 2.81 | 5.80 | 6% | 7.90% | 11.30% |
| 290 | Electrical | | 2.53 | 3.95 | 5.40 | 6.30% | 8% | 10.10% |
| 310 | Total: Mechanical & Electrical | | 5.80 | 11.55 | 17.05 | 15.50% | 24.10% | 31.50% |
| 41-001 | **GARAGES, PARKING** | ↓ | 15.55 | 19.45 | 32.85 | | | |
| 002 | Total project costs | C.F. | 1.34 | 1.77 | 2.92 | | | |
| 272 | Plumbing | S.F. | .28 | .55 | .79 | 2.10% | 2.80% | 3.80% |
| 290 | Electrical | | .65 | .95 | 1.54 | 4.20% | 5.20% | 6.30% |
| 310 | Total: Mechanical & Electrical | ↓ | .97 | 1.53 | 1.95 | 6.80% | 8.30% | 9.40% |
| 900 | Per car, total cost | Car | 4,975 | 6,800 | 9,550 | | | |
| 950 | Total: Mechanical & Electrical | " | 355 | 535 | 660 | | | |
| 43-001 | **GYMNASIUMS** | S.F. | 40.90 | 55 | 70.45 | | | |
| 002 | Total project costs | C.F. | 2.07 | 2.66 | 3.62 | | | |
| 272 | Plumbing | S.F. | 2.52 | 3.43 | 4.34 | 4.80% | 7.20% | 8.50% |
| 277 | Heating, ventilating, air conditioning | | 2.68 | 4.64 | 7.60 | 7.40% | 9.70% | 14% |
| 290 | Electrical | | 3.44 | 4.26 | 6.25 | 6.20% | 9% | 10.70% |
| 310 | Total: Mechanical & Electrical | | 6.50 | 11.55 | 15 | 16.60% | 21.80% | 27% |
| 46-001 | **HOSPITALS** | ↓ | 90.30 | 111 | 147 | | | |
| 002 | Total project costs | C.F. | 6.65 | 7.95 | 11.10 | | | |
| 272 | Plumbing | S.F. | 7.95 | 10.15 | 13.75 | 7.50% | 9.10% | 10.70% |
| 277 | Heating, ventilating, air conditioning | | 8.65 | 14.40 | 19.80 | 8.40% | 13% | 16.60% |
| 290 | Electrical | | 9.30 | 12.65 | 19.20 | 10.30% | 12.30% | 15.20% |
| 310 | Total: Mechanical & Electrical | ↓ | 27.30 | 37.20 | 55.05 | 26.90% | 37.70% | 40.30% |
| 900 | Per bed or person, total cost | Bed | 27,100 | 41,800 | 62,700 | | | |
| 48-001 | **HOUSING** For the Elderly | S.F. | 43.70 | 54.90 | 69.05 | | | |
| 002 | Total project costs | C.F. | 3.08 | 4.28 | 5.60 | | | |
| 272 | Plumbing | S.F. | 3.32 | 4.60 | 6.85 | 8.30% | 9.70% | 10.90% |
| 273 | Heating, ventilating, air conditioning | | 1.49 | 2.30 | 3.24 | 3.20% | 5.60% | 7.10% |
| 290 | Electrical | | 3.20 | 4.51 | 6.35 | 7.50% | 9% | 10.60% |
| 291 | Electrical incl. electric heat | ↓ | 3.69 | 6.95 | 8.25 | 9.60% | 11% | 13.30% |

Figure 2.2

## Square Foot and Cubic Foot Estimates

The use of Square Foot and Cubic Foot Estimates is most appropriate prior to the preparation of plans or preliminary drawings, when budgetary parameters are being analyzed and established. Please refer again to Figure 2.2 and note that costs for each type of project are presented first as "Total project costs" by square foot and by cubic foot. These costs are broken down into different components, and then into the relationship of each component to the project as a whole, in terms of costs per square foot. This breakdown enables the designer, planner or estimator to adjust certain components according to the unique requirements of the proposed project. The costs on this and other pages of *Means Electrical Cost Data* were derived from more than 10,500 projects contained in the Means Data Bank of Construction Costs. These costs include the contractor's overhead and profit but do not include architectural fees or land costs. The 1/4 column shows the value at which 25% of the projects had lower costs, 75% had higher. The 3/4 column value denotes that 75% of the projects had lower costs, 25% had higher. The median column value shows that 50% of the projects had lower costs, 50% had higher.

Historical data for square foot costs of new construction are plentiful (see *Means Electrical Cost Data*, Division 17.1). However, the best source of square foot costs is the estimator's own cost records for similar projects, adjusted to the parameters of the project in question. While helpful for preparing preliminary budgets, Square Foot and Cubic Foot estimates can also be useful as checks against other, more detailed estimates. While slightly more time is required than with Order of Magnitude estimates, a greater accuracy (plus or minus 15%) is achieved due to a more specific definition of the project.

## Systems (or Assemblies) Estimates

One of the primary advantages of systems (or assemblies) estimating is to enable alternate construction techniques to be readily compared for budgetary purposes. Rapidly rising design and construction costs in recent years have made budgeting and cost effectiveness studies increasingly important in the early stages of building projects. Never before has the estimating process had such a crucial role in the initial planning. Unit Price Estimating, because of the time and detailed information required, is not well suited as a budgetary or planning tool. A faster and more cost-effective method is needed for the planning phase of a building project; this is the "Systems", or "Assemblies" Estimate.

The Systems method is a logical, sequential approach that reflects how a building is constructed. Twelve "Uniformat" divisions organize building construction into major components that can be used in Systems estimates. These Uniformat divisions are listed below:

**Systems Estimating Divisions:**

Division 1  — Foundations
Division 2  — Substructure
Division 3  — Superstructure
Division 4  — Exterior Closure
Division 5  — Roofing
Division 6  — Interior Construction
Division 7  — Conveying
Division 8  — Mechanical
Division 9  — Electrical

Division 10 – General Conditions
Division 11 – Special
Division 12 – Site Work

Each division is further broken down into systems. Division 9, which covers electrical construction, is comprised of the following groups of systems: Service, Lighting and Power, and Special.

Each system incorporates several different components into an assemblage that is commonly used in construction. Figure 2.3 is an example of a typical system, in this case "Lighting and Power – Fluorescent Fixture" (from *Means Electrical Cost Data*).

A great advantage of the Systems Estimate is that the estimator/designer is able to substitute one system for another during design development and can quickly determine the relative cost differential. The owner can then anticipate budget requirements before the final details and dimensions are established.

The Systems method does not require the degree of final design details needed for a unit price estimate, but estimators who use this approach must have a solid background knowledge of construction materials and methods, code requirements, design options, and budget considerations.

The Systems Estimate should not be used as a substitute for the Unit Price Estimate. While the Systems approach can be an invaluable tool in the planning stages of a project, it should be supported by Unit Price Estimating when greater accuracy is required.

## Unit Price Estimates

The Unit Price Estimate is the most accurate and detailed of the four estimate types and therefore takes the most time to complete. Detailed working drawings and specifications must be available to the unit price estimator. All decisions regarding the project's materials and methods must have been made in order to complete this type of estimate. Because there are fewer variables, the estimate can be more accurate. Working drawings and specifications are used to determine the quantities of materials, equipment, and labor. Current and accurate unit costs for these items are also necessary. These costs can come from different sources. Wherever possible, estimators should use prices based on experience or developed from actual, similar projects. If these kinds of records are not available, prices may be determined from an up-to-date industry source book, such as *Means Electrical Cost Data*.

Because of the detail involved and the need for accuracy, completion of a Unit Price Estimate entails a great deal of time and expense. For this reason, Unit Price Estimating is best suited for construction bidding. It can also be an effective method for determining certain detailed costs in a conceptual budget or during design development.

Most construction specification manuals and cost reference books, such as *Means Electrical Cost Data*, divide unit price information into the 16 Uniform Construction Index (UCI) divisions as adopted by the Construction Specifications Institute, Inc.

**A. Strip Fixture**    **B. Surface Mounted**

**C. Recessed**    **D. Pendent Mounted**

**Design Assumptions:**

1. A 100 footcandle average maintained level of illumination.
2. Ceiling heights range from 9' to 11'.
3. Average reflectance values are assumed for ceilings, walls and floors.
4. Cool white (CW) fluorescent lamps with 3150 lumens for 40 watt lamps and 6300 lumens for 8' slimline lamps.
5. Four 40 watt lamps per 4' fixture and two 8' lamps per 8' fixture.
6. Average fixture efficiency values and spacing to mounting height ratios.
7. Installation labor is average U.S. rate as of January 1, 1986.

| System Components | QUANTITY | UNIT | COST PER S.F. | | |
|---|---|---|---|---|---|
| | | | MAT. | INST. | TOTAL |
| SYSTEM 09.2-212-0520 | | | | | |
| FLUORESCENT FIXTURE MOUNTED 9'-11' ABOVE FLOOR, 100 FC | | | | | |
| TYPE A, 8 FIXTURES PER 400 S.F. | | | | | |
| Steel intermediate conduit, (IMC) 1/2" diam | .404 | L.F. | .19 | 1.04 | 1.23 |
| Wire, 600V, type THWN-THHN, copper, solid, #12 | .008 | C.L.F. | .03 | .19 | .22 |
| Fluorescent strip fixture 8' long, surface mounted, two 75W SL | .020 | Ea. | .81 | .83 | 1.64 |
| Steel outlet box 4" concrete | .020 | Ea. | .05 | .26 | .31 |
| Steel outlet box plate with stud, 4" concrete | .020 | Ea. | .02 | .06 | .08 |
| TOTAL | | | 1.10 | 2.38 | 3.48 |

| 9.2-212 | Fluorescent Fixtures (by Type) | COST PER S.F. | | |
|---|---|---|---|---|
| | | MAT. | INST. | TOTAL |
| 0520 | Fluorescent fixtures, 9'-11' above floor, 100 FC, type A, 8 fixture per 400 S.F. | 1.10 | 2.38 | 3.48 |
| 0560 | 11 fixtures per 600 S.F. | 1.03 | 2.30 | 3.33 |
| 0600 | 17 fixtures per 1000 S.F. | .98 | 2.26 | 3.24 |
| 0640 | 23 fixtures per 1600 S.F. | .86 | 2.16 | 3.02 |
| 0680 | 28 fixtures per 2000 S.F. | .86 | 2.16 | 3.02 |
| 0720 | 41 fixtures per 3000 S.F. | .82 | 2.15 | 2.97 |
| 0800 | 53 fixtures per 4000 S.F. | .81 | 2.10 | 2.91 |
| 0840 | 64 fixtures per 5000 S.F. | .81 | 2.10 | 2.91 |
| 0880 | Type B, 11 fixtures per 400 S.F. | 2.37 | 3.52 | 5.89 |
| 0920 | 15 fixtures per 600 S.F. | 2.16 | 3.37 | 5.53 |
| 0960 | 24 fixtures per 1000 S.F. | 2.09 | 3.36 | 5.45 |
| 1000 | 35 fixtures per 1600 S.F. | 1.94 | 3.19 | 5.13 |
| 1040 | 42 fixtures per 2000 S.F. | 1.89 | 3.21 | 5.10 |
| 1080 | 61 fixtures per 3000 S.F. | 1.92 | 3.09 | 5.01 |
| 1160 | 80 fixtures per 4000 S.F. | 1.82 | 3.16 | 4.98 |
| 1200 | 98 fixtures per 5000 S.F. | 1.81 | 3.14 | 4.95 |
| 1240 | Type C, 11 fixtures per 400 S.F. | 1.92 | 3.71 | 5.63 |
| 1280 | 14 fixtures per 600 S.F. | 1.64 | 3.46 | 5.10 |
| 1320 | 23 fixtures per 1000 S.F. | 1.64 | 3.44 | 5.08 |
| 1360 | 34 fixtures per 1600 S.F. | 1.55 | 3.40 | 4.95 |
| 1400 | 43 fixtures per 2000 S.F. | 1.59 | 3.36 | 4.95 |
| 1440 | 63 fixtures per 3000 S.F. | 1.53 | 3.32 | 4.85 |
| 1520 | 81 fixtures per 4000 S.F. | 1.48 | 3.27 | 4.75 |
| 1560 | 101 fixtures per 5000 S.F. | 1.48 | 3.27 | 4.75 |
| 1600 | Type D, 8 fixtures per 400 S.F. | 1.96 | 2.82 | 4.78 |
| 1640 | 12 fixtures per 600 S.F. | 1.96 | 2.82 | 4.78 |
| 1680 | 19 fixtures per 1000 S.F. | 1.88 | 2.74 | 4.62 |
| 1720 | 27 fixtures per 1600 S.F. | 1.72 | 2.67 | 4.39 |
| 1760 | 34 fixtures per 2000 S.F. | 1.71 | 2.65 | 4.36 |
| 1800 | 48 fixtures per 3000 S.F. | 1.63 | 2.58 | 4.21 |
| 1880 | 64 fixtures per 4000 S.F. | 1.63 | 2.58 | 4.21 |
| 1920 | 79 fixtures per 5000 S.F. | 1.63 | 2.58 | 4.21 |

*Figure 2.3*

**Uniform Construction Index Divisions:**

Division 1  — General Requirements
Division 2  — Site Work
Division 3  — Concrete
Division 4  — Masonry
Division 5  — Metals
Division 6  — Wood & Plastics
Division 7  — Moisture-Thermal Control
Division 8  — Doors, Windows & Glass
Division 9  — Finishes
Division 10 — Specialties
Division 11 — Equipment
Division 12 — Furnishings
Division 13 — Special Construction
Division 14 — Conveying Systems
Division 15 — Mechanical
**Division 16 — Electrical**

Division 16, Electrical, is further divided into the following subdivisions:

16.0 Raceways
16.1 Conductors and Grounding
16.2 Boxes and Wiring Devices
16.3 Starters, Boards, and Switches
16.4 Transformers and Bus Duct
16.5 Power Systems and Capacitors
16.6 Lighting (fixtures)
16.7 Lighting and Utilities (duct banks and poles)
16.8 Special Systems

This method of organizing the various components provides a standard of uniformity that is widely used by construction industry professionals: contractors, material suppliers, engineers, and architects. A sample unit price page from the 1986 edition of *Means Electrical Cost Data* is shown in Figure 2.4. This page lists various types of cable connectors. (Please note that the heading "16.1 Conductors and Grounding" denotes the UCI subdivision classification for these items in Division 16.) Each page contains a wealth of information useful in Unit Price Estimating. The type of work to be performed is described in detail: typical crew make-ups, unit man-hours, units of measure, and separate costs for material and installation. Total costs are extended to include the installing contractor's overhead and profit.

As a refinement to the unit price estimate, a fifth type of estimate warrants mention. This is the "Scheduling Estimate", which involves the application of realistic manpower allocations. A complete unit price estimate is a prerequisite for the preparation of a scheduling estimate. The purpose of the scheduling estimate is to determine costs based upon actual working conditions. This is done using data obtained from the unit price estimate. A "human factor" can be applied. For example, if a task requires 7.5 hours to complete, based on unit price data, a worker will most likely take 8 hours to complete the work. Costs can be adjusted accordingly. A thorough discussion of scheduling and scheduling estimating is beyond the scope of this book, but a brief overview is included in Chapter Nine; this subject is covered in detail in *Means Scheduling Manual*, 2nd edition, by F. William Horsley.

| 16.1 Conductors & Grounding | | CREW | MAN-HOURS | UNIT | BARE COSTS | | | TOTAL INCL O&P |
|---|---|---|---|---|---|---|---|---|
| | | | | | MAT. | INST. | TOTAL | |
| 050 | 12 wires | 1 Elec | 1.860 | C.L.F. | 59 | 42 | 101 | 125 |
| 060 | 14 wires | | 2.110 | | 64 | 47 | 111 | 140 |
| 070 | 16 wires | | 2.290 | | 74 | 51 | 125 | 155 |
| 080 | 18 wires | | 2.420 | | 84 | 54 | 138 | 170 |
| 081 | 19 wires | | 2.580 | | 87 | 58 | 145 | 180 |
| 090 | 20 wires | | 2.670 | | 92 | 60 | 152 | 185 |
| 100 | 22 wires | ▼ | 2.860 | ▼ | 97 | 64 | 161 | 200 |
| **40-001** | **CABLE CONNECTORS** | | | | | | | |
| 010 | 600 volt, nonmetallic, #14-2 wire | 1 Elec | .050 | Ea. | .20 | 1.12 | 1.32 | 1.83 |
| 020 | #14-3 wire to #12-2 wire | | .060 | | .20 | 1.35 | 1.55 | 2.16 |
| 030 | #12-3 wire to #10-2 wire | | .070 | | .20 | 1.57 | 1.77 | 2.48 |
| 040 | #10-3 wire to #14-4 and #12-4 wire | | .080 | | .20 | 1.79 | 1.99 | 2.80 |
| 050 | #8-3 wire to #10-4 wire | | .100 | | .65 | 2.24 | 2.89 | 3.94 |
| 060 | #6-3 wire | | .200 | | .85 | 4.48 | 5.33 | 7.40 |
| 080 | SER, aluminum 3 #8 insulated + 1 #8 ground | | .250 | | .86 | 5.60 | 6.46 | 9 |
| 090 | 3 #6 + 1 #6 ground | | .333 | | 1.17 | 7.45 | 8.62 | 12 |
| 100 | 3 #4 + 1 #6 ground | | .364 | | 1.17 | 8.15 | 9.32 | 13 |
| 110 | 3 #2 + 1 #4 ground | | .400 | | 2.10 | 8.95 | 11.05 | 15.20 |
| 120 | 3 1/0 + 1 #2 ground | | .444 | | 4.75 | 9.95 | 14.70 | 19.55 |
| 140 | 3 2/0 + 1 #1 ground | | .500 | | 4.75 | 11.20 | 15.95 | 21 |
| 160 | 3 4/0 + 1 # 2/0 ground | | .571 | | 5.82 | 12.80 | 18.62 | 25 |
| 180 | 600 volt, armored , #14-2 wire | | .100 | | .20 | 2.24 | 2.44 | 3.44 |
| 200 | #14-3 and #12-2 wire | | .151 | | .20 | 3.38 | 3.58 | 5.10 |
| 220 | #14-4, #12-3 and #10-2 wire | | .200 | | .20 | 4.48 | 4.68 | 6.65 |
| 240 | #12-4, #10-3 and #8-2 wire | | .250 | | .20 | 5.60 | 5.80 | 8.25 |
| 260 | #8-3 and #10-4 wire | | .308 | | .85 | 6.90 | 7.75 | 10.85 |
| 265 | #8-4 wire | | .364 | | 1.17 | 8.15 | 9.32 | 13 |
| 270 | PVC jacket connector, #6-3 wire, #6-4 wire | | .500 | | 4.05 | 11.20 | 15.25 | 21 |
| 280 | #4-3 wire, #4-4 wire | | .500 | | 4.05 | 11.20 | 15.25 | 21 |
| 290 | #2-3 wire | | .667 | | 4.05 | 14.95 | 19 | 26 |
| 300 | #1-3 wire, #2-4 wire | | .667 | | 6.60 | 14.95 | 21.55 | 29 |
| 320 | 1/0-3 wire | | .727 | | 6.60 | 16.30 | 22.90 | 31 |
| 340 | 2/0-3 wire, 1/0-4 wire | | .800 | | 6.60 | 17.90 | 24.50 | 33 |
| 350 | 3/0-3 wire, 2/0-4 wire | | .889 | | 7.85 | 19.90 | 27.75 | 37 |
| 360 | 4/0-3 wire, 3/0-4 wire | | 1.140 | | 7.85 | 26 | 33.85 | 45 |
| 380 | 250 MCM-3 wire, 4/0-4 wire | | 1.330 | | 11.85 | 30 | 41.85 | 56 |
| 400 | 350 MCM-3 wire, 250 MCM-4 wire | | 1.600 | | 11.85 | 36 | 47.85 | 65 |
| 410 | 350 MCM-4 wire | | 2.000 | | 23 | 45 | 68 | 90 |
| 420 | 500 MCM-3 wire | | 2.000 | | 23 | 45 | 68 | 90 |
| 425 | 500 MCM-4 wire, 750 MCM-3 wire | | 2.290 | | 31 | 51 | 82 | 110 |
| 430 | 750 MCM-4 wire | | 2.670 | | 31 | 60 | 91 | 120 |
| 440 | 5 KV, armored, #4 | | 1.000 | | 24 | 22 | 46 | 59 |
| 460 | #2 | | 1.000 | | 24 | 22 | 46 | 59 |
| 480 | #1 | | 1.000 | | 24 | 22 | 46 | 59 |
| 500 | 1/0 | | 1.250 | | 24 | 28 | 52 | 67 |
| 520 | 2/0 | | 1.510 | | 27 | 34 | 61 | 78 |
| 550 | 4/0 | | 2.000 | | 27 | 45 | 72 | 94 |
| 560 | 250 MCM | | 2.220 | | 33 | 50 | 83 | 110 |
| 565 | 350 MCM | | 2.500 | | 40 | 56 | 96 | 125 |
| 570 | 500 MCM | | 3.200 | | 40 | 72 | 112 | 145 |
| 572 | 750 MCM | | 3.640 | | 51 | 81 | 132 | 175 |
| 575 | 1000 MCM | | 4.000 | | 70 | 90 | 160 | 205 |
| 580 | 15 KV, armored, #1 | | 2.000 | | 33 | 45 | 78 | 100 |
| 590 | 1/0 | | 2.000 | | 33 | 45 | 78 | 100 |
| 600 | 3/0 | | 2.220 | | 40 | 50 | 90 | 115 |
| 610 | 4/0 | | 2.350 | | 40 | 53 | 93 | 120 |
| 620 | 250 MCM | | 2.500 | | 40 | 56 | 96 | 125 |
| 630 | 350 MCM | | 2.960 | | 51 | 66 | 117 | 150 |
| 640 | 500 MCM | ▼ | 4.000 | ▼ | 51 | 90 | 141 | 185 |

Figure 2.4

## Chapter 3

# BEFORE STARTING THE ESTIMATE

The "Invitation to Bid" — To the contractor, this can mean the prospect of weeks of hard work with only a chance of bidding success. It can also mean the opportunity to obtain a contract for a successful and lucrative project. It is not uncommon for the contractor to bid ten or more jobs in order to win just one. The successful bid depends heavily upon estimating accuracy and thus, the preparation, organization and care that go into the estimating process. (If a job is won due to omissions in the estimate, it is likely *not* to be a "successful bid".)

The first step before starting the estimate is to obtain copies of the plans and specifications in *sufficient quantities*. Most estimators mark up plans with colored pencils and make numerous notes and references. For one estimator to work from plans that have been used by another is difficult at best and may easily lead to errors. Most often, only two complete sets are provided by the architect to the general contractor. More sets must be purchased, if needed. When more than one estimator works on a particular project, especially if they are working from the same plans, careful coordination is required to prevent omissions or duplications.

The estimator should be aware of and note any instructions to bidders, which may be included in the specifications or in a separate document. To avoid future confusion, the bid's due date, time and place should be clearly stated and understood upon receipt of the construction documents (plans and specifications). The due date should be marked on a calendar and a schedule. Completion of the takeoff should be made as soon as possible, not two days prior to the bid deadline.

If bid security, or a bid bond, is required, then time must be allowed for arrangements to be made, especially if the bonding capability or capacity of a contractor is limited or has not previously been established.

The estimator should attend all pre-bid meetings with the owner or architect, preferably *after* reviewing the plans and specifications. Important points are often brought up at such meetings and details clarified. Attendance is important, not only to show the owner your interest in the project, but also to assure equal and competitive bidding. For many projects, attendance is required or the bid will not be accepted. It is to the estimator's advantage to examine and review the plans and specifications before any such meetings and before the initial site visit. It is important to become familiar with the project as soon as possible.

All contract documents should be read thoroughly. They exist to protect all parties involved in the construction process. If not provided, the estimator should make a list of all appropriate documents to be checked with the engineer and listed in the bid. The contract documents are written so that the contractors will be bidding equally and competitively, and to ensure that all items in a project are included. The contract documents protect the designer (the architect or engineer) by ensuring that all work is supplied and installed as specified. The owner also benefits from thorough and complete construction documents by being assured of a quality job and a complete, functional project. Finally, the contractor benefits from good contract documents because the scope of work is well defined, eliminating the gray areas of what is implied but not stated. "Extras" are more readily avoided. Change orders, if required, are accepted with less argument if the original contract documents are complete, well stated, and most importantly, read by all concerned parties. The Appendix of this book contains a reproduction of the two electrical pages from SPEC-AID, an R.S. Means Co., Inc. publication. This SPEC-AID was created not only to assist designers and planners when developing project specifications, but also as an aid to the contractor's estimator. The electrical section of the SPEC-AID lists the electrical division (U.C.I. 16) components and variables. The estimator can use this form as a means of outlining a project's requirements, and as a checklist to be sure that all items have been included.

During the first review of the specifications, any items to be priced should be identified and noted. All work to be contracted should be examined for "related work" required from other trades or contractors. Such work is usually referenced and described in a thorough project specification. "Work by others" or "Not in Contract" should be clearly defined on the drawings. Certain materials are often specified by the designer and purchased by the owner to be installed (labor only) by the contractor. These items should be noted and the responsibilities of each party clearly understood. An example of such an item might involve allocating responsibility for receiving, temporary storage, and maintenance of motors furnished by others but installed by electricians.

The "General Conditions", "Supplemental Conditions" and "Special Conditions" sections of the specifications should be examined carefully by *all* parties involved in the project. These sections describe the items that have a direct bearing on the proposed project, but may not be part of the actual, physical installation. Temporary power, lighting, and water are examples of these kinds of items. Also included in these sections is information regarding completion dates, payment schedules (e.g., retainage), submittal requirements, allowances, alternates, and other important project requirements. Each of these conditions can have a significant bearing on the ultimate cost of the project. They must be read and understood prior to performing the estimate.

While analyzing the plans and specifications, the estimator should evaluate the different portions of the project to determine which areas warrant the most attention. The estimator should focus first on those items which represent the largest cost centers of the project or which entail the greatest risk. These cost centers are not always the portions of the job that require the most time to estimate but are those items that will have the most significant impact on the estimate.

| 17.1 | S.F., C.F. and % of Total Costs | UNIT | UNIT COSTS | | | % OF TOTAL | | |
|---|---|---|---|---|---|---|---|---|
| | | | 1/4 | MEDIAN | 3/4 | 1/4 | MEDIAN | 3/4 |
| 01-001 | **APARTMENTS** Low Rise (1 to 3 story) | S.F. | 30.85 | 38.90 | 51.65 | | | |
| 002 | Total project cost | C.F. | 2.77 | 3.57 | 4.49 | | | |
| 272 | Plumbing | S.F. | 2.42 | 3.17 | 4.05 | 6.80% | 9% | 10.20% |
| 277 | Heating, ventilating, air conditioning | | 1.54 | 1.87 | 2.80 | 4.20% | 5.80% | 7.60% |
| 290 | Electrical | | 1.79 | 2.38 | 3.33 | 5.20% | 6.70% | 8.80% |
| 310 | Total: Mechanical & Electrical | ↓ | 5.35 | 6.55 | 8.45 | 15.90% | 18.30% | 22.40% |
| 900 | Per apartment unit, total cost | Apt. | 25,100 | 36,000 | 54,000 | | | |
| 950 | Total: Mechanical & Electrical | " | 4,450 | 64,750 | 8,975 | | | |
| 02-001 | **APARTMENTS** Mid Rise (4 to 7 story) | S.F. | 41.35 | 49 | 60.90 | | | |
| 002 | Total project costs | C.F. | 3.22 | 4.50 | 5.30 | | | |
| 272 | Plumbing | S.F. | 2.42 | 3.05 | 4.19 | 6.20% | 7.40% | 9% |
| 290 | Electrical | | 2.75 | 3.69 | 4.53 | 6.30% | 7% | 8.90% |
| 310 | Total: Mechanical & Electrical | ↓ | 7.60 | 9.40 | 11.40 | 17.10% | 19.70% | 21.50% |
| 900 | Per apartment unit, total cost | Apt. | 38,000 | 49,400 | 66,500 | | | |
| 950 | Total: Mechanical & Electrical | " | 9,100 | 10,600 | 15,600 | | | |
| 03-001 | **APARTMENTS** High Rise (8 to 24 story) | S.F. | 46.70 | 54.35 | 62.75 | | | |
| 002 | Total project costs | C.F. | 3.78 | 5 | 6.25 | | | |
| 272 | Plumbing | S.F. | 2.99 | 4.07 | 5.10 | 6.70% | 9.10% | 10.40% |
| 290 | Electrical | | 3.40 | 4.24 | 5.50 | 7% | 7.70% | 9.10% |
| 310 | Total: Mechanical & Electrical | ↓ | 9.65 | 11.80 | 14.50 | 19.70% | 22.40% | 24.50% |
| 800 | | | | | | | | |
| 900 | Per apartment unit, total cost | Apt. | 42,700 | 50,500 | 55,500 | | | |
| 950 | Total: Mechanical & Electrical | " | 10,400 | 12,000 | 12,700 | | | |
| 04-001 | **AUDITORIUMS** | S.F. | 46.05 | 65.35 | 84.35 | | | |
| 002 | Total project costs | C.F. | 2.89 | 4.23 | 5.85 | | | |
| 272 | Plumbing | S.F. | 3.01 | 4 | 5.20 | 5.70% | 6.80% | 8.40% |
| 277 | Heating, ventilating, air conditioning | | 6.25 | 15.15 | 17.60 | 6.90% | 16% | 19.80% |
| 290 | Electrical | | 3.72 | 5.30 | 6.95 | 6.80% | 8.80% | 10.90% |
| 310 | Total: Mechanical & Electrical | | 7.80 | 9.95 | 16 | 14.40% | 17.80% | 23.30% |
| 05-001 | **AUTOMOTIVE SALES** | ↓ | 32.65 | 41.35 | 53.85 | | | |
| 002 | Total project costs | C.F. | 2.44 | 2.84 | 3.69 | | | |
| 272 | Plumbing | S.F. | 1.63 | 2.29 | 3.11 | 2.80% | 5.90% | 6.40% |
| 277 | Heating, ventilating, air conditioning | | 2.38 | 3.93 | 5.65 | 6.30% | 10.20% | 10.80% |
| 290 | Electrical | | 2.79 | 4.91 | 6.55 | 7.30% | 9.90% | 12.40% |
| 310 | Total: Mechanical & Electrical | | 5.90 | 11.35 | 13.20 | 15.40% | 19.20% | 30.30% |
| 06-001 | **BANKS** | ↓ | 72.05 | 90.25 | 118.45 | | | |
| 002 | Total project costs | C.F. | 5.10 | 6.85 | 9.05 | | | |
| 272 | Plumbing | S.F. | 2.36 | 3.37 | 4.97 | 2.90% | 4% | 5% |
| 277 | Heating, ventilating, air conditioning | | 4.55 | 6.20 | 8.85 | 5.20% | 7.40% | 8.70% |
| 290 | Electrical | | 7 | 9.15 | 12.25 | 8.40% | 10.30% | 12.30% |
| 310 | Total: Mechanical & Electrical | | 11.65 | 16.15 | 23 | 14.20% | 18.10% | 23.50% |
| 13-001 | **CHURCHES** | ↓ | 48.15 | 59.75 | 75.05 | | | |
| 002 | Total project costs | C.F. | 3.04 | 3.80 | 4.95 | | | |
| 272 | Plumbing | S.F. | 1.91 | 2.75 | 4 | 3.60% | 4.90% | 6.30% |
| 277 | Heating, ventilating, air conditioning | | 4.46 | 5.85 | 8.30 | 7.70% | 10% | 12.10% |
| 290 | Electrical | | 3.96 | 5.35 | 7 | 7.30% | 8.80% | 10.90% |
| 310 | Total: Mechanical & Electrical | | 8.60 | 12.35 | 16.30 | 16.10% | 21.80% | 26.50% |
| 15-001 | **CLUBS, COUNTRY** | ↓ | 48.30 | 58 | 75.90 | | | |
| 002 | Total project costs | C.F. | 3.99 | 5 | 6.85 | | | |
| 272 | Plumbing | S.F. | 2.96 | 4.14 | 7.50 | 5.40% | 8.90% | 10% |
| 277 | Heating, ventilating, air conditioning | | 2.85 | 6.15 | 9.15 | 6.70% | 10% | 10.70% |
| 290 | Electrical | | 3.45 | 6.10 | 7.55 | 7.80% | 9.70% | 11% |
| 310 | Total: Mechanical & Electrical | | 9.35 | 14.25 | 20.60 | 17.20% | 24.20% | 30.90% |
| 17-001 | **CLUBS, SOCIAL** Fraternal | ↓ | 41.50 | 57.10 | 74.60 | | | |
| 002 | Total project costs | C.F. | 2.54 | 3.81 | 4.70 | | | |
| 272 | Plumbing | S.F. | 1.94 | 3.08 | 3.52 | 4.90% | 6.70% | 7.80% |
| 277 | Heating, ventilating, air conditioning | | 3.68 | 5.45 | 6.17 | 8.20% | 10.90% | 14.40% |
| 290 | Electrical | | 2.91 | 5.01 | 5.66 | 6.70% | 9.50% | 11.40% |
| 310 | Total: Mechanical & Electrical | ↓ | 9.01 | 11.99 | 16.35 | 18.50% | 28.70% | 33.10% |

Figure 3.1

Figure 3.1 from Division 17 of *Means Electrical Cost Data*, is a chart showing typical percentages of electrical and mechanical work relative to projects as a whole. These charts have been developed to represent the overall average percentages for new construction. Commonly used building types are included.

When the overall scope of the work has been identified, the drawings should be examined to confirm the information in the specifications. This is the time to clarify details while reviewing the general content. The estimator should note which sections, elevations and detail drawings are for which plans. At this point and throughout the whole estimating process, the estimator should note and list any discrepancies between the plans and specifications, as well as any possible omissions. It is often stated in bid documents that bidders are obliged to notify the owner or architect/engineer of any such discrepancies. When so notified, the designer will most often issue an addendum to the contract documents in order to properly notify all parties concerned and to assure equal and competitive bidding. Competition can be fair only if the same information is provided for all bidders.

Once familiar with the contract documents, the estimator should solicit bids by notifying appropriate subcontractors, manufacturers, and vendors. Those whose work may be affected by the site conditions should accompany the estimator on a job site visit (especially in cases of renovation and remodeling, where existing conditions can have a significant effect on the cost of a project).

During a site visit, the estimator should take notes, and possibly photographs, of all situations pertinent to the construction and, thus, to the project estimate. If unusual site conditions exist, or if questions arise during the takeoff, a second site visit is recommended.

In some areas, questions are likely to arise that cannot be answered clearly by the plans and specifications. It is crucial that the owner or responsible party be notified quickly, preferably in writing, so that these questions may be resolved before unnecessary problems arise. Often such items involve more than one contractor and can only be resolved by the owner or construction manager. A proper estimate cannot be completed until all such questions are answered.

# Chapter 4

# THE QUANTITY TAKEOFF

The quantity takeoff is the cornerstone of construction bidding. The amount of detail shown on electrical drawings is perhaps the least of any for the major trades. As a result, the electrical estimator must not only count the items shown on the drawings but also envision the completed design, including fittings, hangers, fasteners, devices, and cover plates. To effectively cover all aspects of the project, certain steps must be followed.

The quantity takeoff should be organized so that the information gathered can be used to future advantage. Scheduling can be made easier if items are taken off and listed by construction phase, or by floor. Material purchasing will similarly benefit.

Units for each item should be used consistently throughout the whole project — from takeoff to cost control. In this way, the original estimate can be equitably compared to progress reports and final cost reports. It will be easier to keep track of a job. (See Chapter 11: Cost Control.)

Part Two of this book is devoted to descriptions of 69 electrical components. In that section, a takeoff procedure is suggested for each component. Typical material and labor units are also given for each component installation. Each material and labor unit consists of a list of items that are generally included in the component. Also given are the units by which the component is measured or counted.

Traditionally, quantities are taken off from the drawings in the same sequence as they are erected or installed. Recently, however, the sequence of takeoff and estimating is more often based on the systems or type of work. In this case, preference in the estimating process is given to certain items or components based on the relative costs.

Quantities should be taken off by one person if the project is not too large and if time allows. For larger projects, the plans are often split and the work assigned to two or more quantity surveyors. In this case, a project leader ought to be assigned to coordinate and assemble the estimate.

When working with the plans during the quantity takeoff, consistency is a very important consideration. If each job is approached in the same manner, a pattern will develop, such as moving from the lower floors to the top, clockwise or counterclockwise. The choice of method is not important, but consistency is. The purpose of being consistent is to avoid duplications as well as omissions and errors. Pre-printed forms provide an excellent means for developing consistent patterns. Figures 4.1 and 4.3 are two examples of takeoff forms. The quantity sheet (Figure 4.1) is designed purely for quantity takeoff. Note that one list of materials can be used for up to four different areas or segments. Part Three of this book (the sample estimate) contains examples of the use of quantity sheets. Figure 4.2 shows a Cost Analysis sheet, which can be used in conjunction with a quantity sheet. Totals of quantities are transferred to this sheet for pricing and extensions. Figure 4.3, a Consolidated Estimate sheet, is designed to be used for both quantity takeoff and pricing on one form. There are many other variations.

Every contractor might benefit from designing custom company forms. If employees of a company all use the same types of forms, then communications and coordination of the estimating process will proceed more smoothly. One estimator will be able to more easily understand the work of another. R.S. Means has published a book completely devoted to forms and their use, entitled *Means Forms for Building Construction Professionals*. Scores of forms, examples, and instructions for use are included.

Appropriate and easy-to-use forms are the first, and most important, of the "tools of the trade" for estimators. Other tools useful to the estimator include scales, rotometers, mechanical counters, and colored pencils.

A number of shortcuts can be used for the quantity takeoff. If approached logically and systematically, these techniques help to save time without sacrificing accuracy. Consistent use of accepted abbreviations saves the time of writing things out. An abbreviations list similar to the one that appears in the Appendix of this book might be posted in a conspicuous place for each estimator to provide a consistent pattern of definitions for use within an office.

All dimensions – whether printed, measured, or calculated – that can be used for determining quantities of more than one item should be listed on a separate sheet and posted for easy reference. Posted gross dimensions can also be used to quickly check for order of magnitude errors.

Rounding off, or decreasing the number of significant digits, should be done only when it will not statistically affect the resulting product. The estimator must use good judgement to determine instances when rounding is appropriate. An overall two or three percent variation in a competitive market can often be the difference between getting or losing a job, or between profit or no profit. The estimator should establish rules for rounding to achieve a consistent level of precision. In general, it is best not to round numbers until the final summary of quantities.

The final summary is also the time to convert units of measure into standards for practical use (linear feet of wire to one hundred linear foot units, for example). This is done to keep the numerical value of the unit cost manageable. The installation of type THW #14 wire requires .00615 man-hours per linear foot, but this information is expressed as .615 M.H./C.L.F.

Be sure to quantify (count) and include "labor only" items that are not shown on plans. Such items may or may not be indicated in the specifications and might include cleanup, special labor for handling materials, testing, etc.

## Means Forms

**QUANTITY SHEET**

| | SHEET NO. | |
|---|---|---|
| PROJECT | ESTIMATE NO. | |
| LOCATION | ARCHITECT | DATE |
| TAKE OFF BY | EXTENSIONS BY: | CHECKED BY: |

| DESCRIPTION | NO. | DIMENSIONS | | | | UNIT | | UNIT | | UNIT | | UNIT |
|---|---|---|---|---|---|---|---|---|---|---|---|---|
| | | | | | | | | | | | | |
| | | | | | | | | | | | | |
| | | | | | | | | | | | | |
| | | | | | | | | | | | | |
| | | | | | | | | | | | | |
| | | | | | | | | | | | | |
| | | | | | | | | | | | | |
| | | | | | | | | | | | | |
| | | | | | | | | | | | | |
| | | | | | | | | | | | | |

*Figure* 4.1

21

**COST ANALYSIS**

| | | | | | | | | | |
|---|---|---|---|---|---|---|---|---|---|
| | | SHEET NO. | | | | | | | |
| PROJECT | | ESTIMATE NO. | | | | | | | |
| ARCHITECT | | DATE | | | | | | | |
| TAKE OFF BY: | QUANTITIES BY: | PRICES BY: | EXTENSIONS BY: | CHECKED BY: | | | | | |

| DESCRIPTION | SOURCE/DIMENSIONS | | | QUANTITY | UNIT | MATERIAL | | LABOR | | EQ./TOTAL | |
|---|---|---|---|---|---|---|---|---|---|---|---|
| | | | | | | UNIT COST | TOTAL | UNIT COST | TOTAL | UNIT COST | TOTAL |

*Figure 4.2*

Figure 4.3

The following list is a summation of the suggestions mentioned above plus a few more guidelines which will be helpful during the quantity take-off:

- Use preprinted forms.
- Transfer carefully when copying numbers from one sheet to the next.
- List dimensions (width, length) in a consistent order.
- Use printed measurements, if available.
- Verify the scale of drawings before using them as a basis for measurement.
- Convert feet and inches to decimal feet.
- Do not round off until the final summary of quantities.
- Mark drawings as quantities are counted.
- Be alert for changes in scale, or notes such as "N.T.S." (not to scale).
- Include required items which may not appear in the plans and specs.

And perhaps the four most important points:

- Write legibly.
- Be organized.
- Use common sense.
- Be consistent.

## Chapter 5

# PRICING THE ESTIMATE

When the quantities have been counted, then values, in the form of unit costs, must be applied and project burdens (overhead and profit) added in order to determine the total selling price (i.e., the quote). Depending upon the chosen estimating method (based on the degree of accuracy required) and the level of detail, these unit costs may be direct or "bare", or may include overhead, profit or contingencies. In Unit Price estimating, the unit costs most commonly used are "bare", or "unburdened". Most contractors extend labor quantities as man-hours and material quantities as dollars. When the total man-hours are determined, it is necessary to multiply these figures by the appropriate hourly rate, before adding mark-ups. Items such as overhead and profit are usually added to the total direct costs at the estimate summary.

## Sources of Cost Information

One of the most difficult aspects of the estimator's job is determining accurate and reliable bare cost data. Sources for such data are varied, but can be categorized in terms of their relative reliability. The most reliable of any cost information is the accurate, up-to-date, well-kept records of completed work by the estimator's own company. There is no better cost for a particular construction item that the *actual* cost to the contractor of that item from another recent job, modified (if necessary) to meet the requirements of the project being estimated.

Quotations by vendors for material costs are also reliable. In this case, however, the estimator must still apply the labor costs.

Whenever possible, all price quotations from vendors or subcontractors should be obtained in writing. Qualifications and exclusions should be clearly stated. The items quoted should be checked to be sure that they are complete and as specified. One way to assure these requirements is to prepare a form on which all subcontractors and vendors must submit their quotations. This form, generally called a "request for quote", can suggest all of the appropriate questions. It also provides a format to organize the information needed by the estimator and, later, the purchasing agent. This technique can be especially useful to the estimator in a smaller organization, since he must often act as purchasing agent as well.

The above procedures are ideal, but in the realistic haste of estimating and bidding, quotations are often received verbally, either in person or by telephone. The importance of gathering all pertinent information is heightened because omissions are more likely. A preprinted form, such as the one shown in Figure 5.1, can be extremely useful to assure that all required information and qualifications are obtained and understood. How often has the subcontractor said, "I didn't know that I was supposed to include that"? With the help of such forms, the appropriate questions may be covered.

If the estimator has no cost records for a particular item and is unable to obtain a quotation, then the next most reliable source of price information is current unit price cost books such as *Means Electrical Cost Data*. R.S. Means presents all such data in the form of national averages; these figures can be adjusted to local conditions, a procedure that will be explained in Part Three of this book. In addition to being a source of primary costs, unit price books can also be useful as a reference or cross-check for verifying costs obtained elsewhere.

Lacking cost information from any of the above-mentioned sources, the estimator may have to rely on experience and personal knowledge of the field to develop costs.

No matter which source of cost information is used, the system and sequence of pricing should be the same as that used for the quantity takeoff. This consistent approach should continue through both accounting and cost control during work on the project.

## Types of Costs

### Types of Costs in a Construction Estimate

| Direct Costs | Indirect Costs |
|---|---|
| Material | Taxes and insurance |
| Labor | Overhead |
| Equipment | Profit |
| Subcontractors | Contingencies |
| Project overhead | |
| Sales tax | |
| Bonds | |

All costs included in a unit price estimate can be divided into two types: direct and indirect. Direct costs are those dedicated solely to the physical construction of a specific project. Material, labor, equipment, and subcontract costs, as well as project overhead costs are all direct.

Indirect costs are usually added to the estimate at the summary stage and are most often calculated as a percentage of the direct costs. They include such items as taxes, insurance, overhead, profit, and contingencies. The indirect costs account for great variation in estimates among different bidders.

A clear understanding of direct and indirect cost factors is a fundamental part of pricing the estimate. Chapters 6 and 7 address the components of direct and indirect costs in detail.

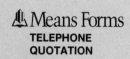 **Means Forms**

**TELEPHONE QUOTATION**

DATE

PROJECT _____ TIME

FIRM QUOTING _____ PHONE ( )

ADDRESS _____ BY

ITEM QUOTED _____ RECEIVED BY

| WORK INCLUDED | AMOUNT OF QUOTATION |
|---|---|
| | |
| | |
| | |
| | |
| | |
| | |
| | |
| | |
| | |
| | |
| | |
| | |
| | |

| DELIVERY TIME | **TOTAL BID** | |
|---|---|---|

DOES QUOTATION INCLUDE THE FOLLOWING:      If ☐ NO is checked, determine the following:

| | | | | |
|---|---|---|---|---|
| STATE & LOCAL SALES TAXES | ☐ YES | ☐ NO | MATERIAL VALUE | |
| DELIVERY TO THE JOB SITE | ☐ YES | ☐ NO | WEIGHT | |
| COMPLETE INSTALLATION | ☐ YES | ☐ NO | QUANTITY | |
| COMPLETE SECTION AS PER PLANS & SPECIFICATIONS | ☐ YES | ☐ NO | DESCRIBE BELOW | |

| EXCLUSIONS AND QUALIFICATIONS | |
|---|---|
| | |
| | |
| | |
| | |
| | |
| | |
| | |
| | |
| | |

| ADDENDA ACKNOWLEDGEMENT | **TOTAL ADJUSTMENTS** | |
|---|---|---|
| | **ADJUSTED TOTAL BID** | |

| ALTERNATES | |
|---|---|
| ALTERNATE NO. | |
| ALTERNATE NO. | |
| ALTERNATE NO. | |
| ALTERNATE NO. | |
| ALTERNATE NO. | |
| ALTERNATE NO. | |
| ALTERNATE NO. | |

*Figure* 5.1

# Chapter 6
# DIRECT COSTS

Direct costs can be defined as those necessary for the completion of the project, in other words, the hard costs. Material, labor, and equipment are among the more obvious items in this category. While subcontract costs include the overhead and profit (indirect costs) of the subcontractor, they are considered to be direct costs to the prime contractor. Also included are certain project overhead costs for items that are necessary for construction. Examples are a storage trailer, tools, and temporary power and lighting. Sales tax and bonds are additional direct costs, since they are essential for the performance of the project.

## Material

When quantities have been carefully taken off, estimates of material cost can be very accurate. For a high level of accuracy, the material unit prices must be reliable and current. The most reliable source of material costs is a quotation from a vendor for the particular job in question. Ideally, the vendor should have access to the plans and specifications for verification of quantities and specified products. Providing quotes is a service which most suppliers will perform on large, bulk procurements, such as lighting fixtures or switchgear.

Material pricing appears relatively simple and straightforward. There are, however, certain considerations that the estimator must address when analyzing material quotations. The reputation of the vendor is a significant factor. Can the vendor "deliver", both figuratively and literally? Estimators may choose not to rely on a "competitive" lower price from an unknown vendor, but will instead use a slightly higher price from a known, reliable vendor. Experience is the best judge for such decisions.

There are many other questions that the estimator should ask. How long is the price guaranteed? At the end of that period, is there an escalation clause? Does the price include delivery charges or sales tax, if required? Where is the point of FOB? (With sensitive electronic or large electrical equipment, this can be an extremely important factor.) Are guarantees and warranties in compliance with the specification requirements? Will there be adequate and appropriate storage space available? If not, can staggered shipments be made? Note that most of these questions are addressed on the form in Figure 5.1. More information should be obtained, however, to assure that a quoted price is accurate and competitive.

The estimator must be sure that the quotation or obtained price is for the materials as per plans and specifications. Architects and engineers may write into the specifications that: a) a particular type or brand of product must be used with no substitution, b) a particular type or brand of product is specified, but alternate brands of equal quality and performance may be accepted *upon approval*, or c) no particular type or brand is specified. Depending upon the options, the estimator may be able to find an acceptable, less expensive alternative. In some cases, these substitutions can substantially lower the cost of a project. Note also that many specification packages will require that "catalogue cuts" be submitted for certain materials as part of the bid proposal. In this case, there is pressure on the estimator to obtain the lowest possible price on materials that he believes will meet the specified criteria.

When the estimator has received material quotations, there are still other considerations which should have a bearing on the final choice of a vendor. Lead time — the amount of time between order and delivery — must be determined and considered. It does not matter how competitive or low a quote is if the material cannot be delivered to the job site on time to support the schedule. If a delivery date is promised, is there a guarantee, or a penalty clause for late delivery?

The estimator should also determine if there are any unusual payment requirements. Cash flow for a company can be severely affected if a large material purchase, thought to be payable in 30 days (90 days 10 years ago!) is delivered C.O.D. Truck drivers may not allow unloading until payment has been received. Such requirements must be determined during the estimating stage so that the cost of borrowing money, if necessary, can be included.

If unable to obtain the quotation of a vendor from whom the material would be purchased, the estimator has other sources for obtaining material prices. These include, in order of reliability:

1. Current price lists from manufacturers' catalogues. Be sure to check that the list is for "contractor prices".
2. Cost records from previous jobs. Historical costs must be updated for present market conditions.
3. Reputable and current annual unit price cost books, such as *Means Electrical Cost Data*. Such books usually represent national averages and must be factored to local markets.

No matter which price source is used, the estimator must be sure to include an allowance for any burdens, such as delivery or finance charges, over the actual cost of the material. (The same kinds of concerns that apply to vendor quotations should be taken into consideration when using these other price sources.)

## Labor

In order to determine the installation cost for each item of construction, the estimator must know two pieces of information: first, the labor rate (hourly wage or salary) of the worker, and second, how much time a worker will need to complete a given unit of the installation- in other words, the productivity or labor units. Wage rates are known going into a project, but productivity may be very difficult to determine.

To estimators working for contractors, the construction labor rates that the contractor pays will be known, well-documented and constantly updated. Estimators for owners, architects, or engineers must determine labor rates from outside sources. Unit price data books, such as *Means Electrical Cost Data* provide national average labor wage rates. The unit costs for labor are based on these averages. Figure 6.1 shows national average *union* rates for the construction industry based on January 1, 1986. Figure 6.2 lists national average *non-union* rates, again based on January 1, 1986.

| Abbr. | Trade | Base Rate Incl. Fringes | | Workers' Comp. Ins. | Average Fixed Overhead | Subs Overhead | Subs Profit | Subs Total Overhead & Profit | | Rate with Subs O & P | |
|---|---|---|---|---|---|---|---|---|---|---|---|
| | | Hourly | Daily | | | | | % | Amount | Hourly | Daily |
| Skwk | Skilled Workers Average (35 trades) | $20.50 | $164.00 | 9.3% | 13.8% | 12.8% | 10% | 45.9% | $ 9.40 | $29.90 | $239.20 |
| | Helpers Average (5 trades) | 15.55 | 124.40 | 9.8 | | 13.0 | | 46.6 | 7.25 | 22.80 | 182.40 |
| | Foremen Average, Inside (50¢ over trade) | 21.00 | 168.00 | 9.3 | | 12.8 | | 45.9 | 9.65 | 30.65 | 245.20 |
| | Foremen Average, Outside ($2.00 over trade) | 22.50 | 180.00 | 9.3 | | 12.8 | | 45.9 | 10.35 | 32.85 | 262.80 |
| Clab | Common Building Laborers | 15.90 | 127.20 | 10.1 | | 11.0 | | 44.9 | 7.15 | 23.05 | 184.40 |
| Asbe | Asbestos Workers | 22.75 | 182.00 | 7.7 | | 16.0 | | 47.5 | 10.80 | 33.55 | 268.40 |
| Boil | Boilermakers | 22.75 | 182.00 | 6.6 | | 16.0 | | 46.4 | 10.55 | 33.30 | 266.40 |
| Bric | Bricklayers | 20.50 | 164.00 | 7.6 | | 11.0 | | 42.4 | 8.70 | 29.20 | 233.60 |
| Brhe | Bricklayer Helpers | 16.00 | 128.00 | 7.6 | | 11.0 | | 42.4 | 6.80 | 22.80 | 182.40 |
| Carp | Carpenters | 20.00 | 160.00 | 10.1 | | 11.0 | | 44.9 | 9.00 | 29.00 | 232.00 |
| Cefi | Cement Finishers | 19.20 | 153.60 | 5.9 | | 11.0 | | 40.7 | 7.80 | 27.00 | 216.00 |
| Elec | Electricians | 22.40 | 179.20 | 4.0 | | 16.0 | | 43.8 | 9.80 | 32.20 | 257.60 |
| Elev | Elevator Constructors | 22.65 | 181.20 | 5.5 | | 16.0 | | 45.3 | 10.25 | 32.90 | 263.20 |
| Eqhv | Equipment Operators, Crane or Shovel | 21.05 | 168.40 | 7.2 | | 14.0 | | 45.0 | 9.45 | 30.50 | 244.00 |
| Eqmd | Equipment Operators, Medium Equipment | 20.60 | 164.80 | 7.2 | | 14.0 | | 45.0 | 9.25 | 29.85 | 238.80 |
| Eqlt | Equipment Operators, Light Equipment | 19.45 | 155.60 | 7.2 | | 14.0 | | 45.0 | 8.75 | 28.20 | 225.60 |
| Eqol | Equipment Operators, Oilers | 17.50 | 140.00 | 7.2 | | 14.0 | | 45.0 | 7.90 | 25.40 | 203.20 |
| Eqmm | Equipment Operators, Master Mechanics | 21.80 | 174.40 | 7.2 | | 14.0 | | 45.0 | 9.80 | 31.60 | 252.80 |
| Glaz | Glaziers | 20.15 | 161.20 | 7.9 | | 11.0 | | 42.7 | 8.60 | 28.75 | 230.00 |
| Lath | Lathers | 20.10 | 160.80 | 6.3 | | 11.0 | | 41.1 | 8.25 | 28.35 | 226.80 |
| Marb | Marble Setters | 20.10 | 160.80 | 7.6 | | 11.0 | | 42.4 | 8.50 | 28.60 | 228.80 |
| Mill | Millwrights | 20.75 | 166.00 | 6.4 | | 11.0 | | 41.4 | 8.60 | 29.35 | 234.80 |
| Mstz | Mosaic and Terrazzo Workers | 19.90 | 159.20 | 5.4 | | 11.0 | | 40.2 | 8.00 | 27.90 | 223.20 |
| Pord | Painters, Ordinary | 19.25 | 154.00 | 7.7 | | 11.0 | | 42.5 | 8.20 | 27.45 | 219.60 |
| Psst | Painters, Structural Steel | 20.00 | 160.00 | 27.0 | | 11.0 | | 61.8 | 12.35 | 32.35 | 258.80 |
| Pape | Paper Hangers | 19.50 | 156.00 | 7.7 | | 11.0 | | 42.5 | 8.30 | 27.80 | 222.40 |
| Pile | Pile Drivers | 20.10 | 160.80 | 17.0 | | 16.0 | | 56.8 | 11.40 | 31.50 | 252.00 |
| Plas | Plasterers | 19.90 | 159.20 | 7.7 | | 11.0 | | 42.5 | 8.45 | 28.35 | 226.80 |
| Plah | Plasterer Helpers | 16.50 | 132.00 | 7.7 | | 11.0 | | 42.5 | 7.00 | 23.50 | 188.00 |
| Plum | Plumbers | 22.55 | 180.40 | 4.6 | | 16.0 | | 44.6 | 10.05 | 32.60 | 260.80 |
| Rodm | Rodmen (Reinforcing) | 21.75 | 174.00 | 16.8 | | 14.0 | | 54.6 | 11.90 | 33.65 | 269.20 |
| Rofc | Roofers, Composition | 18.80 | 150.40 | 18.2 | | 11.0 | | 53.0 | 9.95 | 28.75 | 230.00 |
| Rots | Roofers, Tile & Slate | 18.95 | 151.60 | 18.2 | | 11.0 | | 53.0 | 10.05 | 29.00 | 232.00 |
| Rohe | Roofer Helpers (Composition) | 13.75 | 110.00 | 18.2 | | 11.0 | | 53.0 | 7.30 | 21.05 | 168.40 |
| Shee | Sheet Metal Workers | 22.70 | 181.60 | 6.3 | | 16.0 | | 46.1 | 10.45 | 33.15 | 265.20 |
| Spri | Sprinkler Installers | 23.25 | 186.00 | 5.5 | | 16.0 | | 45.3 | 10.55 | 33.80 | 270.40 |
| Stpi | Steamfitters or Pipefitters | 22.75 | 182.00 | 4.8 | | 16.0 | | 44.6 | 10.15 | 32.90 | 263.20 |
| Ston | Stone Masons | 20.30 | 162.40 | 7.6 | | 11.0 | | 42.4 | 8.60 | 28.90 | 231.20 |
| Sswk | Structural Steel Workers | 21.70 | 173.60 | 19.3 | | 14.0 | | 57.1 | 12.40 | 34.10 | 272.80 |
| Tilf | Tile Layers (Floor) | 19.75 | 158.00 | 5.4 | | 11.0 | | 40.2 | 7.95 | 27.70 | 221.60 |
| Tilh | Tile Layer Helpers | 15.60 | 124.80 | 5.4 | | 11.0 | | 40.2 | 6.30 | 21.90 | 175.20 |
| Trlt | Truck Drivers, Light | 16.35 | 130.80 | 8.6 | | 11.0 | | 43.4 | 7.10 | 23.45 | 187.60 |
| Trhv | Truck Drivers, Heavy | 16.60 | 132.80 | 8.6 | | 11.0 | | 43.4 | 7.20 | 23.80 | 190.40 |
| Sswl | Welders, Structural Steel | 21.70 | 173.60 | 19.3 | | 14.0 | | 57.1 | 12.40 | 34.10 | 272.80 |
| Wrck | *Wrecking | 15.90 | 127.20 | 20.7 | | 11.0 | | 55.5 | 8.80 | 24.70 | 197.60 |

*Not included in Averages.

Figure 6.1

Note the column entitled "Worker's Compensation". Of all the trades, electricians have the lowest rate. This is an average number which varies from state to state (and between crafts). See Figure 6.3. Worker's Compensation is a direct cost when applied to field labor.

If more accurate union labor rates are required, the estimator has a couple of options. Union locals can provide rates (as well as negotiated increases) for a particular location. Employer bargaining groups can usually provide labor cost data as well. R.S. Means Co., Inc. publishes *Labor Rates for the Construction Industry* on an annual basis. This book lists the union labor rates by trade for over 300 U.S. and Canadian cities.

| Abbr. | Trade | Base Rate Incl. Fringes | | Work-ers' Comp. Ins. | Average Fixed Over-head | Subs Over-head | Subs Profit | Subs Total Overhead & Profit | | Rate with Subs O & P | |
|---|---|---|---|---|---|---|---|---|---|---|---|
| | | Hourly | Daily | | | | | % | Amount | Hourly | Daily |
| Skwk | Skilled Workers Average | $10.05 | $ 80.40 | 9.3% | 13.8% | 22.8% | 10% | 55.9% | $ 5.60 | $15.65 | $125.20 |
| | Helpers Average ($2.00 under trade) | 8.05 | 64.40 | 9.8 | | 23.0 | | 56.6 | 4.55 | 12.60 | 100.80 |
| | Foremen Average, ($2.00 over trade) | 12.05 | 96.50 | 9.3 | | 22.8 | | 55.9 | 6.75 | 18.80 | 150.40 |
| Clab | Laborers | 6.95 | 55.60 | 10.1 | | 21.0 | | 54.9 | 3.80 | 10.75 | 86.00 |
| Asbe | Pipe or Duct Insulators | 9.95 | 79.60 | 7.7 | | 26.0 | | 57.5 | 5.70 | 15.65 | 125.20 |
| Boil | Boilermakers | 11.85 | 94.80 | 6.6 | | 26.0 | | 56.4 | 6.70 | 18.55 | 148.40 |
| Bric | Brick or Block Masons | 9.25 | 74.00 | 7.6 | | 21.0 | | 52.4 | 4.85 | 14.10 | 112.80 |
| Carp | Carpenters | 10.30 | 82.40 | 10.1 | | 21.0 | | 54.9 | 5.65 | 15.95 | 127.60 |
| Cefi | Cement Finishers | 9.65 | 77.20 | 5.9 | | 21.0 | | 50.7 | 4.90 | 14.55 | 116.40 |
| Elec | Electricians | 11.15 | 89.20 | 4.0 | | 26.0 | | 53.8 | 6.00 | 17.15 | 137.20 |
| Elev | Elevator Constructors | 11.30 | 90.40 | 5.5 | | 26.0 | | 55.3 | 6.25 | 17.55 | 140.40 |
| Eqhv | Equipment Operators, Crane | 11.75 | 94.00 | 7.2 | | 24.0 | | 55.0 | 6.45 | 18.20 | 145.60 |
| Eqmd | Equipment Operators | 9.45 | 75.60 | 7.2 | | 24.0 | | 55.0 | 5.20 | 14.65 | 117.20 |
| Eqmm | Equipment Mechanics | 11.75 | 94.00 | 7.2 | | 24.0 | | 55.0 | 6.45 | 18.20 | 145.60 |
| Glaz | Glaziers | 9.75 | 78.00 | 7.9 | | 21.0 | | 52.7 | 5.15 | 14.90 | 119.20 |
| Lath | Lathers | 10.30 | 82.40 | 6.3 | | 21.0 | | 51.1 | 5.25 | 15.55 | 124.40 |
| Mill | Millwrights | 10.30 | 82.40 | 6.6 | | 21.0 | | 51.4 | 5.30 | 15.60 | 124.80 |
| Pord | Painters | 9.50 | 76.00 | 7.7 | | 21.0 | | 52.5 | 5.00 | 14.50 | 116.00 |
| Pile | Pile Drivers | 10.30 | 82.40 | 17.0 | | 26.0 | | 66.8 | 6.90 | 17.20 | 137.60 |
| Plas | Plasterers | 9.25 | 74.00 | 7.7 | | 21.0 | | 52.5 | 4.85 | 14.10 | 112.80 |
| Plum | Plumbers | 12.75 | 102.00 | 4.8 | | 26.0 | | 54.6 | 6.95 | 19.70 | 157.60 |
| Rodm | Rodmen (Reinforcing) | 7.75 | 62.00 | 16.8 | | 24.0 | | 64.6 | 5.00 | 12.75 | 102.00 |
| Rofc | Roofers | 10.40 | 83.20 | 18.2 | | 21.0 | | 63.0 | 6.55 | 16.95 | 135.60 |
| Shee | Sheet Metal Workers | 9.95 | 79.60 | 6.3 | | 26.0 | | 56.1 | 5.60 | 15.55 | 124.40 |
| Spri | Sprinkler Installers | 12.90 | 103.20 | 5.5 | | 26.0 | | 55.3 | 7.15 | 20.05 | 160.40 |
| Stpi | Pipefitters | 11.85 | 94.80 | 4.8 | | 26.0 | | 54.6 | 6.50 | 18.35 | 146.80 |
| Ston | Stone Masons | 9.25 | 74.00 | 7.6 | | 21.0 | | 52.4 | 4.85 | 4.10 | 112.80 |
| Sswk | Structural Steel Erectors | 12.25 | 98.00 | 19.3 | | 24.0 | | 67.1 | 8.20 | 20.45 | 163.60 |
| Tilf | Flooring Installers | 10.40 | 83.20 | 5.4 | | 21.0 | | 50.2 | 5.20 | 15.60 | 124.80 |
| Trhv | Truck Drivers | 8.10 | 64.80 | 8.6 | | 21.0 | | 53.4 | 4.35 | 12.45 | 99.60 |
| Wrck | Wreckers | 6.95 | 55.60 | 20.7 | ↓ | 21.0 | ↓ | 65.5 | 4.55 | 11.50 | 92.00 |

*Figure 6.2*

## Table 10.2-203 Workers' Compensation by Trade and State

| STATE | CARPENTRY — 3 stories or less 5651 | CARPENTRY — interior cab. work 5437 | CARPENTRY — general 5403 | CONCRETE WORK—NOC 5213 | CONCRETE WORK — flat (flr, sdwk.) 5221 | ELECTRICAL WIRING — inside 5190 | EXCAVATION — earth NOC 6217 | EXCAVATION — rock 6217 | GLAZIERS 5462 | INSULATION WORK 5479 | LATHING 5443 | MASONRY 5022 | PAINTING & DECORATING 5474 | PILE DRIVING 6003 | PLASTERING 5480 | PLUMBING 5183 | ROOFING 5551 | SHEET METAL WORK (HVAC) 5538 | STEEL ERECTION — door & sash 5102 | STEEL ERECTION — inter. ornam. 5102 | STEEL ERECTION — structure 5040 | STEEL ERECTION — NOC 5057 | TILE WORK — (interior ceramic) 5348 | WATERPROOFING 9014 | WRECKING 5701 |
|---|---|---|---|---|---|---|---|---|---|---|---|---|---|---|---|---|---|---|---|---|---|---|---|---|---|
| AL | 6.32 | 3.56 | 6.49 | 6.53 | 3.67 | 3.00 | 6.05 | 6.05 | 5.61 | 4.85 | 3.51 | 4.27 | 6.46 | 16.85 | 5.66 | 2.61 | 8.74 | 5.68 | 3.87 | 3.87 | 8.60 | 11.69 | 4.33 | 3.18 | 11.69 |
| AK | 12.75 | 6.91 | 10.68 | 10.22 | 7.68 | 7.32 | 7.04 | 7.04 | 9.62 | 10.60 | 8.45 | 3.45 | 9.92 | 19.71 | 10.75 | 7.43 | 19.85 | 7.43 | 12.52 | 12.52 | 34.45 | 19.77 | 6.11 | 3.81 | 34.45 |
| AZ | 8.39 | 7.90 | 19.10 | 8.03 | 6.73 | 4.78 | 6.52 | 6.52 | 11.14 | 8.44 | 7.46 | 11.09 | 7.39 | 18.57 | 16.39 | 4.48 | 19.01 | 6.75 | 11.03 | 11.03 | 8.86 | 12.98 | 6.45 | 4.32 | 19.10 |
| AR | 6.31 | 3.50 | 6.71 | 6.33 | 3.71 | 2.51 | 5.21 | 5.21 | 4.78 | 4.78 | 4.27 | 3.61 | 4.69 | 14.97 | 4.61 | 2.47 | 10.77 | 4.38 | 4.20 | 4.20 | 20.68 | 7.73 | 2.93 | 2.66 | 20.68 |
| CA | 11.39 | 5.44 | 11.46 | 6.08 | 6.08 | 4.54 | 6.38 | 6.38 | 13.00 | 13.81 | 8.01 | 8.62 | 10.61 | 22.68 | 11.70 | 6.29 | 27.70 | 7.24 | 8.71 | 8.71 | 19.30 | 14.97 | 6.48 | 10.61 | 19.83 |
| CO | 7.79 | 5.90 | 10.41 | 8.64 | 6.56 | 3.93 | 8.99 | 8.99 | 9.11 | 8.87 | 7.03 | 12.63 | 7.23 | 31.08 | 12.42 | 5.16 | 25.99 | 6.63 | 6.85 | 6.85 | 15.40 | 18.09 | 3.99 | 3.92 | 18.09 |
| CT | 12.79 | 10.36 | 14.44 | 14.56 | 11.45 | 4.05 | 8.72 | 8.72 | 11.37 | 14.56 | 9.30 | 16.89 | 10.88 | 19.75 | 9.29 | 6.75 | 31.69 | 9.46 | 7.59 | 7.59 | 42.88 | 37.61 | 8.09 | 3.27 | 42.88 |
| DE | 9.39 | 9.39 | 9.39 | 8.44 | 5.76 | 3.28 | 7.51 | 7.51 | 9.36 | 9.39 | 6.96 | 6.04 | 10.20 | 10.90 | 6.96 | 3.35 | 25.31 | 8.00 | 9.19 | 9.19 | 17.99 | 9.19 | 6.43 | 6.04 | 24.81 |
| DC | 8.82 | 7.14 | 12.71 | 20.40 | 8.80 | 10.71 | 12.97 | 12.97 | 11.41 | 11.21 | 8.56 | 17.57 | 9.86 | 30.30 | 10.12 | 12.48 | 22.50 | 9.20 | 15.72 | 15.72 | 40.45 | 32.71 | 18.65 | 4.13 | 40.45 |
| FL | 10.58 | 5.58 | 12.58 | 13.37 | 8.66 | 5.55 | 9.73 | 9.73 | 9.58 | 8.89 | 8.22 | 9.62 | 8.78 | 17.74 | 9.44 | 6.86 | 24.62 | 8.10 | 7.50 | 7.50 | 18.28 | 17.36 | 5.08 | 5.73 | 18.28 |
| GA | 4.43 | 3.10 | 6.70 | 4.75 | 3.14 | 2.53 | 5.58 | 5.58 | 6.52 | 4.33 | 2.50 | 3.67 | 3.63 | 9.63 | 4.33 | 3.44 | 7.70 | 4.68 | 3.63 | 3.63 | 9.96 | 11.50 | 2.35 | 2.24 | 11.50 |
| HI | 15.42 | 15.87 | 54.13 | 27.08 | 13.77 | 13.13 | 27.75 | 27.75 | 33.28 | 16.18 | 15.39 | 18.68 | 16.60 | 44.30 | 22.92 | 11.57 | 44.35 | 11.43 | 11.46 | 11.46 | 37.11 | 42.12 | 13.90 | 11.62 | 37.11 |
| ID | 9.19 | 5.16 | 9.86 | 5.94 | 4.34 | 3.79 | 6.71 | 6.71 | 6.77 | 7.23 | 5.40 | 7.35 | 7.05 | 14.45 | 7.94 | 3.55 | 14.22 | 5.55 | 5.77 | 5.77 | 13.89 | 13.88 | 4.56 | 4.13 | 13.89 |
| IL | 8.76 | 5.99 | 9.61 | 14.06 | 5.83 | 4.76 | 6.40 | 6.40 | 8.51 | 7.19 | 6.08 | 9.34 | 8.96 | 18.88 | 6.14 | 7.12 | 21.60 | 7.26 | 9.23 | 9.23 | 29.05 | 33.16 | 5.81 | 2.92 | 33.16 |
| IN | 2.73 | 1.98 | 2.86 | 2.32 | 1.61 | 1.00 | 2.09 | 2.09 | 3.00 | 2.14 | 1.55 | 1.59 | 2.58 | 3.95 | 2.28 | 1.26 | 3.71 | 1.82 | 1.33 | 1.33 | 3.12 | 7.14 | 1.12 | 1.16 | 3.12 |
| IA | 3.51 | 3.43 | 6.90 | 7.09 | 3.54 | 3.65 | 4.27 | 4.27 | 3.86 | 4.77 | 3.73 | 4.54 | 6.82 | 9.15 | 4.13 | 4.83 | 9.07 | 4.59 | 5.37 | 5.37 | 14.64 | 16.97 | 2.61 | 2.46 | 14.64 |
| KS | 4.82 | 4.20 | 5.40 | 4.02 | 4.15 | 1.93 | 3.17 | 3.17 | 5.64 | 8.25 | 4.53 | 4.87 | 6.75 | 16.77 | 4.13 | 2.85 | 12.72 | 3.85 | 3.86 | 3.86 | 7.97 | 17.39 | 3.11 | 1.87 | 17.39 |
| KY | 6.94 | 3.02 | 6.87 | 3.80 | 3.75 | 2.55 | 4.27 | 4.27 | 4.58 | 5.04 | 4.34 | 4.60 | 5.99 | 9.80 | 4.12 | 2.72 | 11.54 | 3.94 | 4.14 | 4.14 | 10.81 | 7.33 | 3.36 | 4.09 | 10.81 |
| LA | 8.06 | 4.35 | 7.60 | 4.99 | 3.64 | 2.47 | 5.34 | 5.34 | 5.24 | 8.26 | 3.08 | 3.69 | 6.07 | 17.46 | 5.05 | 3.87 | 10.71 | 4.42 | 4.36 | 4.36 | 10.96 | 8.65 | 4.25 | 3.15 | 10.96 |
| ME | 6.59 | 7.64 | 19.90 | 14.58 | 7.71 | 5.73 | 11.59 | 11.59 | 11.18 | 10.36 | 7.54 | 10.04 | 11.33 | 30.32 | 10.32 | 7.39 | 27.20 | 9.41 | 9.07 | 9.07 | 36.50 | 29.85 | 7.24 | 4.96 | 36.50 |
| MD | 10.09 | 8.83 | 10.31 | 22.62 | 7.82 | 7.25 | 10.20 | 10.20 | 13.13 | 12.45 | 8.24 | 9.36 | 11.23 | 26.39 | 9.00 | 9.48 | 26.49 | 11.38 | 11.42 | 11.42 | 37.95 | 28.06 | 8.37 | 4.24 | 37.95 |
| MA | 8.52 | 4.57 | 18.12 | 12.64 | 5.57 | 3.50 | 5.32 | 5.32 | 8.46 | 6.20 | 6.03 | 10.41 | 11.00 | 14.09 | 7.20 | 5.25 | 52.69 | 7.33 | 7.92 | 7.92 | 30.32 | 28.70 | 5.65 | 5.05 | 26.38 |
| MI | 7.54 | 6.27 | 10.24 | 14.50 | 8.88 | 3.52 | 9.97 | 9.97 | 11.79 | 9.40 | 9.08 | 9.60 | 9.71 | 17.68 | 10.09 | 5.01 | 19.17 | 7.52 | 6.74 | 6.74 | 17.29 | 19.78 | 10.12 | NA | 17.29 |
| MN | 10.16 | 10.16 | 20.31 | 13.72 | 8.15 | 4.15 | 8.36 | 8.36 | 9.07 | 8.83 | 9.18 | 10.01 | 10.02 | 23.58 | 9.18 | 8.10 | 28.30 | 8.07 | 11.37 | 11.37 | 27.75 | 30.08 | 6.83 | 5.71 | 27.75 |
| MS | 4.90 | 3.31 | 6.08 | 3.97 | 3.36 | 3.12 | 4.87 | 4.87 | 4.25 | 3.36 | 3.69 | 2.98 | 3.16 | 17.09 | 5.12 | 1.61 | 7.40 | 4.11 | 3.94 | 3.94 | 10.64 | 7.11 | 3.84 | 2.37 | 10.64 |
| MO | 3.67 | 3.73 | 3.83 | 4.79 | 2.91 | 1.75 | 3.47 | 3.47 | 4.62 | 5.47 | 3.44 | 4.09 | 5.81 | 13.56 | 3.85 | 2.49 | 8.22 | 3.49 | 4.00 | 4.00 | 8.19 | 11.36 | 2.40 | 3.05 | 11.36 |
| MT | 9.65 | 6.44 | 13.37 | 5.89 | 5.90 | 4.33 | 9.15 | 9.15 | 7.62 | 10.66 | 6.66 | 12.75 | 10.38 | 27.11 | 9.09 | 7.00 | 25.10 | 7.32 | 7.16 | 7.16 | 32.17 | 20.61 | 5.29 | 5.70 | 32.17 |
| NE | 5.80 | 3.40 | 5.70 | 5.30 | 3.98 | 2.54 | 5.12 | 5.12 | 5.54 | 4.54 | 4.67 | 4.84 | 4.15 | 12.41 | 6.35 | 3.16 | 12.29 | 3.89 | 3.27 | 3.27 | 8.65 | 13.99 | 3.03 | 4.59 | 13.99 |
| NV | 10.82 | 10.82 | 10.82 | 8.51 | 8.51 | 5.48 | 8.51 | 8.51 | 7.34 | 10.82 | 10.82 | 10.24 | 6.92 | 8.51 | 6.92 | 6.74 | 17.31 | 6.74 | 10.82 | 10.82 | 17.11 | 17.11 | 7.34 | 8.51 | 17.11 |
| NH | 11.45 | 5.85 | 11.18 | 15.59 | 7.44 | 5.03 | 15.17 | 15.17 | 8.77 | 10.49 | 9.66 | 8.86 | 10.76 | 26.95 | 12.67 | 6.85 | 36.18 | 6.82 | 8.09 | 8.09 | 24.11 | 19.47 | 6.43 | 4.28 | 24.11 |
| NJ | 5.48 | 4.60 | 5.48 | 5.81 | 4.52 | 2.43 | 5.11 | 5.11 | 4.21 | 5.60 | 6.34 | 6.94 | 8.10 | 13.39 | 3.82 | 8.02 | 8.02 | 14.52 | 6.88 | 6.88 | 14.52 | 19.90 | 2.70 | 2.91 | 19.97 |
| NM | 11.33 | 10.26 | 12.73 | 17.13 | 9.18 | 4.79 | 7.69 | 7.69 | 12.36 | 13.76 | 8.22 | 11.01 | 8.17 | 47.64 | 13.16 | 8.25 | 28.26 | 9.27 | 12.17 | 12.17 | 19.90 | 21.01 | 8.75 | 5.28 | 19.90 |
| NY | 6.63 | 4.03 | 8.89 | 12.97 | 7.04 | 3.92 | 8.40 | 8.40 | 9.92 | 4.73 | 7.20 | 8.65 | 7.28 | 14.54 | 12.46 | 5.39 | NA | 7.00 | 4.19 | 4.19 | 18.60 | 16.60 | 5.70 | 3.33 | 26.03 |
| NC | 3.36 | 2.56 | 5.20 | 3.22 | 2.45 | 2.76 | 3.40 | 3.40 | 3.55 | 3.85 | 2.92 | 2.51 | 4.08 | 13.25 | 3.30 | 2.45 | 8.24 | 3.48 | 2.88 | 2.88 | 16.93 | 5.36 | 2.22 | 1.15 | 16.93 |
| ND | 5.93 | 5.93 | 5.93 | 5.25 | 5.25 | 2.20 | 5.42 | 5.42 | 8.16 | 3.95 | 3.08 | 3.10 | 4.92 | 13.65 | 3.08 | 6.22 | 8.20 | 6.22 | 5.93 | 5.93 | 13.65 | 13.65 | 2.90 | 8.20 | NA |
| OH | 4.77 | 4.77 | 4.77 | 4.81 | 4.81 | 1.98 | 4.81 | 4.81 | 5.38 | 5.01 | 5.01 | 4.40 | 5.38 | 4.81 | 5.01 | 3.26 | 10.95 | 8.26 | NA | NA | 18.31 | 18.31 | 2.32 | 10.95 | NA |
| OK | 7.54 | 4.73 | 7.21 | 6.36 | 4.30 | 2.95 | 5.98 | 5.98 | 6.15 | 6.69 | 4.78 | 4.36 | 4.54 | 16.30 | 7.28 | 4.39 | 19.81 | 5.51 | 5.70 | 5.70 | 23.36 | 12.95 | 3.59 | 3.70 | 23.36 |
| OR | 23.29 | 11.36 | 20.02 | 20.24 | 12.97 | 6.02 | 14.97 | 14.97 | 13.68 | 22.40 | 15.14 | 16.70 | 20.00 | 39.32 | 16.21 | 8.92 | 47.47 | 14.54 | 15.61 | 15.61 | 38.51 | 39.13 | 17.39 | 13.57 | 38.51 |
| PA | 8.64 | 7.65 | 8.64 | 11.55 | 5.70 | 4.25 | 7.95 | 7.95 | 9.67 | 8.64 | 8.49 | 8.84 | 11.84 | 11.24 | 8.49 | 5.01 | 17.57 | 8.21 | 8.87 | 8.87 | 24.68 | 8.87 | 6.33 | 8.84 | 42.10 |
| RI | 8.14 | 5.37 | 9.28 | 10.02 | 8.89 | 5.14 | 7.64 | 7.64 | 9.93 | 7.82 | 5.97 | 8.37 | 9.18 | 24.41 | 8.30 | 3.74 | 24.12 | 6.21 | 7.33 | 7.33 | 40.04 | 38.15 | 4.77 | 4.11 | 40.04 |
| SC | 8.04 | 5.35 | 8.10 | 5.13 | 3.82 | 6.19 | 5.21 | 5.21 | 4.96 | 7.19 | 4.38 | 6.63 | 8.68 | 12.04 | 5.85 | 2.55 | 15.79 | 7.24 | 3.77 | 3.77 | 15.49 | 13.35 | 3.89 | 2.45 | 15.49 |
| SD | 5.04 | 3.70 | 6.91 | 5.66 | 3.96 | 3.20 | 6.57 | 6.57 | 4.85 | 5.21 | 3.85 | 5.00 | 11.20 | 4.69 | 3.39 | 3.55 | 13.55 | 3.83 | 4.03 | 4.03 | 11.08 | 10.81 | 3.27 | 1.94 | 11.08 |
| TN | 4.03 | 3.81 | 6.52 | 4.00 | 3.16 | 2.30 | 4.17 | 4.17 | 4.51 | 4.56 | 3.71 | 4.09 | 4.65 | 8.59 | 3.58 | 2.65 | 7.22 | 4.16 | 3.62 | 3.62 | 9.45 | 6.59 | 2.43 | 1.99 | 9.45 |
| TX | 5.44 | 3.90 | 5.44 | 5.13 | 3.57 | 2.94 | 3.66 | 3.66 | 4.69 | 5.97 | 2.39 | 3.72 | 3.78 | 11.43 | 7.02 | 2.97 | 14.31 | 5.68 | 3.35 | 3.35 | 16.95 | 6.59 | 2.88 | 3.54 | 14.11 |
| UT | NA | NA | 4.97 | 6.15 | 3.10 | 1.72 | 3.74 | 3.74 | 4.22 | 5.47 | 4.42 | 4.97 | 5.12 | 8.67 | 4.94 | 2.65 | 14.99 | 2.93 | 3.64 | 3.64 | NA | 8.17 | 2.64 | 2.44 | 8.17 |
| VT | 4.15 | 3.14 | 5.32 | 6.92 | 3.25 | 3.38 | 5.24 | 5.24 | 5.02 | 5.07 | 3.81 | 5.30 | 4.24 | 12.40 | 4.75 | 2.69 | 11.71 | 4.41 | 4.64 | 4.64 | 13.07 | 10.62 | 3.13 | 2.51 | 13.07 |
| VA | 4.86 | 5.63 | 7.62 | 7.75 | 3.70 | 3.49 | 4.89 | 4.89 | 4.67 | 7.85 | 8.94 | 5.91 | 6.25 | 10.65 | 4.54 | 3.57 | 13.76 | 6.92 | 4.37 | 4.37 | 14.34 | 16.97 | 3.86 | 2.35 | 14.34 |
| WA | 6.66 | 6.66 | 6.66 | 5.38 | 4.61 | 2.31 | 6.25 | 6.25 | 6.99 | 5.35 | 6.66 | 7.43 | 6.17 | 12.85 | 6.76 | 2.44 | 6.38 | 2.83 | 6.17 | 6.17 | 8.58 | 8.58 | 5.16 | 6.97 | 7.64 |
| WV | 9.49 | 9.49 | 9.49 | 11.46 | 11.46 | 3.89 | 9.02 | 9.02 | 4.98 | 4.98 | 12.16 | 6.82 | 12.16 | 7.62 | 12.16 | 3.91 | 10.98 | 4.98 | 6.67 | 6.67 | 9.10 | 6.67 | 6.82 | 2.80 | 9.10 |
| WI | 4.83 | 3.29 | 8.06 | 6.57 | 3.68 | 3.32 | 3.92 | 3.92 | 5.38 | 5.88 | 4.64 | 5.06 | 4.38 | 10.98 | 5.08 | 3.34 | 10.38 | 5.09 | 4.07 | 4.07 | 18.57 | 13.59 | 3.16 | 3.50 | 22.81 |
| WY | 5.00 | 5.00 | 5.00 | 5.00 | 5.00 | 5.00 | 5.00 | 5.00 | 5.00 | 5.00 | 5.00 | 5.00 | 5.00 | 5.00 | 5.00 | 5.00 | 5.00 | 5.00 | 5.00 | 5.00 | 5.00 | 5.00 | 5.00 | 5.00 | 5.00 |
| AVG. | 7.80 | 5.90 | 10.11 | 9.12 | 5.83 | 4.04 | 7.16 | 7.16 | 7.89 | 7.69 | 6.36 | 7.62 | 7.73 | 17.04 | 7.76 | 4.85 | 18.28 | 6.31 | 6.80 | 6.80 | 19.30 | 16.74 | 5.39 | 4.54 | 20.71 |

*Figure 6.3*

Determination of non-union, or "open shop" rates is much more difficult. In larger cities, employer organizations often exist to represent non-union contractors. These organizations may have records of local pay scales, but ultimately, wage rates are determined by each contractor. The national average for electricians' non-union base rate is 49.8% of the union rate. However, due to a higher overhead percentage, the rate with overhead and profit is 53.3% of union shop wages.

Labor units (or man-hours) are the least predictable of all factors for building projects. It is important to determine as accurately as possible prevailing productivity. The best source of labor productivity or labor units (and therefore labor costs) is the estimator's well-kept records from previous projects. If there are no company records for productivity, cost data books, such as *Means Electrical Cost Data*, and productivity reference books, such as *Means Man-Hour Standards*, can be invaluable. Included with the listing for each individual construction item is the designation of a suggested crew make-up (1 electrician in most cases). The crew is the minimum grouping of talent which can be expected to accomplish the task efficiently. Figure 6.4, a typical page from *Means Electrical Cost Data*, includes this data and indicates the required number of man-hours for each "unit" of work. When multiplied by the average union wage rate for electricians (see Figure 6.1), the result is the "bare" installed unit cost, which appears in the "INST." column.

The estimator who has neither company records nor the sources described above must put together the appropriate crews and determine the expected output or productivity. This type of estimating should only be attempted based upon strong experience and considerable exposure to construction methods and practices. There are rare occasions when this approach is necessary to estimate a particular item or a new technique. Even then, the new labor units are often extrapolated from existing figures for similar work, rather than being created from scratch.

## Equipment

In recent years, construction equipment has become more important, not only because of the incentive to reduce labor costs, but also as a response to new, high-technology construction methods and materials. As a result, equipment costs represent an increasing percentage of total project costs in construction. Estimators must carefully address the issue of equipment and related expenses. Equipment costs can be divided into the two following categories:

1. Rental, lease or ownership costs. These costs may be determined based upon hourly, daily, weekly, monthly or annual increments. These fees or payments only buy the "right" to use the equipment (i.e., exclusive of operating costs).
2. Operating Costs. Once the "right" of use is obtained, costs are incurred for actual use or operation. These costs may include fuel, lubrication, maintenance, and parts.

Equipment costs, as described above, do not include the labor expense of operators. However, some cost books and suppliers may include the operator in the quoted price for equipment as an "operated" rental cost. In other words, the equipment is priced as if it were a subcontract cost. In electrical construction, equipment is often procured in this way, via a contract with the general contractor on site. The advantage of this approach is that it can be used on an "as needed" basis and not carried as a weekly or monthly cost.

| 16.3 Starters, Boards & Switches | CREW | MAN-HOURS | UNIT | BARE COSTS MAT. | BARE COSTS INST. | BARE COSTS TOTAL | TOTAL INCL O&P |
|---|---|---|---|---|---|---|---|
| 910          200 amp, to 125 HP motor | 1 Elec | 6.150 | Ea. | 315 | 140 | 455 | 545 |
| 911          400 amp, to 200 HP motor | " | 8.890 | " | 825 | 200 | 1,025 | 1,200 |
| **56-001 TIME SWITCHES** | | | | | | | |
| 010      Single pole, single throw, 24 hour dial | 1 Elec | 2.000 | Ea. | 41 | 45 | 86 | 110 |
| 020          24 hour dial with reserve power | | 2.220 | | 180 | 50 | 230 | 270 |
| 030          Astronomic dial | | 2.220 | | 65 | 50 | 115 | 145 |
| 040          Astronomic dial with reserve power | | 2.420 | | 195 | 54 | 249 | 295 |
| 050          7 day calendar dial | | 2.420 | | 59 | 54 | 113 | 145 |
| 060          7 day calendar dial with reserve power | | 2.500 | | 180 | 56 | 236 | 280 |
| 070          Photo cell 2000 watt | | 1.000 | | 14 | 22 | 36 | 48 |
| 108      Load management device, 2 loads | | 2.000 | | 175 | 45 | 220 | 255 |
| 110      Load management device, 8 loads | ↓ | 8.000 | ↓ | 940 | 180 | 1,120 | 1,300 |
| **60-001 SWITCHBOARDS** Aluminum bus bars incoming section, | | | | | | | |
| 010      not including CT's or PT's | | | | | | | |
| 020      No main disconnect, includes CT compartment | | | | | | | |
| 030          120/208 volt, 4 wire, 600 amp | 1 Elec | 16.000 | Ea. | 1,725 | 360 | 2,085 | 2,425 |
| 040          800 amp | | 18.180 | | 1,900 | 405 | 2,305 | 2,675 |
| 050          1000 amp | | 20.000 | | 2,225 | 450 | 2,675 | 3,100 |
| 060          1200 amp | | 22.220 | | 2,350 | 500 | 2,850 | 3,300 |
| 070          1600 amp | | 24.240 | | 2,575 | 545 | 3,120 | 3,625 |
| 080          2000 amp | | 25.810 | | 2,750 | 580 | 3,330 | 3,850 |
| 100          3000 amp | | 28.570 | | 3,525 | 640 | 4,165 | 4,800 |
| 120          277/480 volt, 4 wire, 600 amp | | 16.000 | | 1,725 | 360 | 2,085 | 2,425 |
| 130          800 amp | | 18.180 | | 1,900 | 405 | 2,305 | 2,675 |
| 140          1000 amp | | 20.000 | | 2,225 | 450 | 2,675 | 3,100 |
| 150          1200 amp | | 22.220 | | 2,350 | 500 | 2,850 | 3,300 |
| 160          1600 amp | | 24.240 | | 2,575 | 545 | 3,120 | 3,625 |
| 170          2000 amp | | 25.810 | | 2,750 | 580 | 3,330 | 3,850 |
| 180          3000 amp | | 28.570 | | 3,525 | 640 | 4,165 | 4,800 |
| 190          4000 amp | ↓ | 30.770 | ↓ | 4,375 | 690 | 5,065 | 5,800 |
| 200      Fused switch & CT compartment | | | | | | | |
| 210          120/208 volt, 4 wire, 400 amp | 1 Elec | 14.290 | Ea. | 2,450 | 320 | 2,770 | 3,150 |
| 220          600 amp | | 17.020 | | 2,850 | 380 | 3,230 | 3,675 |
| 230          800 amp | | 19.050 | | 3,825 | 425 | 4,250 | 4,825 |
| 240          1200 amp | | 23.530 | | 4,800 | 525 | 5,325 | 6,050 |
| 250          277/480 volt, 4 wire, 400 amp | | 14.040 | | 2,450 | 315 | 2,765 | 3,150 |
| 260          600 amp | | 17.020 | | 2,850 | 380 | 3,230 | 3,675 |
| 270          800 amp | | 19.050 | | 3,825 | 425 | 4,250 | 4,825 |
| 280          1200 amp | ↓ | 23.530 | ↓ | 4,800 | 525 | 5,325 | 6,050 |
| 290      Pressure switch & CT compartment | | | | | | | |
| 300          120/208 volt, 4 wire, 800 amp | 1 Elec | 20.000 | Ea. | 5,475 | 450 | 5,925 | 6,675 |
| 310          1200 amp | | 24.240 | | 6,225 | 545 | 6,770 | 7,625 |
| 320          1600 amp | | 25.810 | | 6,925 | 580 | 7,505 | 8,450 |
| 330          2000 amp | | 28.570 | | 8,100 | 640 | 8,740 | 9,825 |
| 331          2500 amp | | 32.000 | | 9,225 | 715 | 9,940 | 11,200 |
| 332          3000 amp | | 36.360 | | 13,800 | 815 | 14,615 | 16,400 |
| 333          4000 amp | | 40.000 | | 18,400 | 895 | 19,295 | 21,500 |
| 340          277/480 volt, 4 wire, 800 amp | | 20.000 | | 7,925 | 450 | 8,375 | 9,350 |
| 360          1200 amp, with ground fault | | 24.240 | | 9,325 | 545 | 9,870 | 11,000 |
| 400          1600 amp, with ground fault | | 25.810 | | 10,000 | 580 | 10,580 | 11,800 |
| 420          2000 amp, with ground fault | ↓ | 28.570 | ↓ | 12,000 | 640 | 12,640 | 14,100 |
| 440      Circuit breaker, molded case & CT compartment | | | | | | | |
| 460          3 pole, 4 wire, 600 amp | 1 Elec | 17.020 | Ea. | 4,050 | 380 | 4,430 | 5,000 |
| 480          800 amp | | 19.050 | | 4,275 | 425 | 4,700 | 5,325 |
| 500          1200 amp | ↓ | 23.530 | ↓ | 5,975 | 525 | 6,500 | 7,325 |
| **61-001 SWITCHBOARD** | | | | | | | |
| 010      Main lugs only, to 600 volt, 3 pole, 3 wire, 200 amp | 1 Elec | 13.330 | Ea. | 1,075 | 300 | 1,375 | 1,600 |
| 011          400 amp | | 13.330 | | 1,075 | 300 | 1,375 | 1,600 |
| 012          600 amp | ↓ | 13.330 | ↓ | 1,175 | 300 | 1,475 | 1,725 |

*Figure 6.4*

Equipment ownership costs apply to both leased and owned equipment. The operating costs of equipment, whether rented, leased or owned, are available from the following sources (listed in order of reliability):

1. The company's own records
2. Annual cost books containing equipment operating costs, such as *Means Electrical Cost Data*
3. Manufacturers' estimates
4. Text books dealing with equipment operating costs

These operating costs consist of fuel, lubrication, expendable parts replacement, minor maintenance, transportation and mobilizing costs. For estimating purposes, the equipment ownership and operating costs should be listed separately. In this way, the decision to rent, lease or purchase can be decided project by project.

There are two commonly used methods for including equipment costs in a construction estimate. The first is to include the equipment as a part of the construction task for which it is used. In this case, costs are included in each line item as a separate unit price. The advantage of this method is that costs are allocated to the division or task that actually incurs the expense. As a result, more accurate records can be kept for each installed component. A disadvantage of this method occurs in the pricing of equipment that may be used for many different tasks. Duplication of costs can occur in this instance. Another disadvantage is that the budget may be left short for the following reason: the estimate may only reflect two hours for a crane truck, when the minimum availability of the truck involves a daily (8-hour) rental charge.

The second method for including equipment costs in the estimate is to keep all such costs separate and to include them in Division 1 as a part of Project Overhead. The advantage of this method is that all equipment costs are grouped together, and that machines used for several tasks are included (without duplication). One disadvantage is that for future estimating purposes, equipment costs will be known only on a job basis and not per installed unit.

Whichever method is used, the estimator must be consistent, and must be sure that all equipment costs are included but not duplicated. The estimating method should be the same as that chosen for cost monitoring and accounting. In this way, the data will be available both for monitoring the project's costs and for bidding future projects.

A final word of caution about equipment is to consider its age and reliability. If an older item, such as a pick-up truck, needs frequent repair, it may cost far more in lost man-hours to the project than is reflected in its calculated cost rate.

## Subcontractors

On the occasions when subcontractors are used, quotations should be solicited and analyzed in the same way as material quotes. A primary concern is that the bid covers the work as per plans and specifications, and that all appropriate work alternates and allowances are included. Any exclusions should be clearly stated and explained. If the bid is received verbally, a form such as that shown in Figure 5.1 will help to assure that it is documented accurately. Any unique scheduling or payment requirements must be noted and evaluated prior to submission of your bid. Such requirements could affect (restrict or enhance) the normal progress of the project, and should therefore be known in advance.

The estimator should note how long the subcontract bid will be honored. This time period usually varies from 30 to 90 days and is often included as a condition in complete bids.

The estimator should question and verify the bonding capability and capacity of unfamiliar subcontractors. Taking such action may be necessary when bidding in a new location. Other than word of mouth, these inquiries may be the only way to confirm subcontractor reliability.

## Project Overhead

Project Overhead represents those construction costs that are usually included in Division One – General Requirements. Site management is covered in this section. Typical items are supervisory personnel, job engineers, daily and final cleanup, and temporary heat and power. While these items may not be directly part of the physical structure, they are a part of the project. Project Overhead, like all other direct costs, can be separated into material, labor, and equipment components. Figures 6.5 and 6.6 are an example of a form that can help ensure that all appropriate costs are included.

Some may not agree that certain items (such as equipment or scaffolding) should be included as Project Overhead, and might prefer to list such items in another division. Ultimately, it is not important *where* each item is incorporated into the estimate but that *every item is included somewhere*.

Project overhead often includes time-related items; equipment rental, supervisory labor, and temporary utilities are examples. The cost for these items depends upon the duration of the project. A preliminary schedule should, therefore, be developed *prior* to completion of the estimate so that time-related items can be properly counted. This will be further discussed in Chapter Nine, "Pre-Bid Scheduling".

## Bonds

Although bonds are really a type of "direct cost", they are priced and based upon the total "bid" or "selling price". For that reason, they are generally figured after indirect costs have been added. Bonding requirements for a project will be specified in Division 1 – "General Requirements", and will be included in the construction contract. Various types of bonds may be required. Listed below are a few common types:

**Bid Bond.** A form of bid security executed by the bidder or principle and by a surety (bonding company) to guarantee that the bidder will enter into a contract within a specified time and furnish any required Performance or Labor and Material Payment bonds.

**Completion Bond.** Also known as "Construction" or "Contract" bond. The guarantee by a surety that the construction contract will be completed and that it will be clear of all liens and encumbrances.

**Labor and Material Payment Bond.** The guarantee by a surety to the owner that the contractor will pay for all labor and materials used in the performance of the contract as per the construction documents. The claimants under the bond are those having direct contracts with the contractor or any subcontractor.

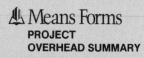

**PROJECT**
**OVERHEAD SUMMARY**

| | | | | | | | | | | |
|---|---|---|---|---|---|---|---|---|---|---|

PROJECT _____  SHEET NO. _____

PROJECT _____  ESTIMATE NO. _____

LOCATION _____  ARCHITECT _____  DATE _____

QUANTITIES BY: _____  PRICES BY: _____  EXTENSIONS BY: _____  CHECKED BY: _____

| DESCRIPTION | QUANTITY | UNIT | MATERIAL/EQUIPMENT | | LABOR | | TOTAL COST | |
|---|---|---|---|---|---|---|---|---|
| | | | UNIT | TOTAL | UNIT | TOTAL | UNIT | TOTAL |
| **Job Organization:** Superintendent | | | | | | | | |
| Project Manager | | | | | | | | |
| Timekeeper & Material Clerk | | | | | | | | |
| Clerical | | | | | | | | |
| Safety, Watchman & First Aid | | | | | | | | |
| | | | | | | | | |
| **Travel Expense:** Superintendent | | | | | | | | |
| Project Manager | | | | | | | | |
| | | | | | | | | |
| **Engineering:** Layout | | | | | | | | |
| Inspection/Quantities | | | | | | | | |
| Drawings | | | | | | | | |
| CPM Schedule | | | | | | | | |
| | | | | | | | | |
| **Testing:** Soil | | | | | | | | |
| Materials | | | | | | | | |
| Structural | | | | | | | | |
| | | | | | | | | |
| **Equipment:** Cranes | | | | | | | | |
| Concrete Pump, Conveyor, Etc. | | | | | | | | |
| Elevators, Hoists | | | | | | | | |
| Freight & Hauling | | | | | | | | |
| Loading, Unloading, Erecting, Etc. | | | | | | | | |
| Maintenance | | | | | | | | |
| Pumping | | | | | | | | |
| Scaffolding | | | | | | | | |
| Small Power Equipment/Tools | | | | | | | | |
| | | | | | | | | |
| **Field Offices:** Job Office | | | | | | | | |
| Architect/Owner's Office | | | | | | | | |
| Temporary Telephones | | | | | | | | |
| Utilities | | | | | | | | |
| Temporary Toilets | | | | | | | | |
| Storage Areas & Sheds | | | | | | | | |
| | | | | | | | | |
| **Temporary Utilities:** Heat | | | | | | | | |
| Light & Power | | | | | | | | |
| Water | | | | | | | | |
| | | | | | | | | |
| **PAGE TOTALS** | | | | | | | | |
| | | | | | | | | |
| | | | | | | | | |

Page 1 of 2

*Figure 6.5*

## ⚓ Means Forms

| DESCRIPTION | QUANTITY | UNIT | MATERIAL/EQUIPMENT | | LABOR | | TOTAL COST | |
|---|---|---|---|---|---|---|---|---|
| | | | UNIT | TOTAL | UNIT | TOTAL | UNIT | TOTAL |
| **Total Brought Forward** | | | | | | | | |
| | | | | | | | | |
| **Winter Protection:** Temp. Heat/Protection | | | | | | | | |
| Snow Plowing | | | | | | | | |
| Thawing Materials | | | | | | | | |
| | | | | | | | | |
| **Temporary Roads** | | | | | | | | |
| | | | | | | | | |
| **Signs & Barricades:** Site Sign | | | | | | | | |
| Temporary Fences | | | | | | | | |
| Temporary Stairs, Ladders & Floors | | | | | | | | |
| | | | | | | | | |
| **Photographs** | | | | | | | | |
| | | | | | | | | |
| **Clean Up** | | | | | | | | |
| Dumpster | | | | | | | | |
| Final Clean Up | | | | | | | | |
| | | | | | | | | |
| **Punch List** | | | | | | | | |
| | | | | | | | | |
| **Permits:** Building | | | | | | | | |
| Misc. | | | | | | | | |
| | | | | | | | | |
| **Insurance:** Builders Risk | | | | | | | | |
| Owner's Protective Liability | | | | | | | | |
| Umbrella | | | | | | | | |
| Unemployment Ins. & Social Security | | | | | | | | |
| | | | | | | | | |
| **Taxes** | | | | | | | | |
| City Sales Tax | | | | | | | | |
| State Sales Tax | | | | | | | | |
| | | | | | | | | |
| **Bonds** | | | | | | | | |
| Performance | | | | | | | | |
| Material & Equipment | | | | | | | | |
| | | | | | | | | |
| **Main Office Expense** | | | | | | | | |
| | | | | | | | | |
| **Special Items** | | | | | | | | |
| | | | | | | | | |
| | | | | | | | | |
| | | | | | | | | |
| | | | | | | | | |
| **TOTALS:** | | | | | | | | |
| | | | | | | | | |
| | | | | | | | | |

*Figure 6.6*

**Performance Bond.** (1) A guarantee that a contractor will perform a job according to the terms of the contracts. (2) A bond of the contractor in which a surety guarantees to the owner that the work will be performed in accordance with the contract documents. Except where prohibited by statute, the performance bond is frequently combined with the labor and material payment bond.

**Surety Bond.** A legal instrument under which one party agrees to answer to another party for the debt, default or failure to perform of a third party.

## Sales Tax

Sales tax varies from state to state and often from city to city within a state (see Figure 6.7). Larger cities may have a sales tax in addition to the state sales tax. Some localities also impose separate sales taxes on labor and equipment.

When bidding takes place in unfamiliar locations, the estimator should check with local agencies regarding the amount and the method of payment of sales tax. Local authorities may require owners to withhold payments to out-of-state contractors until payment of all required sales tax has been verified. Sales tax is often taken for granted or even omitted and, as can be seen in Figure 6.7, can be as much as 7.9% of material costs. Indeed, this can represent a significant portion of the project's total cost. Conversely, some clients and/or their projects may be tax exempt. If this fact is unknown to the estimator, a large dollar amount for sales tax might be needlessly included in a bid.

### Sales Tax Percentages on Materials by State (as of 7/85)

| State | Tax | State | Tax | State | Tax | State | Tax |
|---|---|---|---|---|---|---|---|
| Alabama | 4% | Illinois | 5% | Montana | 0% | Rhode Island | 6% |
| Alaska | 0 | Indiana | 5 | Nebraska | 3.5 | South Carolina | 5 |
| Arizona | 5 | Iowa | 4 | Nevada | 5.75 | South Dakota | 4 |
| Arkansas | 4 | Kansas | 3 | New Hampshire | 0 | Tennessee | 5.5 |
| California | 6 | Kentucky | 5 | New Jersey | 6 | Texas | 4 |
| Colorado | 3 | Louisiana | 4 | New Mexico | 3.75 | Utah | 5.5 |
| Connecticut | 7.5 | Maine | 5 | New York | 4 | Vermont | 4 |
| Delaware | 0 | Maryland | 5 | North Carolina | 3 | Virginia | 4 |
| District of Columbia | 6 | Massachusetts | 5 | North Dakota | 4 | Washington | 7.9 |
| Florida | 5 | Michigan | 4 | Ohio | 5.5 | West Virginia | 5 |
| Georgia | 3 | Minnesota | 6 | Oklahoma | 3 | Wisconsin | 5 |
| Hawaii | 4 | Mississippi | 6 | Oregon | 0 | Wyoming | 3 |
| Idaho | 4 | Missouri | 6.125 | Pennsylvania | 6 | Average | 4.28% |

Figure 6.7

# Chapter 7
# INDIRECT COSTS

Indirect costs are those "costs of doing business" that are incurred by the general staff. These expenses are sometimes referred to as a "burden" to the project. Indirect costs may include certain fixed, or known, expenses and percentages, as well as costs which can be variable and subjectively determined. Government authorities require payment of certain taxes and insurance, usually based upon labor costs and determined by trade. These are a type of fixed indirect cost. Office overhead, if well understood and established, can also be considered as a relatively fixed percentage. Profit and contingencies, however, are more of a variable subjective cost. These figures are often determined based on the judgement and discretion of the person responsible for the company's growth and success.

If the direct costs for the same project have been carefully determined, they should not vary significantly from one estimator to another. It is the indirect costs that are often responsible for variations between bids.

The direct costs of a project must be itemized, tabulated and totalled before the indirect costs can be applied to the estimate. Indirect costs include:

1. Taxes and Insurance (on Office Staff)
2. Office or Operating Overhead (vs. Project Overhead)
3. Profit
4. Contingencies

## Taxes and Insurance

The taxes and insurance included as indirect costs are most often related to the costs of management and/or type of work. This category may include Worker's Compensation, Builder's Risk, and Public Liability insurance, as well as employer-paid social security and unemployment insurance. By law, the employer must pay these expenses. Rates are based on the type and salary of the employees, as well as the location and/or type of business. These are "Home Office expenses".

## Office or Operating Overhead

Office overhead, or the cost of doing business, is perhaps one of the main reasons why so many contractors are unable to realize a profit, or even to stay in business. This effect is manifested in two ways. Either a company does not know its true overhead cost and, therefore, fails to mark up its costs enough to recover them; or management does not restrain or control overhead costs effectively and fails to remain competitive.

If a contractor does not know the costs of operating the business, then, more than likely, these costs will not be recovered. Many companies survive, and even turn a profit, by simply adding an arbitrary percentage for overhead to each job, without knowing how the percentage is derived or what is included. When annual volume changes significantly, whether by increase or decrease, the previously used percentage for overhead may no longer be valid. When such a volume change occurs, the owner often finds that the company is not doing as well as before and cannot determine the reasons. Chances are, overhead costs are not being fully recovered. As an example, Figure 7.1 lists annual office costs and expenses for a "typical" electrical contractor. It is assumed that the anticipated annual volume of the company is $1,500,000. Each of the items is described briefly below:

*Owner*: This includes only a reasonable base salary and does not include profits. An owner's salary is *not* a company's profit.

*Engineer/Estimator*: Since the owner is primarily on the road getting business, this is the person who runs the daily operation of the company and is responsible for estimating. In some operations, the estimator successfully wins a bid, then becomes the "project manager" and is responsible to the owner for its profitability.

*Secretary/Receptionist*: This person manages office operations and handles paperwork. A talented individual in this position can be a tremendous asset.

*Office Worker Insurance & Taxes*: These costs are for main office personnel only and, for this example, are calculated as 37% of the total salaries based on the following breakdown:

| | |
|---|---|
| Worker's Compensation | 6% |
| FICA | 7% |
| Unemployment | 4% |
| Medical & other insurance | 10% |
| Profit sharing, pension, etc. | 10% |
| | 37% |

## ANNUAL MAIN OFFICE EXPENSES

Assume: $1,500,000 Annual Volume in the Field
30% Material, 70% Labor

Office/Operating Expenses:

| | |
|---|---:|
| Owner | $ 60,000 |
| Engineer/Estimator | 44,000 |
| Secretary/Receptionist | 18,000 |
| Personnel Insurance & Taxes | 44,030 |
| Office Rent | 10,000 |
| Utilities | 1,800 |
| Telephone | 6,000 |
| Vehicles (2) | 13,000 |
| Office Equipment | 2,400 |
| Legal/Accounting Services | 5,000 |
| Miscellaneous: | |
|   Advertising | 1,750 |
|   Seminars | 3,000 |
|   Travel & Entertainment | 8,000 |
| Uncollected Receivables | 18,000 |
| Total | $231,980 |

$$\frac{\text{Expenses}}{\text{Labor Volume}} = \frac{\$231,980}{\$1,050,000} = 22.1\%$$

To support this overhead, job site staffing would have to average approximately 23 workers throughout the year.

Figure 7.1

*Physical Plant Expenses*: Whether the office, warehouse and yard are rented or owned, roughly the same costs are incurred. Telephone and utility costs will vary depending on the size of the building and the type of business. Office equipment includes items such as the rental of a copy machine and typewriters.

*Professional Services*: Accountant fees are primarily for quarterly audits. Legal fees go towards collecting and contract disputes. In addition, a prudent contractor will have every contract read by his lawyer prior to signing.

*Miscellaneous*: There are many expenses that could be placed in this category. Included in the example are just a few of the possibilities. Advertising includes the Yellow Pages, promotional materials, etc.

*Uncollected Receivables*: This amount can vary greatly, and is often affected by the overall economic climate. Depending upon the timing of "uncollectables", cash flow can be severely restricted and can cause serious financial problems, even for large companies. Sound cash planning and anticipation of such possibilities can help to prevent severe repercussions.

While the office example used here is feasible within the industry, keep in mind that it is hypothetical and that conditions and costs vary widely from company to company.

In order for this example company to stay in business without losses (profit is not yet a factor), not only must all direct construction costs be paid, but an additional $231,980 must be recovered during the year (as a percentage of volume) in order to operate the office. The percentage may be calculated and applied in two ways:

1. Office overhead applied as a percentage of labor costs only. This method requires that labor and material costs be estimated separately.
2. Office overhead applied as a percentage of total project costs. This is appropriate where material and labor costs are not estimated separately.

Remember that the anticipated volume is $1,500,000 for the year, 70% of which is expected to be labor. Office overhead costs, therefore, will be approximately 22.1% of the labor cost, or 15.5% of annual volume for this example. The most common method for recovering these costs is to apply this percentage to each job over the course of the year.

The estimator must also remember that, if volume changes significantly, then the percentage for office overhead should be recalculated for current conditions. The same is true if there are changes in office staff. Salaries are the major portion of office overhead costs. It should be noted that a percentage is commonly applied to material costs, for handling, regardless of the method of recovering office overhead costs. This percentage is more easily calculated if material costs are estimated and listed separately.

## Profit

Determining a fair and reasonable percentage to be included for profit is not an easy task. This responsibility is usually left to the owner or chief estimator. Experience is crucial in anticipating what profit the market will bear. The economic climate, competition, knowledge of the project, and familiarity with the architect, engineer, or owner all affect the way in which profit is determined. Chapter 10 includes a method to mathematically determine the profit margin based on historical bidding information. As with all facets of estimating, experience is the key to success.

## Contingencies

Like profit, contingencies can also be difficult to quantify. Especially appropriate in preliminary budgets, the addition of a contingency is meant to protect the contractor as well as to give the owner a realistic estimate of potential project costs.

A contingency percentage should be based on the number of "unknowns" in a project, or the level of risk involved. This percentage should be inversely proportional to the amount of planning detail that has been done for the project. If complete plans and specifications are supplied, and the estimate is thorough and precise, then there is little need for a contingency. Figure 7.2, from *Means Electrical Cost Data*, lists suggested contingency percentages that may be added to an estimate based on the stage of planning and development.

If an estimate is priced and each individual item is rounded up, or "padded", this is, in essence, adding a contingency. This method can cause problems, however, because the estimator can never be quite sure of what is the actual cost and what is the "padding", or safety margin, for each item. At the summary, the estimator cannot determine exactly how much has been included as a contingency factor for the project as a whole. A much more accurate and controllable approach is to price the estimate precisely and then add one contingency amount at the bottom line.

| 1.1 Overhead | CREW | MAN-HOURS | UNIT | BARE COSTS | | | TOTAL INCL O&P |
|---|---|---|---|---|---|---|---|
| | | | | MAT. | INST. | TOTAL | |
| 03-001 **BOND PERFORMANCE** see 1.1-34 | | | | | | | |
| 04-001 **CLEANING UP** After job completion, allow | | | Job Cost | | | | .30% |
| 003     Rubbish removal | | | | | | | |
| 005     Cleanup of floor area, continuous, per day | A-5 | 1.500 | M.S.F. | 1.49 | 25 | 26.49 | 38 |
| 010     Final | " | 1.570 | " | 1.58 | 26 | 27.58 | 39 |
| 06-001 **CONSTRUCTION COST INDEX** for 162 major U.S. and | | | | | | | |
| 002     Canadian cities, total cost, min. (Greensboro NC) | | | % | | | | 81 |
| 005 C13.1 -100     Average | | | | | | | 100% |
| 010     Maximum (Anchorage, AK) | | | ↓ | | | | 134.10 |
| 11-001 **CONTINGENCIES** Allowance to add at conceptual stage | | | Project | | | | 15% |
| 005     Schematic stage | | | | | | | 10% |
| 010     Preliminary working drawing stage | | | | | | | 7% |
| 015     Final working drawing stage | | | ↓ | | | | 2% |
| 12-001 C10.3 -300 **CONTRACTOR EQUIPMENT** See division 1.5 | | | | | | | |
| 14-001 **CREWS** For building construction, see foreword p.viii to x | | | | | | | |
| 010 | | | | | | | |
| 15-001 **ENGINEERING FEES** Educational planning consultant, minimum | | | Project | | | | .50% |
| 010     Maximum | | | " | | | | 2.50% |
| 020     Electrical, minimum | | | Contract | | | | 4.10% |
| 030     Maximum | | | | | | | 10.10% |
| 040     Elevator & conveying systems, minimum | | | | | | | 2.50% |
| 050     Maximum | | | | | | | 5% |
| 060     Food service & kitchen equipment, minimum | | | | | | | 8% |
| 070     Maximum | | | | | | | 12% |
| 100     Mechanical (plumbing & HVAC), minimum | | | | | | | 4.10% |
| 110     Maximum | | | ↓ | | | | 10.10% |
| 16-001 C13.1 -100 **HISTORICAL COST INDEXES** Back to 1942 | | | | | | | |
| 18-001 C10.1 -300 **INSURANCE** Builders risk, standard, minimum | | | Job Cost | | | | .10% |
| 005     Maximum | | | | | | | .50% |
| 020     All-risk type, minimum | | | | | | | .12% |
| 025     Maximum | | | ↓ | | | | .68% |
| 040     Contractor's equipment floater, minimum | | | Value | | | | .50% |
| 045     Maximum | | | " | | | | 2.50% |
| 060     Public liability, average | | | Job Cost | | | | .82% |
| 061 | | | | | | | |
| 080 C10.2 -200     Workers' compensation & employer's liability, average | | | | | | | |
| 085     by trade, carpentry, general | | | Payroll | | 10.10% | | |
| 100     Electrical | | | | | 4% | | |
| 115     Insulation | | | | | 7.35% | | |
| 145     Plumbing | | | | | 4.80% | | |
| 155     Sheet metal work (HVAC) | | | ↓ | | 6.30% | | |
| 19-001 **JOB CONDITIONS** Modifications to total | | | | | | | |
| 002     project cost summaries | | | | | | | |
| 010     Economic conditions, favorable, deduct | | | Project | | | | 2% |
| 020     Unfavorable, add | | | | | | | 5% |
| 030     Hoisting conditions, favorable, deduct | | | | | | | 1% |
| 040     Unfavorable, add | | | | | | | 5% |
| 070     Labor availability, surplus, deduct | | | | | | | 1% |
| 080     Shortage, add | | | | | | | 10% |
| 090     Material storage area, available, deduct | | | | | | | 1% |
| 100     Not available, add | | | | | | | 2% |
| 110     Subcontractor availability, surplus, deduct | | | | | | | 5% |
| 120     Shortage, add | | | | | | | 12% |
| 130     Work space, available, deduct | | | | | | | 1% |
| 140     Not available, add | | | ↓ | | | | 4% |
| 20-001 **LABOR INDEX** For 162 major U.S. and Canadian cities | | | | | | | |
| 002     Minimum (Roanoke, VA) | | | % | | 66.90% | | |

Figure 7.2

# Chapter 8
# THE ESTIMATE SUMMARY

At the pricing stage of the estimate, there is typically a large amount of paperwork that must be assembled, analyzed and tabulated. Generally, the information contained in this paperwork is covered by the following major categories:

- Quantity takeoff sheets for all electrical work items (Figure 8.1)
- Material supplier's written quotations (see note below)
- Material supplier's telephone quotations (Figure 8.2)
- Subcontractor's written quotations
- Equipment supplier's quotations
- Cost Analysis or Consolidated Estimate Sheets (Pricing Sheets) (Figures 8.3 and 8.4)
- Recap Summary Sheets

Note: Additional forms, such as Request for Quote forms, are often prepared by the individual electrical contractor.

In the "real world" of estimating, many quotations, especially for bulk materials and for subcontracts, are not received until the last minute before the bidding deadline. Therefore, a system is needed to efficiently handle the paperwork and to ensure that everything will get transferred once (and only once) from the quantity takeoff to the cost analysis sheets. Some general rules for this process are as follows:

- Write on only one side of any document, where possible.
- Code each sheet with a large division number in a consistent place, preferably near one of the upper corners.
- Use Telephone Quotation forms for uniformity in recording prices received by phone.
- Document the source of every quantity and price.
- Keep each type of document in its "pile" (Quantities, Material, Subcontractors, Equipment) filed in order by division number.
- Keep the entire estimate in one or more compartmentalized folders.
- When an item is transferred to the cost analysis sheet, check it off.
- Most importantly, number and code each takeoff sheet and each pricing extension sheet as it is created. At the same time, keep an index list of each sheet by number. If a sheet is to be abandoned, write "VOID" on it, but do not discard it. Keep it until the bid is accepted to be able to account for all pages and sheets.

Note: A helpful technique for organizing these sheets and forms involves the use of pastel colors to code each category of cost sheets. This system makes locating and revising sheets both easier and quicker.

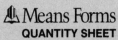

**QUANTITY SHEET**

| | | SHEET NO. |
|---|---|---|
| PROJECT | | ESTIMATE NO. |
| LOCATION | ARCHITECT | DATE |
| TAKE OFF BY | EXTENSIONS BY: | CHECKED BY: |

| DESCRIPTION | NO. | DIMENSIONS | | | | UNIT | | UNIT | | UNIT | | UNIT |
|---|---|---|---|---|---|---|---|---|---|---|---|---|
| | | | | | | | | | | | | |

Figure 8.1

48

 Means Forms

**TELEPHONE**
**QUOTATION**

| | | |
|---|---|---|
| | DATE | |
| PROJECT | TIME | |
| FIRM QUOTING | PHONE ( ) | |
| ADDRESS | BY | |
| ITEM QUOTED | RECEIVED BY | |

| WORK INCLUDED | AMOUNT OF QUOTATION |
|---|---|
| | |
| | |
| | |
| | |
| | |
| | |
| | |
| | |
| | |
| | |
| | |
| | |
| | |

| | | | | |
|---|---|---|---|---|
| DELIVERY TIME | | | **TOTAL BID** | |
| DOES QUOTATION INCLUDE THE FOLLOWING: | | | If ☐ NO is checked, determine the following: | |
| STATE & LOCAL SALES TAXES | ☐ YES | ☐ NO | MATERIAL VALUE | |
| DELIVERY TO THE JOB SITE | ☐ YES | ☐ NO | WEIGHT | |
| COMPLETE INSTALLATION | ☐ YES | ☐ NO | QUANTITY | |
| COMPLETE SECTION AS PER PLANS & SPECIFICATIONS | ☐ YES | ☐ NO | DESCRIBE BELOW | |

| EXCLUSIONS AND QUALIFICATIONS | |
|---|---|
| | |
| | |
| | |
| | |
| | |
| | |
| | |
| | |

| | | |
|---|---|---|
| ADDENDA ACKNOWLEDGEMENT | **TOTAL ADJUSTMENTS** | |
| | **ADJUSTED TOTAL BID** | |

| ALTERNATES | |
|---|---|
| ALTERNATE NO. | |
| ALTERNATE NO. | |
| ALTERNATE NO. | |
| ALTERNATE NO. | |
| ALTERNATE NO. | |
| ALTERNATE NO. | |
| ALTERNATE NO. | |

*Figure 8.2*

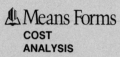

**COST**
**ANALYSIS**

SHEET NO.

PROJECT

ESTIMATE NO.

ARCHITECT

DATE

| TAKE OFF BY: | QUANTITIES BY: | | PRICES BY: | | EXTENSIONS BY: | | CHECKED BY: | |
|---|---|---|---|---|---|---|---|---|

| DESCRIPTION | SOURCE/DIMENSIONS | | | QUANTITY | UNIT | MATERIAL | | LABOR | | EQ./TOTAL | |
|---|---|---|---|---|---|---|---|---|---|---|---|
| | | | | | | UNIT COST | TOTAL | UNIT COST | TOTAL | UNIT COST | TOTAL |

Figure 8.3

**Means Forms**
**CONSOLIDATED ESTIMATE**

| | | |
|---|---|---|
| PROJECT | CLASSIFICATION | SHEET NO. |
| LOCATION | ARCHITECT | ESTIMATE NO. |
| TAKE OFF BY | QUANTITIES BY | PRICES BY | EXTENSIONS BY | DATE |
| | | | | CHECKED BY |

| DESCRIPTION | NO. | DIMENSIONS | QUANTITIES | | EXTENSIONS | | | | | | | | | | | |
|---|---|---|---|---|---|---|---|---|---|---|---|---|---|---|---|---|
| | | | | UNIT | MATERIAL | | LABOR | | EQUIPMENT | | TOTAL | | | | | |
| | | | | | UNIT COST | TOTAL | UNIT COST | TOTAL | UNIT COST | TOTAL | UNIT COST | TOTAL | | | | |

Figure 8.4

51

All subcontract costs should be properly noted and listed separately. These costs contain the subcontractor's markups and may be treated differently from other direct costs when the estimator calculates the general contractor's overhead, profit and contingency allowance.

After all the unit prices, subcontractor prices, and allowances have been entered on the pricing sheets, the costs are extended. In making the extensions, ignore the cents column and round all totals to the nearest dollar. In a column of figures, the cents will average out and will not be of consequence. Finally, each subdivision is added and the results checked, preferably by someone other than the person doing the extensions.

It is important to check the larger items for order of magnitude errors. If the total subdivision costs are divided by the building area, the resulting square foot cost figures can be used to quickly pinpoint areas that are out of line with expected square foot costs. These cost figures should be recorded for comparison to past projects and as a resource for future estimating.

The takeoff and pricing method, as discussed, has been to utilize a Quantity Sheet for the material takeoff (see Figure 8.1), and to transfer the data to an analysis form for pricing the material, labor, and subcontractor items (see Figure 8.3).

An alternative to this method is a consolidation of the takeoff task and pricing on a single form. This approach works well for smaller bids and for change orders. An example, the Consolidated Estimate Form, is shown in Figure 8.4. The same sequences and recommendations used to complete the Quantity Sheet and Cost Analysis form are to be followed when using the Consolidated Estimate form to price the estimate.

When the pricing of all direct costs is complete, the estimator has two choices:   1) to make all further price changes and adjustments on the Cost Analysis or Consolidated Estimate sheets, *or* 2) to transfer the total costs for each subdivision to an Estimate Summary sheet so that all further price changes, until bid time, will be done on one sheet. Any indirect markups and burdens will be figured on this sheet.

Unless the estimate has a limited number of items, it is recommended that costs be transferred to an Estimate Summary sheet. This step should be double-checked since an error of transposition may easily occur. Pre-printed forms can be useful, although a plain columnar form may suffice. This summary with page numbers from each extension sheet can also serve as an index.

A company that repeatedly uses certain standard listings can save valuable time by having a custom Estimate Summary sheet printed with these items listed. The printed UCI division and subdivision headings serve as another type of checklist, ensuring that all required costs are included. Appropriate column headings or categories for any estimate summary form could be as follows:

1. Material
2. Labor
3. Equipment
4. Subcontractor
5. Total

As items are listed in the proper columns, each category is added and appropriate markups applied to the total dollar values. Different percentages may be added to the sum of each column at the estimate summary. These percentages may include the following items, as discussed in Chapter 7:

1. Taxes and Insurance
2. Overhead
3. Profit
4. Contingencies

# Chapter 9
# PRE-BID SCHEDULING

The need for planning and scheduling is clear once the contract is signed and work commences on the project. However, some scheduling is also important during the bidding stage for the following reasons:

1. To determine if the project can be completed in the allotted or specified time using normal crew sizes
2. To identify potential overtime requirements
3. To determine the time requirements for supervision
4. To anticipate possible temporary heat and power requirements
5. To price certain general requirement items and overhead costs
6. To budget for equipment usage
7. To anticipate and justify material and equipment delivery requirements. (An awareness of these requirements may, for example, justify using a more expensive vendor quote with better delivery terms.)

The schedule produced prior to bidding may be a simple bar chart or network diagram that includes overall quantities, probable delivery times and available manpower. Network scheduling methods, such as the Critical Path Method (CPM) and the Precedence Chart simplify pre-bid scheduling because they do not require time-scaled line diagrams.

In the CPM Diagram, the activity is represented by an arrow. Nodes indicate start/stop between activities. The Precedence Diagram, on the other hand, shows the activity as a node with arrows used to denote precedence relationships between the activities. The precedence arrows may be used in different configurations to represent the sequential relationships between activities. Examples of CPM and Precedence diagrams are shown in Figures 9.1 and 9.2, respectively. In both systems, duration times are indicated along each path. The sequence (path) of activities requiring the most total time represents the shortest possible time (critical path) in which those activities may be completed.

For example, in both Figure 9.1 and Figure 9.2, activities A, B, and C require 20 successive days for completion before activity G can begin. Activity paths for D and E (15 days), and for F (12 days) are shorter and can easily be completed during the 20-day sequence. Therefore, this 20-day sequence is the shortest possible time (i.e., the "critical path") for the completion of these activities — before activity G can begin.

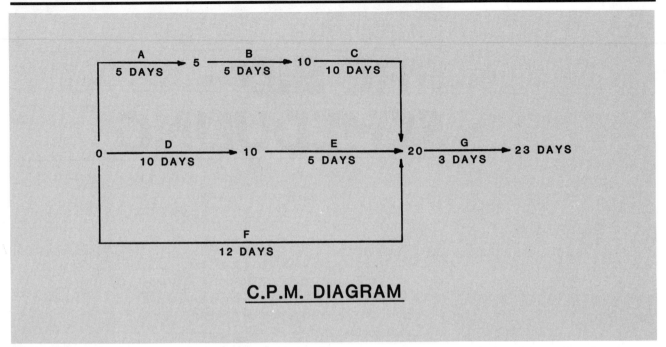

## C.P.M. DIAGRAM

Figure 9.1

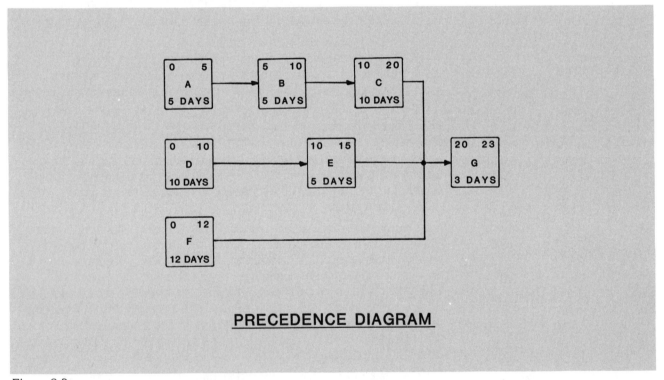

## PRECEDENCE DIAGRAM

Figure 9.2

56

Past experience or a prepared rough schedule may suggest that the time specified in the bidding documents is insufficient to complete the required work. In such cases, a more comprehensive schedule should be produced prior to bidding; this schedule will help to determine the added overtime or premium time work costs required to meet the completion date.

A three-story office building project is used for the Sample Estimate in Part Three of this book. A preliminary schedule is needed to determine the supervision and manning requirements of the job. The specifications state that the building must be completed within one year. Normally, the excavation, foundations and superstructure can be completed in approximately one half of the construction period. Therefore, the major portion of the electrical work (assuming exposed conduit and undercarpet power and phone systems) is restricted to the last six months of the one year allotted. The service and distribution equipment installation might be restrained by the delivery of the equipment.

A rough schedule for the electrical work might be produced as shown in Figure 9.3. The man-days used to develop this schedule are derived from the man-hour figures in the estimate. Man-hours can be determined based on the figures in *Means Electrical Cost Data*. Man-days can also be figured by dividing the total labor cost shown on the estimate by the cost per man-day for an electrician.

As shown, the preliminary schedule can be used to determine supervision requirements, to develop appropriate crew sizes, and as a basis for ordering materials. All of these factors must be considered at this preliminary stage in order to determine how to meet the required one-year completion date.

A pre-bid schedule can provide much more information than simple job duration. It can be used to refine the estimate by introducing realistic manpower projections. The schedule may also help the contractor to adjust the structure and size of the company based on projected requirements for months, even years ahead. A schedule can also become an effective tool for negotiating contracts.

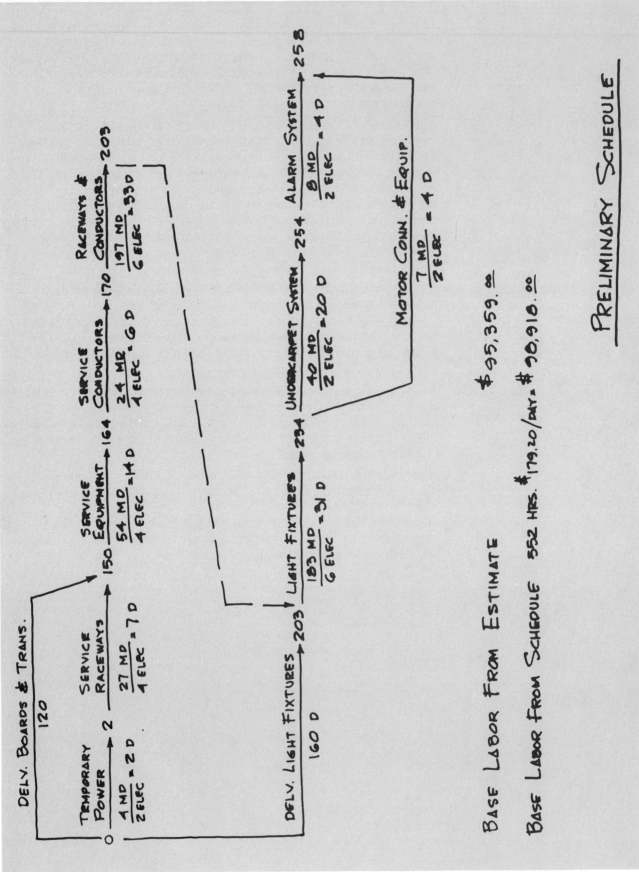

Figure 9.3

58

# Chapter 10
# BIDDING STRATEGIES

The goal of most contractors is to make as much money as possible on each job, but more importantly, to maximize return on investment on an annual basis. Often, this can be done by taking *fewer* jobs at a *higher* profit and by limiting bidding to the jobs which are most likely to be successful for the company.

## Resource Analysis

Since a contractor cannot physically bid every job in a geographic area, a selection process must determine which projects to bid. This process should begin with an analysis of the strengths and weaknesses of the contractor. The following items must be considered as objectively as possible:

- Individual strengths of the company's top management
- Management experience with the type of construction involved, from top management to project superintendents
- Cost records adequate for the appropriate type of construction
- Bonding capability and capacity
- Size of projects with which the company is "comfortable"
- Geographic area that can be managed effectively
- Unusual corporate assets such as:
    - Specialized equipment availability
    - Reliable and timely cost control systems
    - Strong balance sheet

## Market Analysis

Most contractors tend to concentrate on a few particular kinds of projects. From time to time, the company should step back and examine the portion of the industry they are serving. During this process, the following items should be carefully analyzed:

- Historical trend of the market segment
- Expected future trend of the market segment
- Geographic expectations of the market segment
- Historical and expected competition from other contractors
- Risk involved in the particular market segment
- Typical size of projects in this market
- Expected return on investment from the market segment

If several of these areas are experiencing a downturn, then it is definitely appropriate to examine an alternate market. On the other hand, many managers may feel that "bad times" are the times when they should consolidate and narrow their market. These managers are only likely to expand or broaden their market when current activities are especially strong.

## Bidding Analysis

Certain steps should be taken to develop a bid strategy within a particular market. The first is to obtain the bid results of jobs in the prospective geographic area. These results should be set up on a tabular basis. This is fairly easy to do in the case of public jobs since the bid results are normally published (or at least available) from the agency responsible for the project. In private work, this step is more difficult, since the bid results are not normally divulged by the owner.

### Determining Risk in a New Market Area

One way to measure success in bidding is how much money is "left on the table", the difference between the low bid and next lowest bid. The contractor who consistently takes jobs by a wide margin below the next bidder is clearly not making as much money as possible. Information on competitive public bidding is used to determine the amount of money left on the table; this information serves as the basis for fine-tuning a future bidding strategy.

For example, assume a public market where all bid prices and the total number of bidders are known. For each "type" of market sector, create a chart showing the percentage left on the table versus the total number of bidders. When the median figure (percent left on the table) for each number of bidders is connected with a smooth curve, the usual shape of the curve is shown in Figure 10.1.

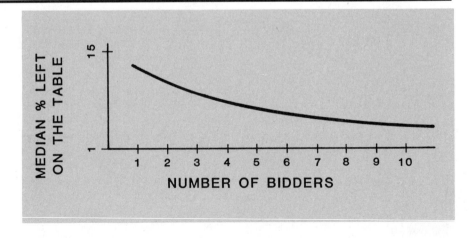

Figure 10.1

The exact shape and magnitude of the amounts left on the table will depend on how much risk is involved with that type of work. If the percentages left on the table are high, then the work can be assumed to be very risky — with a high profit or loss potential if the award is won. If the percentages are low, the work is probably neither as risky nor as potentially profitable.

## Analyzing the Bid's Risk

If a company has been bidding in a particular market, certain information should be collected and recorded as a basis for a bidding analysis. First, the percentage left on the table should be tabulated (as shown in Figure 10.1), along with the number of bidders for the projects in that market on which the company was the low bidder. By probability, half the bids should be above the median line and half below. If more than half are below the line, the company is doing well; if more than half are above, the bidding strategy should be examined.

## Maximizing the Profit-to-Volume Ratio

Once the bidding track record for the company has been established, the next step is to reduce the historical percentage left on the table. One method is to create a chart showing, for instance, the last ten jobs on which the company was low bidder, and the dollar spread between the low and second lowest bid. Next, rank the percentage differences from one to ten (one being the smallest and ten being the largest left on the table). An example is shown in Figure 10.2. This example is for a larger general contractor, but the principles apply to any size or type of contractor involved in bidding.

| Job No. | "Cost" | Low Bid | Second Bid | Difference | % Diff | % Rank | Profit (Assumed at 10%) |
|---|---|---|---|---|---|---|---|
| 1 | $ 918,000 | $ 1,009,800 | $1,095,000 | $ 85,200 | 9.28 | 10 | $ 91,800 |
| 2 | 1,955,000 | 2,150,500 | 2,238,000 | 87,500 | 4.48 | 3 | 195,500 |
| 3 | 2,141,000 | 2,355,100 | 2,493,000 | 137,900 | 6.44 | 6 | 214,100 |
| 4 | 1,005,000 | 1,105,500 | 1,118,000 | 12,500 | 1.24 | 1 | 100,500 |
| 5 | 2,391,000 | 2,630,100 | 2,805,000 | 174,900 | 7.31 | 8 | 239,100 |
| 6 | 2,782,000 | 3,060,200 | 3,188,000 | 127,800 | 4.59 | 4 | 278,200 |
| 7 | 1,093,000 | 1,202,300 | 1,282,000 | 79,700 | 7.29 | 7 | 109,300 |
| 8 | 832,000 | 915,200 | 926,000 | 10,800 | 1.30 | 2 | 83,200 |
| 9 | 2,372,000 | 2,609,200 | 2,745,000 | 135,800 | 5.73 | 5 | 237,200 |
| 10 | 1,681,000 | 1,849,100 | 2,005,000 | 155,900 | 9.27 | 9 | 168,100 |
| | $17,170,000 | $18,887,000 | | $1,008,000 | | | $1,717,000 = 10% of Cost |

*Figure* 10.2

The company's "costs" ($17,170,000) are derived from the company's low bids ($18,887,000) assuming a 10% profit ($1,717,000). The "second bid" is the next lowest bid. The "difference" is the dollar amount between the low bid and the second bid (money "left on the table"). The differences are then ranked based on the percentage of job "costs" left on the table for each. Figure the median difference by averaging the two middle percentages — that is, the fifth and sixth ranked numbers.

$$\text{Median \% Difference} = \frac{5.73 + 6.44}{2} = 6.09\%$$

From Figure 10.2, the median percentage left on the table is 6.09%. To maximize the potential returns on a series of competitive bids, a useful formula is needed for pricing profit. The following formula has proven effective.

$$\text{Normal Profit\%} + \frac{\text{Median \% Difference}}{2} = \text{Adjusted Profit \%}$$

$$10.00 + \frac{6.09}{2} = 13.05\%$$

Now apply this adjusted profit percentage to the same list of ten jobs as shown in Figure 10.3. Note that the job "costs" remain the same, but that the low bids have been revised. Compare the bottom line results of Figure 10.2 to those of Figure 10.3 based on the two profit margins, 10% and 13.05%, respectively.

Total volume *drops* from $18,887,000 to $17,333,900.

Net profits *rise* from $1,717,000 to $2,000,900.

| Job No. | Company's "Cost" | Revised Low Bid | Second Bid | Adj. Diff. | Profit [10% + 3.05%] | | Total |
|---|---|---|---|---|---|---|---|
| 1 | $ 918,000 | $ 1,037,800 | $1,095,000 | $ 57,200 | $ 91,800 + | $28,000 | $ 119,800 |
| 2 | 1,955,000 | 2,210,100 | 2,238,000 | $ 27,900 | $195,500 + | 59,600 | 255,100 |
| 3 | 2,141,000 | 2,420,400 | 2,493,000 | $ 72,600 | 214,100 + | 65,300 | 279,400 |
| 4 | (1,005,000) | (1,136,100) | 1,118,000 (L) | — | 100,500 + | 30,600 | 0 |
| 5 | 2,391,000 | 2,703,000 | 2,805,000 | 102,000 | 239,100 + | 72,900 | 312,000 |
| 6 | 2,782,000 | 3,145,100 | 3,188,000 | 42,900 | 278,200 + | 84,900 | 363,100 |
| 7 | 1,093,000 | 1,235,600 | 1,282,000 | 46,400 | 109,300 + | 33,300 | 142,600 |
| 8 | (832,000) | (940,600) | 926,000 (L) | — | 83,200 + | 25,400 | 0 |
| 9 | 2,372,000 | 2,681,500 | 2,745,000 | 63,500 | 237,200 + | 72,300 | 309,500 |
| 10 | 1,681,000 | 1,900,400 | 2,005,000 | 104,600 | 168,100 + | 51,300 | 219,400 |
| | $15,333,000 | $17,333,900 | | $517,100 | | | $2,000,900 |

*Figure* 10.3

Profits rise while volume drops! If the original volume is maintained or even increased, profits rise even further. Note how this occurs. By determining a reasonable increase in profit margin, the company has, in effect, raised all bids. By doing so, the company loses two jobs to the second bidder (jobs 4 and 8 in Figure 10.3).

A positive effect of this volume loss is reduced exposure to risk. Since the profit margin is higher, the remaining eight jobs collectively produce more profit than the ten jobs based on the original, lower profit margin. From where did this money come? The money "left on the table" has been reduced from $1,008,000 to $517,100. The whole purpose is to systematically lessen the dollar amount difference between the low bids and the second low bids. Caution: This is a hypothetical approach based upon the following assumptions.

- Profit must be assumed to be the bid or estimated profit, not the actual profit when the job is over.
- Bidding must be done with the same market in which data for the analysis was gathered.
- Economic conditions should be stable from the time the data is gathered until the analysis is used in bidding. If conditions change, use of such an analysis should be reviewed.
- Each contractor must make roughly the same number of bidding mistakes. For higher numbers of jobs in the sample, this requirement becomes more probable.
- The company must bid additional jobs if total annual volume is to be maintained or increased. Likewise, if net total profit margin is to remain constant, even fewer jobs need be won.
- Finally, the basic cost numbers and sources must remain constant. If a new estimator is hired or a better cost source is found, this technique cannot work effectively until a new track record has been established.

The accuracy of this strategy depends upon the criteria listed above. Nevertheless, it is a valid concept that can be applied, with appropriate and reasonable judgement, to many bidding situations.

# Chapter 11
# COST CONTROL AND ANALYSIS

An internal accounting system should be used by contractors and construction managers to logically gather and track the costs of a construction project. With this information, a cost analysis can be made about each activity — both during the installation process and at its conclusion. This information or "feedback" becomes the basis for management decisions through the duration of the project. This cost data is also helpful for future bid proposals.

The categories for an electrical project are major items of construction (e.g., lighting) which can be subdivided into component activities (e.g., wire, conduit, fixtures, etc.). These activities should coincide with the system and methods of the quantity takeoff. Uniformity is important in terms of the units of measure and in the grouping of components into cost centers. For instance, cost centers for an electrical project might include categories such as lighting, switchgear, branch power, MCC's and motors, and alarm systems. Activities within one of these categories — branch power, for example, might include such items as receptacles, boxes, conduit, and wire.

The major purposes of cost control and analysis are as follows:
- To provide management with a system to monitor costs and progress
- To provide cost feedback to the estimator(s)
- To determine the costs of change orders
- To be used as a basis for progress payment requisitions to the owner, the general contractor, or his representative
- To manage cash flow
- To identify areas of potential cost overruns to management for corrective action

It is important to establish a cost control system that is uniform — both throughout the company and from job to job. Such a system might begin with a uniform Chart of Accounts, a listing of code numbers for work activities. The Chart of Accounts is used to assign time and cost against work activities for the purpose of creating cost reports. A Chart of Accounts should have enough scope and detail so that it can be used for any of the projects that the company may win. Naturally, an effective Chart of Accounts will also be flexible enough to incorporate new activities as the company takes on new or different kinds of projects. Using a cost control system, the various costs can be consistently allocated. The following information should be recorded for each cost component:

- Labor charges in dollars and man-hours are summarized from weekly time cards and distributed by code.
- Quantities completed to date must also be recorded in order to determine unit costs.
- Equipment rental costs are derived from purchase orders or from weekly charges issued by an equipment company.
- Material charges are determined from purchase orders.
- Appropriate subcontractor charges are allocated.
- Job overhead items may be listed separately or by component.

Each component of costs — labor, materials and equipment — is now calculated on a unit basis by dividing the quantity installed to date into the cost to date. This procedure establishes the actual installed unit costs to date.

At this point, it is also useful to calculate the percent complete to date. This is done in two steps, or levels. First, the percent complete for each activity is calculated based on the actual quantity installed to date divided by the total quantity estimated for the activity. Second, the project total percent complete is calculated. To arrive at this number, multiply the percent complete of each activity times the total estimated cost for that activity. Then sum these results to a total (sometimes called "earned" dollars) and divide this sum by the estimated total. The result is the project's overall percent complete.

The quantities that remain to be installed for each activity should be estimated based on the actual unit cost to date. This is the projected cost to complete each activity. Due to inefficiencies and mobilization costs as the work begins, it is not practical to use the actual units for projecting costs until an activity is 20% complete. Below 20%, the costs as originally estimated should be used. The actual costs to date are added to the projected costs to obtain the anticipated costs at the end of the project.

Typical forms that may be used to develop a cost control system are shown in Figures 11.1 to 11.6.

The analysis of categories serves as a useful management tool, providing information on a constant, up-to-date basis. Immediate attention is attracted to any center that is projecting a loss. Management can concentrate on this item in an attempt to make it profitable or to minimize the expected loss.

The estimating department can use the unit costs developed in the field as background information for future bidding purposes. Particularly useful are unit labor costs and unit man-hours (productivity) for the separate activities. This information should be integrated into the accumulated historical data. A particular advantage of unit man-hour records is that they tend to be constant over time. Current unit labor costs can be figured simply by multiplying these man-hour standards times the current labor rate per hour.

# Means Forms

**DAILY TIME SHEET**

PROJECT _____

FOREMAN _____

WEATHER CONDITIONS _____

TEMPERATURE _____

DATE _____

SHEET NO. _____

DESCRIPTION OF WORK

| NO. | NAME | | TOTALS HOURS REG-ULAR | OVER-TIME | COST INFORMATION RATE | TOTAL | UNIT COST |
|-----|------|--|------|------|------|-------|------|
| | | HOURS | | | | | |
| | | UNITS | | | | | |
| | | HOURS | | | | | |
| | | UNITS | | | | | |
| | | HOURS | | | | | |
| | | UNITS | | | | | |
| | | HOURS | | | | | |
| | | UNITS | | | | | |
| | | HOURS | | | | | |
| | | UNITS | | | | | |
| | | HOURS | | | | | |
| | | UNITS | | | | | |
| | | HOURS | | | | | |
| | | UNITS | | | | | |
| | | HOURS | | | | | |
| | | UNITS | | | | | |
| | | HOURS | | | | | |
| | | UNITS | | | | | |
| | | HOURS | | | | | |
| | | UNITS | | | | | |
| | | HOURS | | | | | |
| | | UNITS | | | | | |
| | | HOURS | | | | | |
| | | UNITS | | | | | |
| | | HOURS | | | | | |
| | | UNITS | | | | | |
| | | HOURS | | | | | |
| | | UNITS | | | | | |
| | TOTALS | HOURS | | | | | |
| | | UNITS | | | | | |
| | EQUIPMENT | | | | | | |
| | | | | | | | |
| | | | | | | | |
| | | | | | | | |
| | | | | | | | |
| | | | | | | | |

*Figure* 11.1

**Means Forms**

**WEEKLY TIME SHEET**

| EMPLOYEE | | EMPLOYEE NUMBER | | WEEK ENDING | |
|---|---|---|---|---|---|
| FOREMAN | | HOURLY RATE | | PIECE WORK RATE | |

| PROJECT | NO. | DESCRIPTION OF WORK | DAY | | | | | | TOTAL | HOURS | RATE | TOTAL COST | UNIT COST |
|---|---|---|---|---|---|---|---|---|---|---|---|---|---|
| | | | HOURS | | | | | | | REG. | | | |
| | | | UNITS | | | | | | | O.T. | | | |
| | | | HOURS | | | | | | | REG. | | | |
| | | | UNITS | | | | | | | O.T. | | | |
| | | | HOURS | | | | | | | REG. | | | |
| | | | UNITS | | | | | | | O.T. | | | |
| | | | HOURS | | | | | | | REG. | | | |
| | | | UNITS | | | | | | | O.T. | | | |
| | | | HOURS | | | | | | | REG. | | | |
| | | | UNITS | | | | | | | O.T. | | | |
| | | | HOURS | | | | | | | REG. | | | |
| | | | UNITS | | | | | | | O.T. | | | |
| | | | HOURS | | | | | | | REG. | | | |
| | | | UNITS | | | | | | | O.T. | | | |
| | | | HOURS | | | | | | | REG. | | | |
| | | | UNITS | | | | | | | O.T. | | | |
| | | TOTAL | HOURS | | | | | | | REG. | | | |
| | | | UNITS | | | | | | | O.T. | | | |

| PAYROLL DEDUCTIONS | F.I.C.A. | INCOME TAXES | | | HOSPITAL-IZATION | | | TOTAL DEDUCTIONS | | LESS DEDUCTIONS |
|---|---|---|---|---|---|---|---|---|---|---|
| | | FEDERAL | STATE | CITY/COUNTY | | | | | | |
| | $ | $ | $ | $ | $ | $ | $ | $ | | NET PAY |

Figure 11.2

## Means Forms

**LABOR
COST RECORD**

SHEET NO.

DATE FROM:

PROJECT

DATE TO:

LOCATION

BY:

| DATE | CHARGE NO. | DESCRIPTION | HOURS | RATE | AMOUNT | HOURS | RATE | AMOUNT | HOURS | RATE | AMOUNT |
|------|-----------|-------------|-------|------|--------|-------|------|--------|-------|------|--------|
|      |           |             |       |      |        |       |      |        |       |      |        |
|      |           |             |       |      |        |       |      |        |       |      |        |
|      |           |             |       |      |        |       |      |        |       |      |        |
|      |           |             |       |      |        |       |      |        |       |      |        |
|      |           |             |       |      |        |       |      |        |       |      |        |
|      |           |             |       |      |        |       |      |        |       |      |        |
|      |           |             |       |      |        |       |      |        |       |      |        |
|      |           |             |       |      |        |       |      |        |       |      |        |
|      |           |             |       |      |        |       |      |        |       |      |        |
|      |           |             |       |      |        |       |      |        |       |      |        |

*Figure* 11.3

## Means Forms

**MATERIAL COST RECORD**

SHEET NO. _____

DATE FROM _____

PROJECT _____   DATE TO _____

LOCATION _____   BY _____

| DATE | NUMBER | VENDOR/DESCRIPTION | QTY. | UNIT PRICE | | | QTY. | UNIT PRICE | | | QTY. | UNIT PRICE | | |
|------|--------|--------------------|------|-----------|---|---|------|-----------|---|---|------|-----------|---|---|
| | | | | | | | | | | | | | | |
| | | | | | | | | | | | | | | |
| | | | | | | | | | | | | | | |
| | | | | | | | | | | | | | | |
| | | | | | | | | | | | | | | |
| | | | | | | | | | | | | | | |
| | | | | | | | | | | | | | | |
| | | | | | | | | | | | | | | |
| | | | | | | | | | | | | | | |
| | | | | | | | | | | | | | | |
| | | | | | | | | | | | | | | |
| | | | | | | | | | | | | | | |
| | | | | | | | | | | | | | | |
| | | | | | | | | | | | | | | |
| | | | | | | | | | | | | | | |
| | | | | | | | | | | | | | | |
| | | | | | | | | | | | | | | |
| | | | | | | | | | | | | | | |

*Figure 11.4*

70

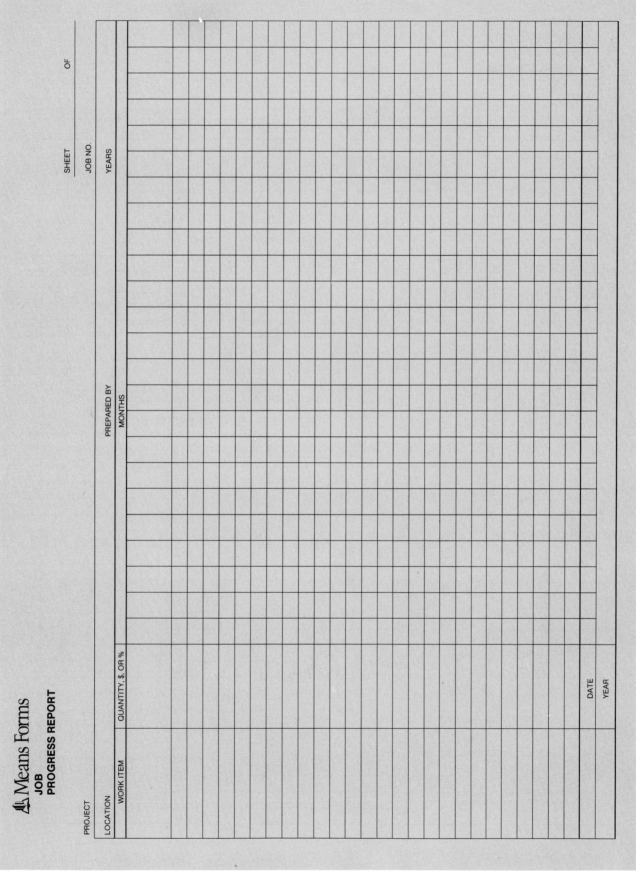

*Figure* 11.5

71

## ⚑ Means Forms
**PERCENTAGE**
**COMPLETE ANALYSIS**

PAGE

PROJECT

DATE

ARCHITECT

BY

FROM

TO

| NO. | DESCRIPTION | ACTUAL OR ESTIMATED | TOTAL PROJECT | THIS PERIOD | | PERCENT TOTAL TO DATE | | | | | | | | | | |
|---|---|---|---|---|---|---|---|---|---|---|---|---|---|---|---|---|
| | | | | QUANTITY | % | QUANTITY | 10 | 20 | 30 | 40 | 50 | 60 | 70 | 80 | 90 | 100 |
| | | ACTUAL | | | | | | | | | | | | | | |
| | | ESTIMATED | | | | | | | | | | | | | | |
| | | ACTUAL | | | | | | | | | | | | | | |
| | | ESTIMATED | | | | | | | | | | | | | | |
| | | ACTUAL | | | | | | | | | | | | | | |
| | | ESTIMATED | | | | | | | | | | | | | | |
| | | ACTUAL | | | | | | | | | | | | | | |
| | | ESTIMATED | | | | | | | | | | | | | | |
| | | ACTUAL | | | | | | | | | | | | | | |
| | | ESTIMATED | | | | | | | | | | | | | | |
| | | ACTUAL | | | | | | | | | | | | | | |
| | | ESTIMATED | | | | | | | | | | | | | | |
| | | ACTUAL | | | | | | | | | | | | | | |
| | | ESTIMATED | | | | | | | | | | | | | | |
| | | ACTUAL | | | | | | | | | | | | | | |
| | | ESTIMATED | | | | | | | | | | | | | | |
| | | ACTUAL | | | | | | | | | | | | | | |
| | | ESTIMATED | | | | | | | | | | | | | | |
| | | ACTUAL | | | | | | | | | | | | | | |
| | | ESTIMATED | | | | | | | | | | | | | | |
| | | ACTUAL | | | | | | | | | | | | | | |
| | | ESTIMATED | | | | | | | | | | | | | | |
| | | ACTUAL | | | | | | | | | | | | | | |
| | | ESTIMATED | | | | | | | | | | | | | | |
| | | ACTUAL | | | | | | | | | | | | | | |
| | | ESTIMATED | | | | | | | | | | | | | | |
| | | ACTUAL | | | | | | | | | | | | | | |
| | | ESTIMATED | | | | | | | | | | | | | | |
| | | ACTUAL | | | | | | | | | | | | | | |
| | | ESTIMATED | | | | | | | | | | | | | | |
| | | ACTUAL | | | | | | | | | | | | | | |
| | | ESTIMATED | | | | | | | | | | | | | | |
| | | ACTUAL | | | | | | | | | | | | | | |
| | | ESTIMATED | | | | | | | | | | | | | | |
| | | ACTUAL | | | | | | | | | | | | | | |
| | | ESTIMATED | | | | | | | | | | | | | | |
| | | ACTUAL | | | | | | | | | | | | | | |
| | | ESTIMATED | | | | | | | | | | | | | | |
| | | ACTUAL | | | | | | | | | | | | | | |
| | | ESTIMATED | | | | | | | | | | | | | | |
| | | ACTUAL | | | | | | | | | | | | | | |
| | | ESTIMATED | | | | | | | | | | | | | | |

*Figure 11.6*

Frequently, items are added to or deleted from the contract either via field change orders or through contract amendments. Accurate cost records are an excellent basis for determining the cost changes that will result.

As discussed above, the determination of completed quantities is necessary in order to calculate unit costs. These quantities are used to figure the percent complete in each category. These percentages can, in turn, be used to calculate the billing for progress payment requisitions (invoices).

A cost system is only as good as the people responsible for coding and recording the required information. Simplicity is the key word. Do not try to break down the code into very small items unless there is a specific need.

Continuous updating of reports is important so that operations which are not in control can be immediately brought to the attention of management.

Also, be sure to draft clear directions and instructions for each phase of the process. Adequate time must be spent to ensure that all who are involved (especially the foremen and supervisors) clearly understand the program.

## Productivity and Efficiency

When using a cost control system such as the one described above, the unit costs should reflect standard practices. Productivity should be based on a five-day, eight-hour-per-day (during daylight hours) work week. Exceptions can be made if a company's requirements are particularly and most often unique. Installation costs should be derived using normal minimum crew sizes, under normal weather conditions, during the normal construction season.

All unusual costs incurred or expected should be recorded separately for each category of work. For example, an overtime situation might occur on every job and in the same proportion. In this case, it would make sense to carry the unit price adjusted for the added cost of premium time. Likewise, unusual weather delays, strike activity, owner/architect delays, or contractor interference should have separate, identifiable cost contributions; these are applied as isolated costs to the activities affected by the delays. This procedure serves two purposes:

- To identify and separate the cost contribution of the delay so that future job estimates will not automatically include an allowance for these "non-typical" delays, and
- To serve as a basis for an extra compensation claim and/or as justification for reasonable extension of the job.

## Overtime Impact

The use of long-term overtime is counter-productive on almost any construction job; that is, the longer the period of overtime, the lower the actual production rate. There have been numerous studies conducted which come up with slightly different numbers, but all reach the same conclusion. Figure 11.7 tabulates the effects of overtime work on efficiency.

As illustrated in Figure 11.7, there can be a difference between the *actual* payroll cost per hour and the *effective* cost per hour for overtime work. This is due to the reduced production efficiency with the increase in weekly hours beyond 40. This difference between actual and effective cost results from overtime work over a prolonged period. Short-term overtime work does not result in as great a reduction in efficiency, and in such cases, effective cost may not vary significantly from the actual payroll cost. As the total hours per week are increased on a regular basis, more time is lost because of fatigue, lowered morale, and an increased accident rate.

As an example, assume a project where workers are working 6 days a week, 10 hours per day. From Figure 11.7 (based on productivity studies), the effective productive hours are 52.5 hours. This represents a theoretical production efficiency of 52.5/60 or 87.5%.

| Days per Week | Hours per Day | Production Efficiency | | | | | Payroll Cost Factors | |
|---|---|---|---|---|---|---|---|---|
| | | 1 Week | 2 Weeks | 3 Weeks | 4 Weeks | Average 4 Weeks | @ 1-1/2 times | @ 2 times |
| 5 | 8 | 100% | 100% | 100% | 100% | 100% | 100% | 100% |
| | 9 | 100 | 100 | 95 | 90 | 96.25 | 105.6 | 111.1 |
| | 10 | 100 | 95 | 90 | 85 | 91.25 | 110.0 | 120.0 |
| | 11 | 95 | 90 | 75 | 65 | 81.25 | 113.6 | 127.3 |
| | 12 | 90 | 85 | 70 | 60 | 76.25 | 116.7 | 133.3 |
| 6 | 8 | 100 | 100 | 95 | 90 | 96.25 | 108.3 | 116.7 |
| | 9 | 100 | 95 | 90 | 85 | 92.50 | 113.0 | 125.9 |
| | 10 | 95 | 90 | 85 | 80 | 87.50 | 116.7 | 133.3 |
| | 11 | 95 | 85 | 70 | 65 | 78.75 | 119.7 | 139.4 |
| | 12 | 90 | 80 | 65 | 60 | 73.75 | 122.2 | 144.4 |
| 7 | 8 | 100 | 95 | 85 | 75 | 88.75 | 114.3 | 128.6 |
| | 9 | 95 | 90 | 80 | 70 | 83.75 | 118.3 | 136.5 |
| | 10 | 90 | 85 | 75 | 65 | 78.75 | 121.4 | 142.9 |
| | 11 | 85 | 80 | 65 | 60 | 72.50 | 124.0 | 148.1 |
| | 12 | 85 | 75 | 60 | 55 | 68.75 | 126.2 | 152.4 |

Figure 11.7

Depending upon the locale and day of week, overtime hours may be paid at time and a half or double time. For time and a half, the overall (average) *actual* payroll cost (including regular and overtime hours) is determined as follows:

**For time and a half:**

$$\frac{40 \text{ reg. hrs.} + (20 \text{ overtime hrs. x } 1.5)}{60 \text{ hrs.}} = 1.167$$

Based on 60 hours, the payroll cost per hour will be (on average) 116.7% of the normal rate at 40 hours per week. However, because the effective production (efficiency) for 60 hours is reduced to the equivalent of 52.5 hours, the effective cost of overtime is calculated as shown below:

**For time and a half:**

$$\frac{40 \text{ reg. hrs.} + (20 \text{ overtime hrs. x } 1.5)}{52.5 \text{ hrs.}} = 1.33$$

Installed cost will be 133% of the normal rate (for labor).

Thus, when figuring overtime, the actual cost per unit of work will be higher than the apparent overtime payroll dollar increase, due to the reduced productivity of the longer work week. These calculations are true only for those cost factors determined by hours worked. Costs that are applied weekly or monthly, such as equipment rentals, will not be similarly affected.

## Retainage and Cash Flow

The majority of construction projects have some percentage of retainage held back by the owner until the job is complete. This retainage can range from 5% to as high as 15% or 20% in unusual cases. The most typical retainage is 10%. Since the profit on a given job may be less than the amount of withheld retainage, the contractor must wait longer before a positive cash flow is achieved than if there were no retainage.

Figures 11.9 and 11.10 are graphic and tabular representations of the projected cash flow for a small project. With this kind of projection, the contractor is able to anticipate cash needs throughout the course of the job. Note that at the eleventh of May, before the second payment is received, the contractor has paid out about $25,000 more than has been received. This is the maximum amount of cash (on hand or financed) that is required for the whole project. At an early stage of planning, the contractor can determine if there will be adequate cash available or if a loan is needed. In the latter case, the expense of interest could be anticipated and included in the estimate. On larger projects, the projection of cash flow becomes crucial, because unexpected interest expense can quickly erode profit margin.

A note on the subject of financing may be helpful here. In the above example, assume that the contractor has adequate cash resources to finance the project. Does this mean that the project need not bear any interest charges? No. At the very least, money withdrawn from savings or investments will result in unrealized interest (effective losses). These "losses" should be included in the project records for the purpose of determining actual job profit. For this reason, many companies assign interest charges to their jobs for negative cash flow. Likewise, the project is credited if a positive cash flow is realized. In this way, an owner or manager can readily assess the contribution (or liability) of each job to the company's "financial health". For example, if a project has a 5% profit at completion but has incurred 10% in "finance charges", it certainly has not helped the company to stay in business.

The General Conditions section of the specifications usually explains the responsibilities of both the owner and the contractor with regard to billing and payments. Even the best planning and projections are contingent upon the owner paying requisitions as anticipated. There is an almost unavoidable adversary relationship between contractor and owner regarding payment during the construction process. However, it is in the best interest of the owner that the contractor be solvent so that delays, complications, and financial difficulties can be avoided prior to final completion of the project. The interest of both parties is best served if information is shared and communication is open. Both are working toward the same goal: the timely and successful completion of the project.

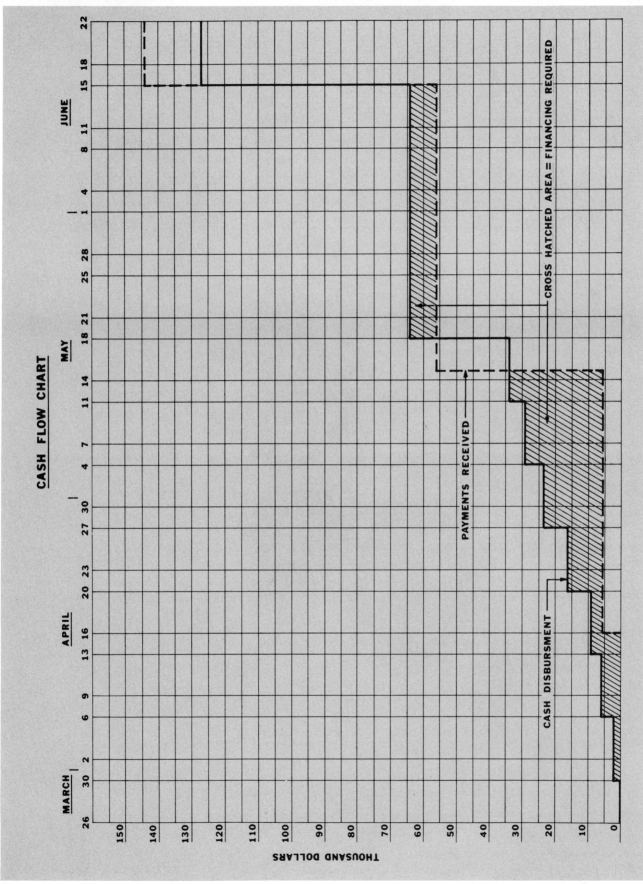

*Figure* 11.9

# REPAIR GARAGE PROJECTED CASH FLOW

| Date | 1 Payroll Incl Taxes | 2 Workers Comp | 3 Monthly Billing | 4 Retainage | 5 Subs Billing + Payment | 6 Retainage | 7 Material Incl Taxes | 8 Equip Incl Taxes | 9 Payments | 10 Accumulated Costs $ |
|---|---|---|---|---|---|---|---|---|---|---|
| 3-30 Payroll | 1926 | 148 | | | | | | | | -1926 |
| 3-30 monthly Billing | | | 4558 | 506 | 708 | 77 | | | | |
| 4-6 Payroll | 3246 | 353 | | | | | | | | -5172 |
| 4-13 Payroll | 4197 | 311 | | | | | | | | -9369 |
| 4-16 Payment | | | | | | | | | 4558 | -4811 |
| 4-20 Pay Sub | | | | | 708 | | | | | -5519 |
| 4-20 Pay material | | | | | | | 515 | | | -6304 |
| 4-20 Pay Equipment | | | | | | | | 965 | | -6999 |
| 4-20 Payroll | 3837 | 275 | | | | | | | | -10816 |
| 4-27 Payroll | 4525 | 333 | | | | | | | | -15351 |
| 4-30 monthly Billing | | | 51330 | 5704 | 5245 | 582 | | | | -20876 |
| 5-4 Payroll | 5525 | 508 | | | | | | | | -24407 |
| 5-11 Payroll | 3531 | 370 | | | | | | | | -26495 |
| 5-15 Payment | | | | | | | | | 61830 | +24810 |
| 5-18 Payroll | 2613 | 243 | | | | | | | | +19065 |
| 5-18 Pay Sub | | | | | 15245 | | | | | +3053 |
| 5-18 Pay material | | | | | | | 30118 | | | +3835 |
| 5-18 Pay Equipment | | | | | | | | 779 | | |
| 6-1 Final Billing | | | 88683 | | 48915 | 5435 | | | | +85051 |
| 6-15 Payment | | | | | | | | | 88683 | +36136 |
| Pay Subs | | | | | 48915 | | | | | +23451 |
| Pay material | | | | | | | 12685 | | | +22713 |
| Pay Equipment | | | | | | | | 938 | | |
| | $ 29390 | $ 2441 | $ 144771 | $ 16080 | $ 54868 | $ 6094 | 85318 | $ 2482 | $ 144771 | |

General Conditions
Permit 150 $3.00
Supervision - Carpenter Foreman 35 days @ $170
Temp Power & Water
Temp Office & Storage (works @ $30)
Clean Up Laborer 2 days @ $132

Cash
Retainage
Pay Workers Comp
Pay Subs Retainage
Gen. Conditions

Figure 11.10

# Life-Cycle Costs

Life-cycle costing is a valuable method of evaluating the total costs of an economic unit during its entire life. Regardless of whether the unit is a piece of excavating machinery or a manufacturing building, life-cycle costing gives the owner an opportunity to look at the economic consequences of a decision. Today, the initial cost of a unit is often not the most important cost factor; the operation and maintenance costs of some building types far exceed the initial outlay. Hospitals, for example, may have operating costs within a three-year period that exceed the original construction costs.

Estimators are in the business of initial costs. But what about the other costs that affect any owner: taxes, fuel, inflation, borrowing, estimated salvage at the end of the facility's lifespan, and expected repair or remodeling costs? These costs that may occur at a later date need to be evaluated in advance. The thread that ties all of these costs together is the time-value of money. The value of money today is quite different from what it will be tomorrow. One thousand dollars placed in a savings bank at 5% interest will, in four years, have increased to $1,215.50. Conversely, if $1,000 is needed four years from now, then $822.70 should be placed in an account today. Another way of saying the same thing is that at 5% interest, the value of $1,000 four years from now is only worth $822.70 today.

Using interest and time, future costs are equated to the present by means of a present worth formula. Standard texts in engineering economics have outlined different methods for handling interest and time. A present worth evaluation could be used, or all costs might be converted into an equivalent, uniform annual cost method.

The present worth of a piece of equipment (such as a trenching machine) can be figured, based on a given lifespan, anticipated maintenance cost, operating costs, and money borrowed at the current rate. Having these figures can help in determining whether to rent or purchase. The costs between two different machines can also be analyzed in this manner to determine which is a better investment.

# Part II
# COMPONENTS OF ELECTRICAL SYSTEMS

## Chapter 12
# RACEWAYS

Raceways are channels constructed to house and protect electrical conductors. This chapter contains descriptions of 11 different types of raceway, fittings, and associated installation activities. The appropriate units for measure are provided for each of these components, along with material and labor requirements and a step-by-step takeoff procedure. Cost modification factors are given when economy of scale or difficult working conditions may affect the cost of the project.

## Cable Tray

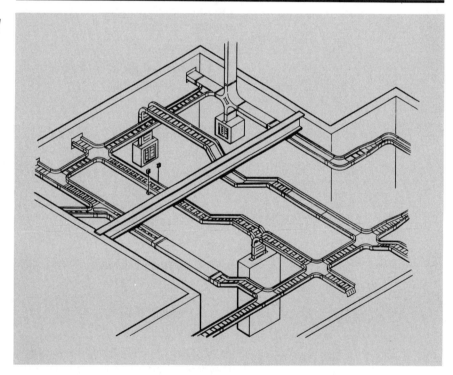

A cable tray system is a prefabricated metal raceway structure consisting of lengths of tray with associated fittings. Together, these components form a continuous rigid open support for cables. Cable tray is usually made of aluminum or galvanized steel but is also available in PVC-coated steel, fiberglass and stainless steel.

There are three basic types of tray: ladder, trough, and solid bottom. Trays have an overall nominal depth of 4 through 7 inches and are supplied in standard lengths of 12 and 24 feet. Standard widths are 6, 9, 12, 18, 24, 30, and 36 inches.

Fittings are the components which provide changes of both direction and size that are needed to bypass obstructions and vary run direction, while providing a continuous support path. Horizontal and vertical bends are offered with a selection of radii from 12″ to 48″ to match the bending capability of the cable being supported.

*Horizontal fittings*: Horizontal bends are tray fittings which provide a change of direction in a horizontal plane (i.e., left or right). Bends are offered in 30, 45, 60 and 90 degree configurations.

*Vertical fittings*: Vertical bends are tray fittings which provide a change in direction in a vertical plane (i.e., up or down). Bends are offered in 30, 45, 60, and 90 degree configurations.

*Miscellaneous fittings*: T's, Y's and X's are examples of other fittings available in both vertical and horizontal configurations.

Cable tray is used for many types of power and control needs. Making the tray system adaptable to the various situations is the purpose of accessories. The linear foot cost of cable tray and fittings, in most estimating standards, does not include accessories, since including them would limit cost flexibility. Accessories should be thought of as providing the following functions: a means of exit, a means of dividing, and a means of support.

*Exit accessories*:

> Trough drop-out bushings
> Drop-out fittings
> Cable ties and clamps
> Conduit to tray clamps
> Cable grips

*Dividing accessories*:

> Barrier Strips
>> Straight
>> Vertical inside and outside
>> Horizontal

*Support accessories*:

> Cable tray brackets
> Trapeze hangers
> Hold down clamps and hanger clamps
> Threaded rod
> Beam clamps

Accessories are, in most cases, not shown on the electrical print. Usually, however, a reference is made to a detail that shows a typical configuration.

There are two basic types of cable tray covers — solid and vented. Solid covers are used when maximum protection of cables is required and no accumulation of heat is expected. Vented covers provide cable protection while allowing heat to dissipate from the cables. Both ventilated and solid covers protect cables from damage. In many instances, however, protection is only required in certain areas or on vertical risers subject to physical damage, or for personnel protection.

Covers are available in either aluminum or steel, and finishes are offered to match the tray construction. Aluminum and pregalvanized covers 24″ wide or less come in 6′ and 12′ lengths. Covers wider than 24″ come in 6′ lengths only.

For straight cover, the material cost per unit includes the cover in 6' lengths, each one including 4 cover clamps. The material cost for cable tray covers for horizontal and vertical bends includes 4 cover clamps; for tees, 6 cover clamps; and for crosses, 8 cover clamps. Fitting covers come in sizes and widths that conform to the arrangements of the fittings.

**Units for Measure**: Cable tray is normally measured in linear feet (L.F.). The total linear feet must be figured for each size and depth of tray. Each unique fitting (EA.) must be counted and tabulated. The same applies to each hanger style and/or support (EA.). Related items, such as welded joints or field fabrication must be identified and listed (EA.).

**Material Units**: Various assumptions are made in all estimating standards as to what is included per unit (L.F., EA., etc.) Being familiar with these assumptions and the standard being used will help you to adjust material and labor prices to suit your own project. In turn, estimates will be both more reliable and competitive.

The following items are generally included per L.F. of cable tray:

> Tray in standard lengths (i.e., 12 feet)
> One pair of connector plates every 12 L.F. (including bolts and nuts)
> One pair clamp-type hangers every 12 L.F.
> One pair 3/8" threaded rods

**Labor Units**: The following procedures are generally included per L.F. of cable tray:

> Receiving and uncrating
> Measuring and marking
> Installing 1 pair beam clamp hangers every 12 L.F.
> Installing cable tray straight sections, using and including the
>     setting up of rolling stages
> Bolting sections of tray together
> Adjusting and leveling of tray

These additional procedures are listed and extended:

> Fittings
> Field fabrication
> Hanger supports
> Welded joints
> Cable

Labor adjustments are made for work over 15' high.

**Typical Job Conditions**: Productivity is based on the following conditions:

> New installation
> Work plane to 15' using rolling staging
> Unrestricted area
> Material storage within 100' of the installation
> Installation on the first three floors
> Minimum of 100 L.F. to be installed

**Takeoff Procedure**: First, consult architectural prints for area construction, marking on the electrical print this information: height of ceiling, size of area, construction type. Note on the print the typical hanger arrangement for each area being taken off: clamp type or beam type, trapeze, etc. Mark the parameters of like types and sizes of cable tray.

Next, set up the quantity takeoff sheet — typical for straight sections only. Note the scale, and be careful of reduced prints. Using a scale or rotometer, work from the origin of service out to the end or point of use. Mark each section of tray measured with a colored pencil. For better results and to leave an audit trail, mark different sizes with different colors.

Then transfer each section measured to the takeoff sheet, noting the height on the right hand corner. Go back over the print and mark any changes in elevation or risers. Transfer the riser quantities to the takeoff sheet.

The previous procedure should show how to take off straight lengths. Now go back to the print and list all fittings by their respective sizes and types on the worksheet. Mark all fittings on the print to complete the audit trail.

## Cost Modifications:

When tray is installed at heights over 15', add the following percentages to labor:

| | | | | |
|---|---|---|---|---|
| 15' | to | 20' | high | +10% |
| 20' | | 25' | | +20% |
| 25' | | 30' | | +25% |
| 30' | | 35' | | +30% |
| 35' | | 40' | | +35% |
| Over | | 40' | | +40% |

Check the prints for hole-cutting requirements in walls to allow the tray to go through. Adjust the estimates according to the following figures:

Up to 4 S.F. opening:

| | | |
|---|---|---|
| 2" Gypsum block | = | .50 M.H. EA. |
| 4" | = | .57 |
| 8" | = | .73 |
| 12" Brick wall | = | 4.00 |
| 12" Concrete block | = | 1.67 |
| 8" Brick wall | = | 2.22 |
| 5/8" Thick drywall | = | .33 |

Note: On masonry walls, use a minimum charge of 2 M.H.

To substitute the hanger type (if included in L.F. cost of cable tray) with a particular type as specified for the project, deduct the cost of the hangers used and add the type required. For example: Deduct the following from the L.F. material cost and man-hours for the price of cable tray without hangers or supports.

| | Material | Man-hours |
|---|---|---|
| 6" Tray | −12% | −17% |
| 9" | −17% | −17% |
| 12" | −17% | −17% |
| 18" | −16% | −17% |
| 24" | −17% | −17% |
| 30" | −15% | −16% |
| 36" | −16% | −16% |

Add the cost of each new type of hanger selected.

## Conduit (to 15' High)

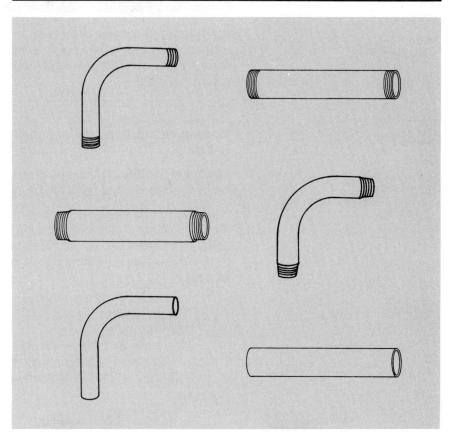

Conduit should be divided into the following three categories: power distribution, branch lighting, and branch power. Using these distinctions, the estimator does not take off all quantities of pipe at one time. Instead, he can concentrate on the quantities in one particular system. In this way, he can associate the conduit with the other components in the system to arrive at the appropriate man-hours.

The most common types of overhead conduit systems are as follows:

>   Aluminum
>   Rigid galvanized steel
>   Steel intermediate conduit (IMC)
>   Rigid steel, plastic coated
>   Electric metal tubing (EMT)

In many instances, architectural prints are simply marked "conduit", with a given size. It is important that the estimator find out the type of conduit specified, as the costs can vary widely.

Fittings are used to connect, change direction, or support conduit runs. The complexity and time required to perform particular pipe installations increase by the number of fittings required. A certain number of fittings is assumed for each 100 L.F. of conduit installed. If the quantities exceed those in the linear foot models, then the excess material and labor units must be included.

An alternative method of conduit pricing does not include fittings, terminations, or hangers. It is designed to give the estimator greater flexibility when adapting his estimate to unique needs.

For example, if hangers to be used support multiple conduit runs, the estimator can estimate the cost of each hanger and not have to bother "backing out" the cost of hangers included in the L.F. cost of the previous section.

This method is also useful when there are many straight conduit runs. The estimator does not have to pay attention to terminations, bushings, locknuts, etc., that are built into the cost per 100 L.F. of conduit in the first method.

*Units for Measure:* Conduit is measured in linear feet (L.F.). Fittings, hangers, supports and racks are counted as each (EA.).

*Material Units:* The estimator must be aware of what is included in the L.F. cost of conduit before taking it off. The following materials are normally included per L.F. of Aluminum, Rigid galvanized, steel-intermediate, and steel-plastic coated conduit on a per 100 linear foot basis.

> 11 Threaded steel couplings
> 11 Beam hangers
> 2 Elbows (factory sweeps)
> 4 Locknuts and 2 fiber bushings
> 2 Field thread pipe terminations
> 2 Concentric knockouts removed

For EMT:

> 11 Set screw steel couplings
> 11 Beam clamps
> 2 Field bends on 1/2 and 3/4"
> 2 Elbows (Factory sweeps) for 1' and above
> 2 Set screw steel box connectors
> 2 Concentric knockouts removed

Note: For congested areas or complex runs, it will be necessary to augment these basic assumptions by adding items on a unit basis.

*Labor Units:* The following procedures are generally included per L.F. of conduit:

> Unloading, uncrating, and hauling
> Set up of rolling stage
> Laying out
> Installing conduit and fittings

These additional procedures are listed and extended:

> Staging (rental or purchase)
> Structural modifications
> Wire
> Junction boxes
> Additional fittings
> Painting of conduit

*Takeoff Procedure:* Set up the quantity takeoff sheet by size and type of conduit. Use a rotometer and start from the panel or source, working out to the point of use. Mark each run with a colored pencil. Then mark changes of level and drops. Measure each size before going to another stage. Transfer the measurements to a cost sheet and price.

*Cost Modifications*: Add the following percentages to labor according to the height of these elevated installations:

| | | | | |
|---|---|---|---|---|
| 15' | to | 20' | high | + 10% |
| 20' | | 25' | | + 20% |
| 25' | | 30' | | + 25% |
| 30' | | 35' | | + 30% |
| 35' | | 40' | | + 35% |
| 40' | and | over | | + 40% |

Add these percentages to the L.F. labor cost, but not to the material cost of the fittings. Add these percentages only to the quantities exceeding the different height levels, not to the total pipe quantities.

Where conduit runs through walls of different types of material, add appropriate cutting costs. Figure the cost of each penetration by the size of pipe, type of wall, and thickness of wall. (See "Cutting and Drilling" in this chapter.)

# Conduit Fittings

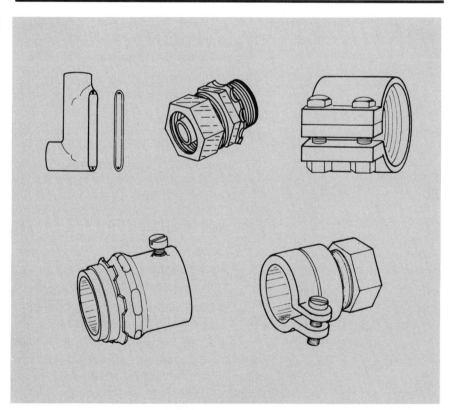

For estimating purposes, we will define fittings as any material item that must be added to our 100 L.F. cost models (as described in the previous chapter). By adding the appropriate fittings, we are also beginning to define the complexity of the conduit. For example, a run requiring the installation of four LB's and one T fitting per 100 L.F. can certainly be considered a more complex installation than one containing only two field bends.

Our approach will be to analyze some common fittings and to explain the logic behind each type. The estimator can apply this logic to any type of fitting, including those that are not listed in any standard, but follow the same installation procedure. In such cases, the estimator must still find out the material price but will be able to figure reasonable man-hours.

Locknuts are used to terminate and secure pipe endings. This procedure involves screwing the locknuts onto an existing pipe thread. Only a material price is needed. Note: Some locknuts also have a sealing ring insert for wet areas.

Bushings are used to protect wires from damage when a conduit terminates in an enclosure. There are several material variations for bushings. Among them are plastic, steel and steel grounding bushings. Although the installation method is the same as for a locknut, several labor assumptions have been added for estimating ease. Since it is unlikely that a 10′ conduit length will end at an exact location, the man-hour cost includes field cutting and threading one conduit end.

When field threading is not practical, threadless couplings are used to join two pieces of galvanized steel conduit, and threadless connectors are used to terminate a rigid galvanized steel conduit run into an enclosure. Threaded LB's and T's are used to change the direction or split a conduit run when field bends are not practical or allowed by code.

Nipples (manufactured type) are used to join electrical enclosures or raceway where there is not enough room for standard conduit techniques. There are many variations of nipples, such as chase and offset types. Nipples are available in the same diameters as conduit. Generally, nipples are under 6″ in length.

Expansion couplings are used where the expansion and contraction of a building (or bridge) will cause damage or a loss of continuity of the conduit run. They are also used if a run exceeds a certain length – to compensate for the thermal expansion and contraction of the conduit material.

*Units for Measure*: Fittings are counted as each (EA.) and listed by size and type.

*Material Units*: There is a great variety of fittings available for each type of conduit. The most common include:

      Locknuts
      Bushings
      Threadless couplings
      Threadless connectors
      Threaded LB's and T's
      Nipples of various lengths
      Expansion couplings
      Unions

Note: Fittings are manufactured in several types, such as explosion-proof, PVC-coated and malleable iron. Be sure your fittings conform to the installation.

*Labor Units*: Different fitting types require different types of labor units:

> **Bushings:** generally include installing locknuts, installing bushings, and cutting and threading one conduit end.

> **Threadless couplings:** generally include cutting to fit two ends of rigid galvanized steel conduit (when field threading is not practical) and installing the coupling.

> **Threadless connectors:** generally include cutting to fit one end of a galvanized steel conduit and installing the connector.

> **Threaded LB's and T's:** generally include cutting and threading two ends of a conduit for LB's and three ends for T's, installing the fitting, and installing the fitting cover.

> **Nipples:** generally include removal of two concentric KO's and installing the nipple (including locknuts).

> **Expansion couplings:** generally include field threading of one conduit end, installing a bushing on one conduit end, and installing the fitting.

These basic operations can be assumed for all fittings. Explosion-proof, seal-off fittings, for example, will include the above approach plus time for packing and pouring the sealing compound.

*Takeoff Procedure*: In most instances, a fitting will not be shown on a print for standard conduit installations. Many of the applications of particular fittings are dictated by legal and national codes, and/or by the project's own specifications. An estimator must be aware of these electrical codes. He must also be aware of the location of restrictions requiring the use of fittings instead of field bends.

If possible, list the fittings on the same quantity sheet as the conduit. List by type and size of fitting. Count each fitting and put a check mark on the print. (It is easy to lose track of changes in direction of conduit runs.) Transfer the quantities to a cost analysis sheet and extend.

# Conduit in Concrete Slab

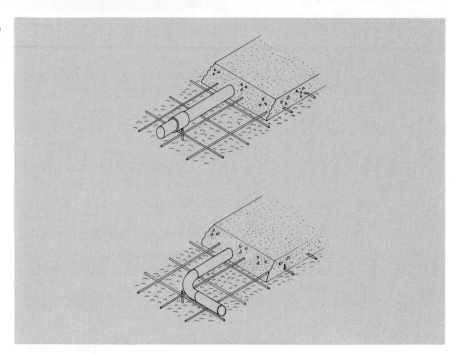

Conduit in concrete slab is used for both branch circuit piping and power distribution where the locations of end use will not change. Embedded conduit is usually a very cost-effective method of raceway installation because of savings in support costs and reduced run lengths. The most common types of pipe used in "slab work" are PVC schedule 40 (plastic), rigid galvanized steel, and intermediate metallic conduit (IMC). These types of conduit are most resistant to corrosion and oxidation. Aluminum conduit is not recommended for use in concrete that contains chlorides, due to its potential for oxidation and expansion.

**Units for Measure:** Conduit in slab is measured in linear feet for estimating. (Do not combine these footages with exposed raceways.) Special fittings, such as union-type (Erickson) couplings, are counted as each (EA.).

**Material Units:** The following items are generally included per L.F. of Rigid galvanized conduit, IMC conduit or PVC conduit:

    Conduit
    Ties to slab reinforcing
    Couplings
    Elbows (factory sweeps)
    Locknuts and 2 fiber bushings
    Removal of concentric knockouts

**Labor Units:** The following procedures are generally included per L.F. of conduit:

    Unloading, uncrating, and hauling from the loading dock
    Laying out
    Installing conduit (including the material items above)

These additional procedures are listed and extended:

    Wire
    Additional bends (over 2 per 100 ft.)

*Takeoff Procedure*: Set up the quantity takeoff sheet by size and type of conduit. Start from the panel or source and work out to the point of use. Mark each run with colored pencil. Measure each size before going to another stage.

*Cost Modifications*: When multiple conduits are run parallel in slab at the same time, apply the following percentage deductions to the man-hours:

| | |
|---|---|
| 2 runs | – 5% |
| 3 | –10% |
| 4 | –15% |
| 5 | –20% |
| Over 5 | –25% |

Example # 1:

6 runs of 100' long 2" PVC, run parallel at the same time by the same crew:

| | |
|---|---|
| 600 L.F.  x  .067 M.H./L.F. | 40.20 M.H. |
| Less 25% | –10.05 |
| | 30.15 M.H. |

Example # 2:

PVC Conduit run in slab involves "stub ups" in 5 locations for equipment, etc. The total length of the runs is 100'. To adjust material and labor, it is important to remember what was included in the linear foot estimate explained in the material and labor assumptions. Next, take off the number of additional field bends or factory sweeps. Add the resulting figures to the man-hours and material.

Total length involving "stub ups" is 100 L.F.
5 stubs require 10 manufactured elbows
Total 2" pipe runs = 100 L.F.

| | |
|---|---|
| Piping 100 L.F. x .067 M.H./L.F. | 6.70 M.H. |
| Elbows 8 EA. x .444 M.H./EA. | 3.55 M.H. |
| | 10.25 M.H. |

Note: Although there are 10 elbows in the model, only 8 are included in the adjustment calculation. This is because 2 elbows were previously listed in the material per 100' of conduit. When an elbow is added, a coupling must also be added.

# Conduit in Trench

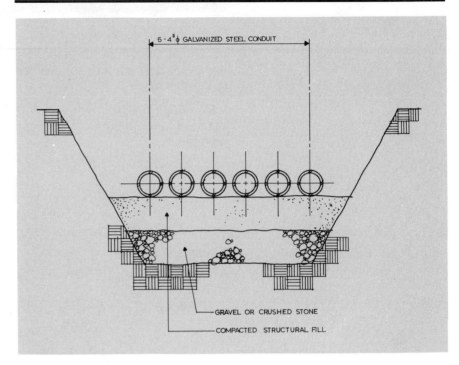

6 - 4"φ GALVANIZED STEEL CONDUIT

GRAVEL OR CRUSHED STONE

COMPACTED STRUCTURAL FILL

Conduit in trench is used for power distribution and for communications. Also, it is commonly used for outdoor lighting applications, such as roadways or parking lots. In this case, the conduit installation may be either direct burial or concrete encased. Rigid galvanized steel and rigid PVC conduit are usually used in trenches. For corrosive soil applications, PVC-coated rigid steel may be specified. Fiber duct or PVC (plastic) conduit may also be used for the installation of a concrete duct bank. Transite duct (asbestos cement) is no longer used.

*Units for Measure*:  Direct burial conduit is measured in linear feet (L.F.). Bends, fittings, spacers, etc. are counted as units, each (EA.). Trenching, sand, backfill and concrete are all calculated separately (usually on a per cubic yard basis) (C.Y.). Man holes are not included in the conduit units and must also be counted as units (EA.).

*Material Units*:  The following items are generally included for rigid galvanized steel pipe:

> Conduit
> Threaded steel couplings
> Elbows (factory sweeps)
> Pipe terminations
> Locknuts and fiber bushings

For PVC conduit, include these items:

> Conduit
> Field cemented couplings
> Elbows (factory sweeps)
> Terminal adapters (male)

*Labor Units*:  The following procedures are generally included per L.F. of conduit:

> Unloading, uncrating, and hauling
> Laying out
> Installing conduit

These additional procedures are listed and extended:

Excavation
Bedding
Concrete
Backfill
Wire

**Typical Job Conditions**: Productivity is based on a new installation in a dry prepared trench to 4' deep. Material staging area within 200' of installation.

**Takeoff Procedure**: Set up the quantity takeoff sheet by size and type of conduit. Start from the source and work out to the point of use. Mark each run with colored pencil. Measure each size before going to another stage.

# Cutting and Drilling

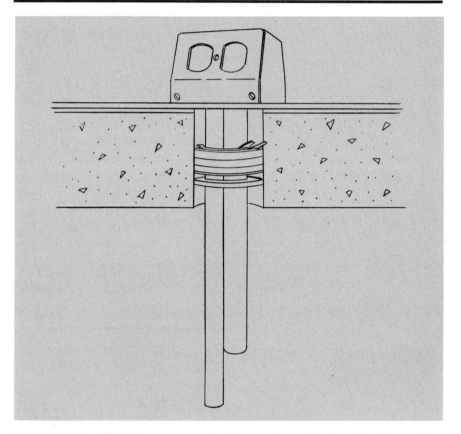

Cutting and core drilling are required in certain restrictive situations. The most practical approach to this kind of work is to identify the physical restriction and price it in as if it were part of the installation.

Walls are the most common restriction. The types that hinder conduit runs are usually concrete or masonry unit. Their thickness ranges from 8" to 24". Since the penetration must accommodate conduit sizes from 1/2" to 6", the core drill must be sized larger than the conduit's O.D. (outside diameter).

**Units for Measure**: Penetration holes (EA.) are taken off by the size of the penetration and for each different wall thickness. Also itemize for elevations over 10'.

## Material and Labor Units:
The following items are generally included per hole:

Measuring and marking
Set up of drill equipment with available power
Core drilling
Clean up

These additional items are listed and extended:

Cost of drill bits
Dust protection
Patching and grouting

## Typical Job Conditions:
Productivity is usually based on an unrestricted area, drilling to a height of 10' using a rotary drill.

## Takeoff Procedure:
Set up the work sheet by the type of wall. Mark each penetration on the print. Transfer the quantities to the work sheet.

## Cost Modifications:
When drilling at a level above 10', add the following percentages to the man-hours:

| | | | | |
|---|---|---|---|---|
| 10' | to | 15' high – add | 5% |
| 16' | | 20' | 10% |
| 21' | | 25' | 15% |
| 26' | | 30' | 20% |
| Over 30' | | | 30% |

Example: Twenty 2" holes are needed in an 8" thick concrete wall. Five conduits are at a height of 12', 10 are at 24' and 5 are at 35'. Adjust the labor accordingly.

| | | | | |
|---|---|---|---|---|
| 5 holes | 2" @ 12' = | 1.8 M.H. x 1.05 = | 9.45 M.H. |
| 10 | 2" @ 24' = | 1.8 M.H. x 1.15 = | 20.70 |
| 5 | 2" @ 35' = | 1.8 M.H. x 1.35 = | 12.20 |
| | | Total | 42.35 M.H., |

Note: If only 1 hole is to be drilled, a 2-hour minimum charge should be applied for setting up, mobilization, etc.

# Wire Duct – Plastic

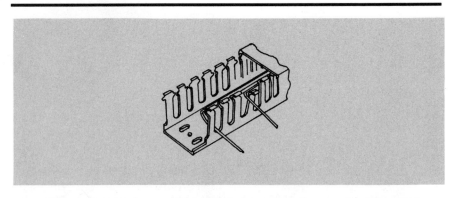

Wire duct is designed to organize both branch and control wires within panels and enclosures. Slotted sides for wire ins/outs make this raceway accessible at 1/2" intervals, and snap-on covers make it easy to add or change wire. Wire duct is available in standard 6' lengths; its cross-sectional dimensions range from .5" x .5" to 4" x 5". Most wire duct is PVC and has an adhesive backing. This backing is used to hold the duct temporarily in place until it is riveted or screwed to a panel mounting plate.

*Units for Measure*: Wire duct is tabulated on a linear foot (L.F.) basis for each size of duct.

*Material and Labor Units*: The following items are generally included per L.F. of wire duct:

> Measuring and marking
> Duct
> Mounting screws, drilled and tapped
> Cutting to length
> Field installation

These additional items are listed and extended:

> Cover
> Wire
> Marking or labeling duct
> Panel or enclosure

*Takeoff Procedure*: To take off, first set up the work sheet by the dimensions of the wire duct. Include the duct cover on the work sheet. Measure the total linear footage and round to the next 6 linear feet. Mark the different sizes of duct with a colored pencil. Transfer the lengths to a quantity sheet.

*Cost Modifications*: If wire duct is not field-installed in cabinets after setting, but is instead field installed assembly-line fashion before the cabinets are mounted, then deduct the percentages below from the total man-hours, according to the number of cabinets being worked.

| | | | |
|---|---|---|---|
| 1 | to | 3 | – 0% |
| 4 | | 6 | –15% |
| 7 | | 9 | –30% |
| 10 | | 15 | –25% |
| Over 16 | | | –30% |

For example, 18 boxes are to be fabricated at benches. Thirty feet of 4" W. x 5" H. duct and cover are to be installed in each cabinet:

| | |
|---|---|
| 18 boxes x 30 L.F. duct/box | = 540 L.F. of 4" W. x 5" H. duct |
| | = 540 L.F. of cover |
| .160 M.H. x 540 L.F. | = 86 M.H. duct |
| .08 M.H. x 540 L.F. | = 43 M.H. cover |
| | 129 M.H. Total |
| | 129 x .70 = 90.3 Adjusted M.H. |

# Trench Duct

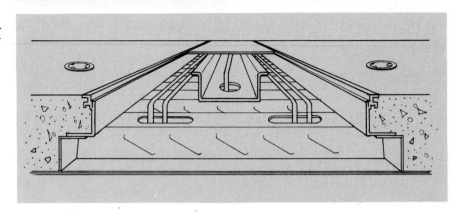

Trench duct is a steel trough system set into a concrete floor, with a top cover fitted flush with the finished floor. Trench duct is used as a raceway system in concrete slabs and to feed underfloor raceways. It is also used as a flush raceway for such situations as computer rooms, language labs, shop areas and x-ray rooms. A trench duct system is made up of individual components which allow flexibility for specific needs. The basic components are as follows:

> U-trough
> Side rail assembly (standard 10' lengths)
> Bottom plates (standard 5' long by cover plate width)
> Cover plate of 1/4" steel

Standard widths are 9", 12", 18", 24", 30", and 36", with standard depths of 2-3/8" to 3-3/8". All fittings come in standard duct widths. Changes in direction are made by an assortment of fittings, such as horizontal elbows, vertical elbows, crosses, and tees. The start and end accessories are end closures, risers and cabinet connectors. All fittings must be added to material and labor calculations.

*Units for Measure*: Trench duct is measured on a linear foot (L.F.) basis. Fittings, supports, boxes, etc. are counted as each (EA.).

*Material and Labor Units*: Rather than pricing each individual component, most estimating standards have put the required components together to come up with a per L.F. cost rate. For example, the following items are generally included per single compartment, 36" wide trench duct:

> Two side rails
> One 36" wide bottom plate
> One 36" wide cover plate
> Two tack welds to a cellular steel floor
> Adjusting and leveling

These additional items are listed and extended:

> Wire
> Fittings
> Access holes from duct to cellular grid
> Cutting and patching

*Takeoff Procedure*: Set up the quantity takeoff sheet by width and number of compartments of duct. Measure from the source of power to the end, rounding the measurement to the next 5'. Transfer quantities from the print to the work sheet. Go back over the print and mark the fittings, identifying by type of fitting, width, and number of compartments. Take off these accessories by type and width.

**Cost Modifications**: If man-hours are based on the installation of up to 100 linear feet of trench duct, you can adjust for quantities over 100'. This can be done by decreasing the man-hours by the following percentages. For lengths of:

| | | | |
|---|---|---|---|
| 100' | to | 200' | −10% |
| 200' | | 300' | −20% |
| Over 300' | | | −30% |

Deduct for these labor factors only if the lengths are located in the same general area on the same floor, and are being installed by the same crew. This applies to the labor for linear feet of trench duct and associated fittings.

For example, 350 L.F. of 24" wide trench duct is to be installed on a first floor area. There are 5 horizontal elbows within this system. Adjust the man-hours for these quantities:

$$350 \text{ L.F.} \times .720 \text{ M.H./L.F.} = 252 \text{ M.H.}$$
$$5 \text{ EA.} \times 5 \text{ M.H./EA.} = \underline{25}$$
$$277 \text{ M.H.}$$

Adjustment:

$$277 \text{ M.H. } (1.00 - .30) = 277 \text{ M.H. } (.70) = 194 \text{ M.H.}$$

# Underfloor Duct

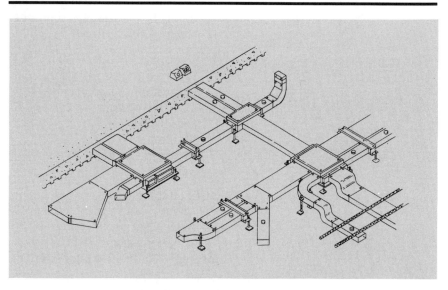

The purpose of underfloor duct is to make power and communication wiring available at numerous locations within a room. Used almost exclusively in office areas, the duct system is set before the concrete floor is poured. Wire or cable is accessed by hand holes spaced 24" O.C. There are two standard types of ducts; both come blank or with hand holes and are manufactured in 10' lengths. The sizes are standard duct and, for larger wire capacities, super duct. Underfloor duct systems are available in two basic designs, single level and two level.

*The single-level duct system* involves ducts that are all mounted at the same elevation. Special junction boxes are required which accommodate one, two or three ducts on each side. If power and communication cables run through a common junction box, a divider system must be installed to separate the two types of cables.

In *the two-level duct system*, feeder ducts are run below branch distribution ducts. Power and communication wiring can be separated more easily with this system and junction boxes are less crowded.

**Units for Measure**: Underfloor duct is estimated per linear foot (L.F.) and rounded upward to the next 10' length. All fittings, junction boxes, supports, etc. are listed as each (EA.).

**Material Units**: The following items are generally included per 10 linear feet of underfloor duct:

> One section duct
> One coupling

These additional items are listed and extended:

> Fittings and junction boxes
> Supports
> Adapters
> Outlets
> Wire and cable

**Labor Units**: The following procedures are generally included per 10 linear feet of underfloor duct:

> Unloading and uncrating
> Hauling from the loading dock
> Measuring and marking
> Setting the raceway and fittings in slab or on grade
> Leveling the raceway

These additional procedures are listed and extended:

> Floor cutting
> Excavation or backfill
> Concrete pour
> Grouting or patching
> Wire pulls
> Conduit

**Takeoff Procedure**: Because underfloor duct systems can vary from simple to complex, most standards require the estimator to list all components, and to price material and labor individually. Because of the wide range and cost of fittings, junctions, and accessories, it is not safe to assume that there will be a certain number of fittings or junction boxes per 100 L.F. of underfloor duct.

To take off, first set up the work sheet by type of duct (super or standard). Note whether or not the duct is insert type, and set it up by the on-center distance between inserts. Record the fittings separately from straight lengths of underfloor duct. Identify junction boxes as single level or two level. Rather than trying to describe all the different types, sizes and partition arrangements of each junction box, list them by the manufacturer's catalogue number. Most underfloor duct systems have been pre-engineered and contain a schedule of part numbers and descriptions.

Set up a third quantity sheet heading for "Accessories". This sheet will be used to list such items as 90-degree bends, cabinet connectors, supports, offset elbows, and conduit adapters.

Measure all straight lengths of underfloor duct and mark each type with a colored pencil. Transfer the lengths to a quantity sheet and round to the next 10' length. Go back over the prints and list and mark all junction boxes and associated fittings; then count and transfer this information to the quantity sheet. Total the quantities on all sheets and transfer these figures to the cost analysis sheet.

*Cost Modifications*: Man-hours are based on the installation of up to 100 linear feet of underfloor duct. Decrease the man-hours by the following percentages to adjust for higher quantities.

| | | | |
|---|---|---|---|
| 100' | to | 200' | – 5% |
| 200' | to | 500' | –10% |
| Over 500' | | | –15% |

For example, 1500 L.F. of 1-3/8" x 7-1/4" blank super duct is to be installed in the same area of an office building; adjust the man-hours for this installation:

| Total length | | M.H. per L.F. | | Modifier | | |
|---|---|---|---|---|---|---|
| 1500 | x | .130 | x | .85 | = | 165.8 M.H. |

Use these percentages for straight lengths only; do not deduct from the man-hours for junctions, fittings or accessories.

Note: It is prudent to get material costs from a vendor or manufacturer whenever possible.

## Surface Metal Raceways

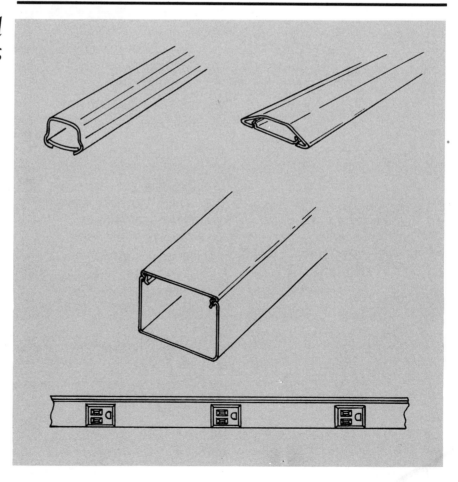

Surface metal raceway (wiremold) is installed on walls or floors in existing buildings. It is used in situations where it would be too costly or difficult to install raceway within the wall. Surface metal raceway is available in many sizes and configurations. The smaller sizes are one-piece units and fasten to a surface with snap-on clips or straps. Wire must then be pulled in. The larger raceway consists of a base with a separate cover. The base is attached to the surface, and the cover snaps over the wires after they are put into place. Some larger sizes can be used with a divider, allowing power and telephone or other signal systems to share the same raceway.

Certain types of raceway are prewired with receptacles at regular spaces; in others, an assembled harness of prewired receptacles is installed in the raceway to match pre-cut holes in the cover.

A special type is available for use on floors. It is shallow, flat on top, and slanted on the sides to minimize tripping. Most raceway is available in 5' or 10' lengths, while most covers come in 5' lengths. One-piece raceways are sold with couplings included.

Many different types of elbows, bushings, couplings, clips, adapters, boxes and other fittings are manufactured. A catalogue, readily available at most electrical supply stores, is helpful for both takeoff and installation.

Units for Measure:  Wiremold is taken off in linear feet (L.F.) and rounded upward to the nearest 10'. Fittings, clips, boxes, adapters, etc. are counted as each (EA.).

Material Units:  The following items are generally included per 10' of raceway:

>  Section(s) of straight raceway and covers
>  Coupling(s)
>  Supporting clip(s)

These additional items must be listed and extended:

>  All other fittings and boxes

Labor Units:  The following procedures are generally included in the man-hours:

>  Unloading and uncrating
>  Hauling from the loading dock
>  Installing raceway and fastening every 10'
>  Measuring and marking

These additional procedures must be listed and extended:

>  Installing wire, fittings and boxes
>  Installing device covers
>  Installing on walls, floors or ceilings over 10' high

Takeoff Procedure:  First, label the work sheet according to style and size, fittings and boxes. Then, take off the straight lengths of raceway for each size.

Next, take off the fittings for each size and run, being careful to note the correct elbow designation — horizontal, vertical or offset. Finally, boxes are taken off. All of the above information is then transferred to a pricing sheet.

# Wireways

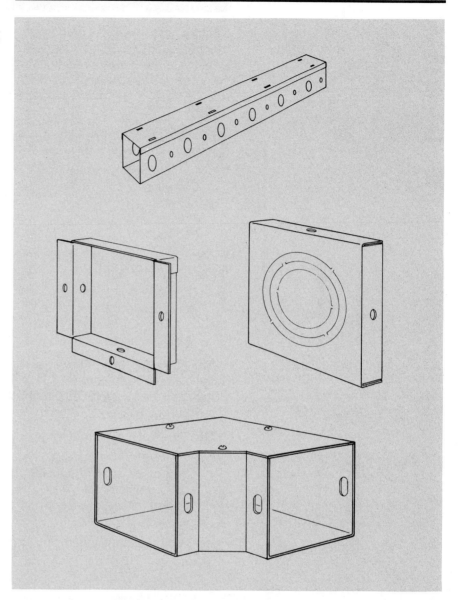

Wireways are sheet-metal troughs with hinged or removable covers. Wireways are used to house and protect electrical wires and cable. Conductors are placed after the wireway system is completed. There are two basic types of wireway. The first is "NEMA 1", which provides protection for wiring installations where oil, water or dust are not a problem. The second is "oiltight". This latter type ("NEMA 12"), with flanged, gasketed ends and hinged, gasketed covers is used in areas that are exposed to oil, dust, coolants, and water.

Generally, wireway must be supported every five feet. However, supports every ten feet are permitted for wireway which is designed specifically for that purpose. (Vertical runs may be supported at 15′ intervals.) Wireways can only be used for exposed situations.

NEMA 1 lay-in systems come in standard sizes of 2-1/2", 4", 6", 8", 10", and 12" square. Standard lengths are 1' through 5', and 10'. Ten-foot lengths are the most common and can be assumed for estimating purposes unless otherwise specified. Oiltight wireways come in standard sizes of 2-1/2", 4", 6" and 8" square. Standard lengths are 1' through 5', and 10'.

For both NEMA 1 and oiltight wireway, fittings come in the same standard widths as the tray itself. Fittings allow changes of direction for straight lengths of wireway and are grouped into four classes:

   Couplings
   Elbows
   Crosses and "T" boxes
   Reducers

Wireway accessories are used to terminate or support wireway installations and consist of items such as:

   End caps
   Panel connectors
   U connectors
   Special hangers

## Units for Measure: Wireway is measured in linear feet (L.F.). Standard lengths are 10', but lengths of 1' to 5' are available to suit the installation. Fittings, brackets, reducers, and panel connectors must be listed as each (EA.).

## Material Units: The following items are generally included per 10 linear feet of wireway:

   One section straight raceway
   One coupling or flange kit
   Two wall bracket hangers
   One section cover

These additional items are listed and extended:

   Wire
   Fittings
   Accessories

## Labor Units: The following procedures are generally included per 10 L.F. of wireway:

   Receiving
   Material handling
   Installing raceway on wall mounted hangers up to 10 feet high and
      spaced every 5 linear feet
   Installing cover

## Takeoff Procedure: To take off, first set up the work sheet by type and size of tray. List the fittings below straight sections of raceway and identify by type, size, and configuration. Measure from the source and work to the termination point. Many estimators like to measure total runs and transfer size quantities as they go. Because of short sections (manufactured sizes other than 10' long) and different sizes and types, it may be in the best interest of the estimator to measure all of one size first before proceeding to another. This is not a dramatic change of

approach, since most wireway sizes remain constant or decline in size from source to termination. Go back over the print and list all fittings; be especially careful to distinguish between vertical and horizontal fittings. Review the print once more and list the accessories (hangers, etc.). Many areas require a variety of support or hanger arrangements. Transfer the totals to the cost analysis sheet.

## Cost Modifications:
If the specs call for fabricated hanger and support systems, the estimator must make an adjustment. To do so, deduct the following percentages from the L.F. cost for material and labor to arrive at a cost for wireway without hangers. Then add the cost of the hanger system specified.

| Standard wireway | | Oiltight | |
|---|---|---|---|
| Material | -8% | Material | -4% |
| Labor | -4% | Labor | -4% |

For example: Adjust material and labor for 1000 L.F. of 6" x 6" standard type wireway so that hangers are not included.

|  | L.F. | x | Material $ | x | Adjustment factor | = | Price without hangers |
|---|---|---|---|---|---|---|---|
| Material: | 1000 | x | $7.70 | x | .92 | = | $7084.00 |
|  |  |  | Unit M.H. |  |  |  | Man-Hours |
| Labor: | 1000 | x | .260 | x | .96 | = | 249.6 |

Man-hours are based on installing wireway to a height of 10'. To adjust man-hours for installations over 10', add these percentages:

| 11' | to | 15' high | +10% |
|---|---|---|---|
| 16' |  | 20' | +15% |
| 21' |  | 25' | +20% |
| Over 25' |  |  | +25% |

Add these percentages to labor only for those lengths that are installed at the specified heights; do not add to the material price.

# Chapter 13
# CONDUCTORS AND GROUNDING

A conductor is a wire or metal bar with a low resistance to the flow of electric current. Grounding is accomplished by a conductor connected between electrical equipment or a circuit and the earth. In this chapter, there are ten sections on various types of wiring and terminations, and a section on grounding. Each section begins with a definition of the component, followed by units for measure, material and labor requirements, and a takeoff procedure.

## Flexible Metallic Conduit

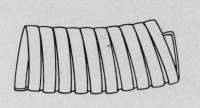

Flexible metallic conduit, sometimes referred to as "Greenfield", or "Flex" is a single strip of aluminum or galvanized steel, spiral wound and interlocked to provide a circular cross section of high strength and flexibility. Its flexibility, together with a continuous length, make it more cost effective than rigid conduit for use in finished frame buildings. Flexible metallic conduit can be purchased in lengths ranging from 25' to 250', depending on its diameter. It requires no elbow fittings or field bends for changes in direction.

Flexible metallic conduit does, however, have restrictions. It cannot be used in wet locations, hoistways, acid storage areas, hazardous locations, underground or embedded in concrete. Sizes of flexible metallic conduit range from 3/8" to 4" diameter. The minimum size allowed by the N.E.C. (National Electrical Code) is 1/2" diameter except in the case of motor wiring, or for a run of less than 72". Flexible metallic conduit must be secured at intervals not exceeding 4-1/2', as well as within 12" of every outlet or fitting. Conductors are pulled into flexible conduit with fish tape in much the same manner as rigid conduit.

107

"Liquidtight" ("Sealtight") is similar to flexible metallic conduit except that it is covered with a liquid-tight, plastic covering. It is used most often for conduit connections to motors and raceways in machines where protection from liquids is required.

Because of the flexibility of Greenfield and Sealtite, only a limited selection of fittings is necessary. The most common fittings are box connectors with either non-insulated or insulated throats. Both are available in straight and 90 degree versions. Rigid to flexible conduit couplings are also very common. Forty-five degree connectors and swivel adapters are also available for Sealtite.

*Units for Measure*: Flexible metallic conduit is measured in linear feet (L.F.) and the connectors are counted as units (EA.).

*Material Units*: The following items are generally included in the material cost per linear foot of Greenfield and Sealtite:

> The flexible metallic conduit
> The cost of a 1-hole strap every 4-1/2'

These additional items must be listed and extended:

> Flex fittings and connectors
> Wire

*Labor Units*: The following procedures are generally included per linear foot of flexible metallic conduit:

> Material handling on site
> Installing conduit
> Securing conduit every 4-1/2 linear feet on masonry wall

Material and labor is based on an average of 50' runs.

These additional items are listed and extended:

> Installing wire
> Installing fittings

*Takeoff Procedure*: To take off, first set up the work sheet by type and size of flexible metallic conduit. List the fittings on the same work sheet. There are many installation methods for flexible conduit. For example, flex used to feed branch circuits involves a takeoff procedure similar to that for rigid conduit.

This is a good time to calculate the wire quantity, since wire is not included in the installation price of flex. This tabulation is done by multiplying the quantities of conduit times the number of conductors. Add 10% for waste. Transfer the quantities from the takeoff sheet to the cost analysis sheet and extend.

*Cost Modifications*: If the installation is done above suspended ceilings, the conduits may be tied in place rather than fastened with straps. This method results in a savings of both material and labor costs. Deduct 7% from the material cost and 35% from the labor in this instance.

# Wire

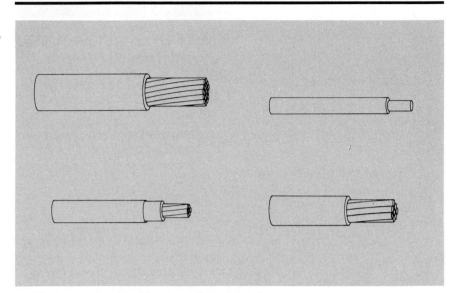

Wire is used to conduct current from electrical source to electrical use. Wire is made of either copper or aluminum conductors with an insulating jacket. Copper is normally used for sizes smaller than #6, while copper or aluminum is used on #6 and larger. Wire comes with various voltage ratings and insulation materials. Some types of insulation are thicker than others and require a larger size conduit. Aluminum conductors of equal ampacity are larger in diameter than copper and may require larger conduit.

The cost impact of using different types of wire goes beyond the cost of the wire itself. For example, typical installed costs of several common wire types (of 600 volt wire per C.L.F. ) are as follows:

| | |
|---|---|
| #6 type THW copper stranded | $42.00 |
| #6 THW aluminum stranded | 30.10 |
| #6 THHN-THWN copper | 42.40 |
| #6 THHN-THWN alum. | 30.80 |
| #6 XHHW copper | 42.00 |
| #6 XHHW alum. | 30.80 |
| #4 XHHW alum. | 38.30 |
| #4 THW alum. | 37.70 |
| #4 THHN-THWN alum. | 38.80 |

#6 copper wire is rated at 65 Amps. #6 aluminum is rated at 50 Amps. #4 aluminum is rated at 65 Amps. #6 THHN-THWN and XHHW, 4 wires, requires 3/4" conduit. #6 THW, 4 wires, requires 1" conduit. #4 THHN-THWN, 4 wires, requires 1". #4 THW, 4 wires, requires 1-1/4". *Thus, both the cost of the wire and the installation must be considered in arriving at the true cost.*

*Units for Measure*: Wire quantity is usually expressed in units of 100 linear feet (C.L.F.). Care must be taken to multiply the run length by the total number of individual wires to get the total length. The ground wire is not always shown on the drawings, but it may be required by specification or code.

Note: Cable and wire are not always treated the same way. Cable is an assembly of two or more insulated wires laid up together, usually with an overall jacket. While some cables are estimated in C.L.F. units, heavier types may be measured in linear feet (L.F.). The estimator must be vigilant to avoid costly errors.

**Material Units:** Generally, the material costs include only the wire and cable. Such items as dunnage and deposits for reels must be figured separately.

**Labor Units:** The price per C.L.F. of wire generally includes these procedures:

> Setting up wire coils or spools on racks
> Attaching wire to pulling line
> Measuring and cutting wire
> Pulling wire into raceway
> Identifying and tagging

These additional items are listed and extended:

> Terminations to breakers, panelboards or equipment
> Splices
> Reel storage and handling

**Takeoff Procedure:** Wire should be taken off in the following categories:

> Feeders & Service Entrance
> Branch lighting
> Branch power

Feeder wire is usually taken off from the conduit feeder schedule sheet, using conduit lengths and an additional allowance for terminations at each end. As an example, switchboard-to-distribution panel would add 12 L.F. to each wire. Feeders are considered to be wire size #6 or larger. Branch circuits are #8, #10 or #12 wire size.

Wire quantities are taken off in one of two ways: by measuring each cable run, or by extending the conduit and raceway quantities times the number of conductors in the raceway. Ten percent should be added for waste and tie-ins. Keep in mind that the unit of measure of wire is C.L.F., not L.F. as in raceways. The formula reads:

$$\frac{\text{L.F. Raceway x No. of Conductors x 1.10}}{100} = \text{C.L.F.}$$

**Cost Modifications:** If more than one wire is being pulled at one time, the following percentages can be deducted from labor for that grouping:

| | |
|---|---|
| 3-5 Wires | |
| 5-10 | −30% |
| 10-15 | −35% |
| Over 15 | −40% |

If wire pull is less than 100' in length and is interrupted several times by boxes, lighting outlets, etc., it may be necessary to add the following lengths to each wire being pulled:

| | |
|---|---|
| Junction box to junction box | 2 L.F. |
| Lighting panel to junction box | 6 L.F. |
| Distribution panel to sub panel | 8 L.F. |
| Switchboard to distribution panel | 12 L.F. |
| Switchboard to motor control center | 20 L.F. |
| Switchboard to cable tray | 40 L.F. |

*Measure of drops and risers:* It is important when taking off wire quantities to include the wire for drops to electrical equipment. If the heights of various electrical equipment items are not clearly stated, use the following guide:

|  | Riser (Bottom Feed) | Drop (Top Feed) |
|---|---|---|
| Safety switch to 100 A | 5' | 6' |
| Safety switch 400 to 600 A | 4' | 6' |
| 100A panel 12 to 30 circuit | 4' | 6' |
| 42 circuit panel | 3' | 6' |
| Switch box | 3' | 3'6" |
| Switchgear | 0' | 8' |
| Motor control centers | 0' | 8' |
| 1-35 KVA transformers | 0' | 2' |
| 45 KVA transformers | 0' | 4' |

# Armored Cable

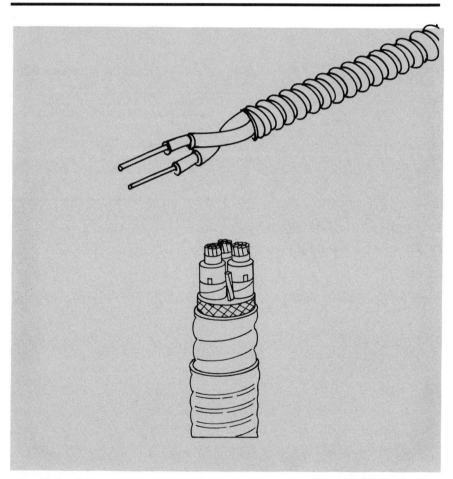

Armored cable is a fabricated assembly of approved cable with an aluminum or galvanized metal enclosure. Type AC (also called BX) is available in wire sizes from #12 aluminum or #14 copper through #1 A.W.G. Type AC is rated for 600 volts or less. Type MC (metal clad) is allowed for applications at 5,000 volts or less. It may have a plastic overall jacket.

Both types of armored cable are used in applications where physical protection is needed and conduit is not practical. Armored cable is not installed in conduit, underground, or in wet locations.

One of the advantages of armored cable is protection against vermin.

*Units for Measure:* Armored cable is measured in linear feet (L.F.) of cable.

*Material Units:* Only the cable itself is included in the takeoff.

For large size armored cable, box connectors must be added. When using BX, the cable is connected to boxes with built-in clamps designed to hold BX. No connectors are required in this case, and labor is included in the box installation.

*Labor Units:* These procedures are generally included for installation of BX cable:

> Field preparation
> Pulling cable
> Identification and tagging of the wire

These additional procedures must be listed and extended:

> Drilling of studs
> Wire terminations to equipment and breakers

*Takeoff Procedure:* First, measure the linear feet of each cable run, and tabulate according to size and type. Be sure to measure drops and risers. Total each column. Allow 10% for waste. Then transfer to a cost analysis sheet and extend. Refer to "Shielded Cable" (in this chapter) for special notes.

## Cable Terminations (to 600 Volt)

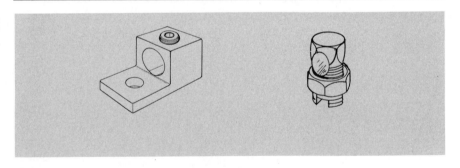

Any point at which a wire or cable starts or ends can be described as a "termination". The estimator must determine the number of individual cable runs and provide terminations for each conductor at each end. When equipment is of a nature that it must be set prior to pulling wire, cable terminations are not included. Switchgear, MCC's, control panels, etc. do not include terminations in their installation cost.

When working with lighting and wiring devices, such as switches and receptacles, the wires and cables are in place before the fixture or device is installed. Therefore, the installation price of a receptacle, for example, includes stripping, marking, and installing wire to the terminals on the receptacle. The installation of breakers, panels, and load centers all include terminations for branch wiring.

*Units for Measure*: Cable terminations are counted for each conductor of a cable at each end (EA.).

*Material Units*: The following items are generally included per unit:

Termination lug
Split bolt connector and tape (splice)
Two way connector and tape (splice)
Tape (if required)

*Labor Units*: The following procedures are generally included per unit:

Stripping of insulation from cable
Cleaning and application of deoxidizing compound on aluminum cable
Installation of terminal lug or connector on wire
Installation of terminal lug to equipment
Taping as required

*Takeoff Procedure*: First, list each cable size and type of lug. Provide for conductors for each cable end. Tabulate and add to get totals. Then transfer to cost analysis sheets and extend.

Note: Do not figure terminations for lighting and branch wiring devices, but do provide terminations for splices and jumpers as needed.

# Shielded Power Cables

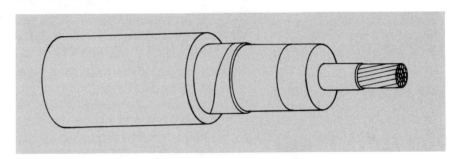

Shielded Power Cables are required in systems where the voltage is 5 KV or above. In the range of 2 KV to 5 KV, the use of shielded cables is a design option that depends upon the intended use or application (such as in damp conduits). Below 2 KV, shielded power cable is not required.

Shielding is the practice of wrapping semiconducting layers – either closely fitting or bonded to the inner insulation layers with no voids. The purpose is to equalize the stress within the insulation and to confine the dielectric field within the cable. Shielding extends the life of the cable, limits electromagnetic interference, minimizes surface discharges, and reduces the hazard of shock.

Voltage ratings may be from 5 KV to 69 KV, and conductor sizes range from #6 to 1250 MCM with copper, or to 2000 MCM with aluminum.

As with any cable, shielded cable must be handled carefully. Particular attention must be paid to the manufacturer's specified minimum bending radius in order to avoid damaging the shield layers. Such damage would lead to premature cable failure.

**Units for Measure:** Shielded cable is measured in units of 100 linear feet (C.L.F.).

**Material Units:** The following items are generally included per C.L.F. of cable:

    Cable
    Wire ties

These additional items must also be listed and extended:

    Terminations

**Labor Units:** The following procedures are generally included for the installation of shielded cable:

    Receiving
    Material handling
    Setting up pulling equipment
    Measuring and cutting cable
    Pulling cable into conduit
    Preparation of conduit for wire

These additional procedures must be listed and extended:

    Conduit or cable tray
    Wire terminations and splices

**Takeoff Procedure:** First, list the sizes, voltage rating, and types of cables on a takeoff sheet. Then measure the length of each run and list in the appropriate column. Total each column and provide + 10% for waste and contingency. Finally, transfer to the cost analysis sheets and extend.

When taking off the cable for procurement purposes, particular attention must be paid to the shipping lengths ordered. Most engineers will not permit splices. For example, a standard reel length is 1000 feet. If a cable run of 400 feet and 3 conductors is needed (i.e., 1200 feet total), it cannot be made from two reels, one with 1000 feet and the other with 200 feet. In this case, it is necessary to order 800 feet on one reel and 400 feet on another, or, better still, 3 reels with 400 feet on each.

Another option is to order the cable from the factory as 3 parallel 400-foot lengths on one reel. This reduces labor costs for handling in the field.

Whenever possible, order cable based upon actual field-measured lengths of raceways. If the length measured from a drawing turns out to be 5 feet short, then the cable ordered in that too-short length would probably be scrap, and a new length would have to be ordered.

# Cable Terminations (High Voltage)

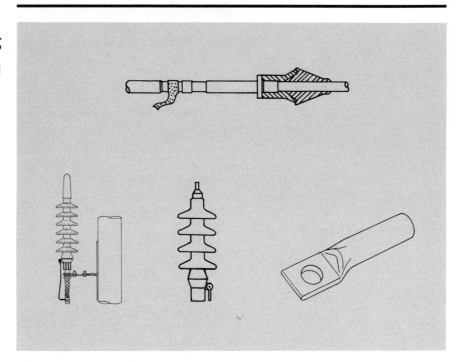

A cable termination is a means by which the conductor of a wire is fastened to the devices or equipment it serves. Many low-voltage (600 V and less) devices have wire attachment terminals with built-in lugs. A screw, nut, bolt, or compression device may be used to firmly and directly attach the wire. Other equipment (motors, for example) must be terminated by attaching a termination lug to the wire, usually by crimping it to the wire. The lug is then bolted or screwed to the device's terminal. Tape or other insulation may be required. Splices are made in a similar manner.

For cables rated at 5 KV and above, the termination process is more complex. A termination lug is compressed onto the conductor, but special attention must be paid to the insulation. If the conductor's shield insulation was abruptly stopped, an electrostatic field would concentrate at the spot and, in time, cause the cable to fail. By tapering the insulation for several inches and applying layers of tape and other insulating materials, a "stress cone" is formed, which gradually equalizes the electrostatic potential over a long surface area.

In addition to stress cones, a series of skirts may be used for outdoor applications. These are intended to prevent leakage from the termination along the damp surface of the cable terminations by effectively increasing the surface distance or length. Splices in high-voltage cables involve the same concerns and precautions as stress cones.

Both splices and terminations require a great deal of time and careful attention. A number of manufacturers supply kits for preforming terminations.

*Units for Measure*:  Each end of each conductor is counted as one termination (EA.).

*Material Units*:  The following items are generally included per unit:

> The termination lug
> A prepackaged kit to form a stress cone

***Labor Units:*** The following procedures are generally included per unit:

Cutting the cable to length
Stripping the insulation
Forming and shaping the stress cone
Attaching the completed lug

These additional procedures must be listed and extended:

Testing

***Takeoff Procedure:*** The cable size, as well as the type of insulation and jacket, must be carefully noted when doing the takeoff. Also allow time for coordinating with others and checking phase. Take into account the possibility of overtime. (When a stress cone is started, it should be completed without interruption). Count each cable by size and multiply by the number of conductors, allowing for waste. List under the appropriate heading. Total and transfer the information to the cost analysis form and extend.

For low-voltage (under 600 V) devices that have built-in terminals, such as outlets and switches, the labor for termination is included with the device's installation unit.

When purchasing termination lugs, it is important to assure that they are compatible with the conductor material. Although some lugs are made to be suitable for either copper conductors (CU) or aluminum conductors (AL), many are not and must be purchased accordingly.

## Mineral Insulated Cable

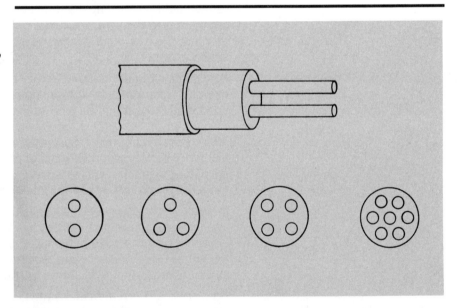

Mineral insulated cable, type "M.I.", is an assembly of one or more bare copper conductors, insulated with a highly compressed refractory mineral insulation and enclosed in a seamless metallic sheath (which is both liquid- and gas-tight). The mineral insulation is magnesium oxide, and the sheath is phosphorous deoxidized copper. This type of cable can be used in hot or cold, wet or dry locations, and in hazardous areas. It can be exposed, embedded in concrete, or buried.

Mineral insulated cable is rated for 600 V. Sizes range from #16 to 250 MCM single conductor, up to #4-3 conductor, to #6-4 conductor, and #10-7 conductor. M.I. cable is also available in heating versions, in 318 stainless jackets, or #16 wire rated at a maximum of 300 volts. Special fittings are used to terminate, a procedure which requires particular care. When M.I. cable is stripped, it must be sealed with an adaptor immediately to prevent moisture from entering the insulation. Special tools are available for use with mineral insulated cable. While M.I. cable *does not* need to be installed in conduit, conduit is sometimes used as a sleeve to protect the cable where it comes out of a slab. M.I. cable must be supported every 6 feet.

*Units for Measure*: Count M.I. cable in units of 100 linear feet (C.L.F.). Count each type of end termination (EA.).

*Material Units*: The following items are generally included per C.L.F. of M.I. cable:

> Cable (with clips and fasteners every 6 feet)

These additional items must be listed and extended:

> Each terminal

*Labor Units*: The following procedures are generally included per C.L.F. of M.I. cable:

> Installing the M.I. wire
> Fastening with bolts and clips

These additional procedures must be listed and extended:

> Terminal kits

*Takeoff Procedure*: Cable should be taken off by size and number of conductors and by type of termination, (hazardous or non-hazardous). Allow 10" or *more* for making terminations and connections at each end of run. Supports must be installed at intervals of not more than six feet per National Electrical Code. Keep in mind that the unit of measure of cable is C.L.F. List the lengths of each type of cable on the takeoff sheets. Total and add 10% for waste. Count the terminations required. Transfer the figures to a cost analysis sheet and extend.

# Nonmetallic Sheathed Cable

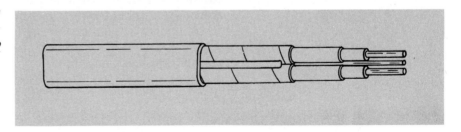

Nonmetallic sheathed cable is factory-constructed of two, three, or four insulated conductors enclosed in an outer sheath. This covering is a plastic or fibrous material. Nonmetallic sheathed cable comes with or without a bare ground wire and is made with either aluminum or copper conductors. The conductor sizes range from #14 to 4/0.

Nonmetallic sheathed cable is a category that includes a variety of individual types. Each has specific uses and code restrictions. Some types of nonmetallic sheathed cable are listed below:

**Type NM:** Nonmetallic sheathed cable used for dry areas in residential wiring and in certain commercial buildings up to 3 stories high.

**Type NMC:** Similar to type NM but with a corrosion-resistant sheath. It can be used in damp areas.

**Type SNM:** Shielded type NM cable. It is used in cable trays and raceways.

**Type SE:** Service entrance cable. It has a flame-retardant, moisture-resistant covering.

**Type USE:** Underground service entrance cable. Its covering is moisture-resistant but not flame-retardant.

*Units for Measure:* Nonmetallic cable is measured in units of 100 linear feet (C.L.F.).

*Material Units:* The following items are generally included when taking off nonmetallic sheathed cable:
Cable

  Staples, clips, or fasteners

These additional items must be listed and extended:

  Box connectors

*Labor Units:* The following procedures are generally included in the man-hours for nonmetallic cable:

  Drilling wood (or bushings in metal) studs
  Measuring
  Running the cable
  Staples and clips as required

These additional procedures must be listed and extended:

  Knock-outs
  Box connections
  Terminations

*Takeoff Procedure:* When measuring, allow for drops, risers, and the amount used within boxes. List each type and size to be used. Measure and list under the appropriate column. Next, count the box connectors, grommets, etc. Total and add 10% for waste. Transfer the quantities to a cost analysis sheet and extend.

# Special Wires

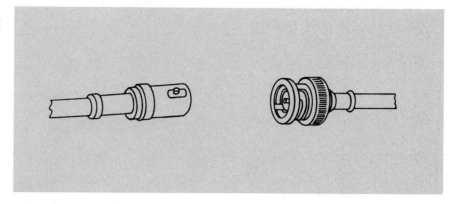

This category includes a number of distinct wire types, such as fixture wires, low-voltage instrumentation wire, signal cables, telephone cables, tray cables, and coaxial cables. Each of these cables is designed to meet unique performance criteria for specific applications. Fixture wire, for example, has insulation rated for the high temperatures created in lighting fixtures. Coaxial cables are designed to have specific impedances that will match the input and output equipment's impedance; this arrangement results in the maximum energy transfer of signals.

*Units for Measure*: These wires are measured in linear feet and extended in units of 100 linear feet (C.L.F.). Special connectors for these cable(s), when required, are counted as one for each cable end (EA.).

*Material Units*: Wire only is generally included for special wires.

These additional items must be listed and extended:

> Special terminators (such as coaxial cable connectors)
> Strain relief grips, etc.
> Special tools for connectors or stripping

*Labor Units*: The following procedures are generally included for special wires:

> Cable pulled and supported as needed

These additional procedures must be listed and extended:

> Coax connector installed on cable with special crimp tool
> Strain relief
> Terminations

*Takeoff Procedure*: List the cables by type and size on the takeoff sheet. Then carefully measure cable runs for one system at a time. (Include drop and riser lengths.) Mark runs with a colored pencil as they are taken off. Count fittings, terminators, or special support devices and list on the same sheet. Next, total the columns and allow an extra 10% for wire footages for waste. Finally, transfer the totals to a cost analysis sheet, paying attention to the unit of measure, and extend.

# Grounding

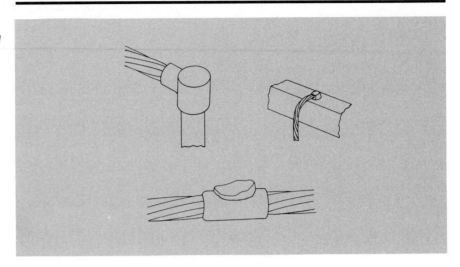

In most distribution systems, one conductor of the supply is grounded. This conductor is called the "neutral wire". In addition, NEC requires that a grounding conductor be supplied to connect non-current-carrying, conductive parts to ground. This distinction between the "grounded conductor" and the "grounding conductor" is important.

Grounding protects persons from injury in the event of an insulation failure within equipment; it also stabilizes the voltage with respect to ground and prevents surface potentials between equipment — which could harm both people and equipment (in a hospital, for example).

Lightning protection systems are a separate concern but are closely related to grounding in intent and practice. Lightning poses two kinds of danger. The first is the lightning strike itself. This can damage structures and distribution systems by passing a very high current for a brief time — causing heating, fire, and/or equipment failure. The second danger is from induced voltages in lines running near to the lightning's path. These pulses can be very high and can damage electrical equipment or cause injury to people.

In both systems, it is essential that a good, low-resistance ground path be provided. This can be accomplished via the use of ground rods, a ground grid, or attachment to metal pipes in contact with the earth (such as water pipes). Note: In communities where plastic water pipes are used between the street main and a building, this grounding method is *not* suitable — even though copper pipes may be used for the building's interior.

Connections between ground wires to pipes, conduits, or boxes are often made with cable clamps. Permanent connections between heavy ground wires to ground rods, building structural steel, other ground wires, and lightning rods are usually made with an exothermic weld process.

**Units for Measure:** Ground wire is measured in linear feet (L.F.) and extended in units of 100 linear feet (C.L.F.). Ground clamps and exothermic welds are counted as individual units (EA.).

**Material Units:** The following items are generally listed in the grounding takeoff:

> Ground rods
> Bare or insulated ground wire
> Ground clamps
> Exothermic weld metal (less molds and accessories)

Note: A different type of mold is necessary for each different type of connection.

These additional items must be listed and extended:

> Sleeves or raceways to protect ground wires
> Lightning rods and devices

**Labor Units:** The following procedures are generally listed for grounding takeoffs:

> Running ground wire and supports as required
> Ground clamps or lugs
> Exothermic weld

These additional procedures must be listed and extended:

> Excavation
> Backfill
> Sleeves or raceways used to protect grounding wires
> Wall penetrations
> Floor cutting
> Core drilling

**Takeoff Procedure:** First, list the sizes and types of ground wires on the takeoff sheets. List ground rods by diameter and length. Identify and list each unique type of ground connection. Each shape and size of exothermic weld connection will require a separate mold. List each mold required on the takeoff sheet for material only. Next, measure the length of the cables by size and allow for drops and risers. Count the ground rods, clamps, and exothermic welds. Total and add 10% to wire for waste. Also, for estimating purposes, add 5–10% to exothermic welds. Finally, transfer the quantities to a cost analysis sheet and extend.

# Undercarpet Wiring

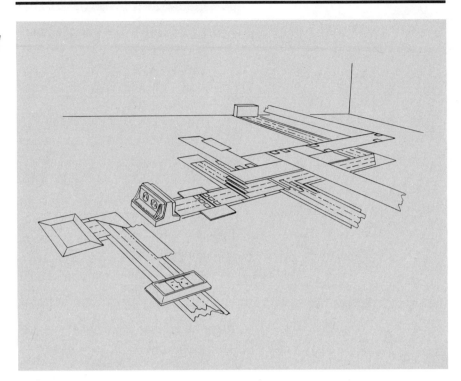

Branch Circuits are those conductors used between the circuit protection (fuse or breaker) and plug outlets for power and lighting. Flat Conductor Cable — type FCC — (also referred to as "Undercarpet Power Systems") is designed to be an accessible, flexible system for branch circuits.

Flat Conductor Cable consists of three or more flat copper conductors assembled edge-to-edge in an insulating jacket. Both a top and a bottom shield are required. All components must be firmly anchored (usually with adhesives). Crossings of two types of FCC, such as telephone (cable), are permitted, provided metal shield is placed between them. Transition assemblies are used to connect the Flat Conductor Cable to conventional wiring for the home run to a breaker panel.

Type FCC cable is used in many existing buildings to rework obsolete wiring systems, and in new buildings to allow customizing flexibility to the tenants and users of office areas.

As the FCC cable system has gained popularity, a wide spectrum of complementary products has also become available. These products include telephone systems, flat coaxial data cable systems, and fiber optic data link systems.

Type FCC cable must be installed under carpet squares not larger than 30" by 30". It is permitted on hard, smooth, continuous floor surfaces and on wall surfaces in metal raceways. It is not permitted outdoors or in wet locations; in corrosive atmospheres; or in residential, school, or hospital buildings.

*Units for Measure*:  The cable and shield are measured per linear foot (L.F.). Fittings, taps, splices, bends, etc. are counted as each unit (EA.).

*Material Units*: The following items are generally included for installation of undercarpet power systems:

**Cable:** Includes flat conductor cable with bottom shield, top shield, adhesive and tape.

**Transition fitting:** Includes 1 base, 1 cover, and 1 transition block.

**Floor fitting:** Includes 1 frame/base kit, 1 transition block, and 2 covers (duplex/blank).

**Tap:** Includes 1 tap connector for each conductor, 2 insulating patches, and 2 top shield connectors.

**Splice:** Includes 1 splice connector for each conductor, 2 insulating patches, and 2 top shield connectors.

**Cable bend:** Includes 2 top shield connectors.

**Cable dead end (outside of transition block):** Includes 2 insulating patches.

A special connecting tool is also required.

*Labor Units*: The following procedures are generally included in the installation of undercarpet power systems:

**Flat conductor cable:** Includes placing the cable only. Additional procedures to be listed and extended are top shield, mark floor, tape primer, and hold down tape, patching or leveling uneven floors, filling in holes or removing projections from concrete slabs, sealing porous floors, sweeping and vacuuming floors, removal of existing carpeting, carpet square cut-outs, and installation of carpet squares.

**Transition fitting:** Includes assembling the transition block and attaching wires. Additional procedures to be listed and extended are wall outlet box for the transition fitting with cover, and home run wires and conduit.

**Floor fitting:** Includes placing the receptacle device and terminating to the cable.

**Tap:** Includes terminating the conductors with connectors. Additional procedures are listed and extended are insulating patches and top shield.

**Splice:** Includes terminating the conductors with connectors. Additional procedures to be listed and extended are insulating patches and top shield.

**Cable bend:** Includes folding the cable at 90 degrees to change direction. An additional item to be listed and extended is top shield.

**Cable dead end:** Includes cutting the cable to length and insulating patches.

*Takeoff Procedure for Power Systems*: List the cable by number of conductors. List components for each fitting type, tap, splice, and bend on your quantity takeoff sheet. Each component must be priced separately.

Start at the power supply transition fittings and survey each circuit for the components needed. Tabulate the quantities of each component under a specific circuit number. Use the floor plan layout scale to get the cable footage. Combine the count for each component in each circuit and list the total quantity in the last column.

Allow approximately 5% waste on items such as cable, top shield, tape, and spray adhesive.

Suggested guidelines:

> Equal amounts of cable and top shield should be included.
> For each roll of cable, include a set of cable splices.
> For every 1' of cable, include 2-1/2' of hold down tape.
> For every 3 rolls of hold down tape, include 1 can of spray adhesive.

Adjust the final figures wherever possible to accommodate standard packaging of the product. (This information is available from the distributor.)

## Undercarpet Telephone Systems

An undercarpet telephone system is designed to connect telephone devices to a distribution closet using undercarpet cabling. This method provides flexibility in open office situations.

*Units for Measure for Undercarpet Telephone Systems:* The cable is measured per linear foot (L.F.). Fittings, taps, splices, bends, etc. are counted as each unit (EA.).

*Material Units for Undercarpet Telephone Systems:* The following items are generally included:

**Cable runs:** Include cable as with power systems. Additional items to be listed and extended are tape and adhesive.

**Transition fittings:** Include 1 base plate, 1 cover, and 1 transition block.

**Floor fittings:** Include 1 frame/base kit, 2 covers, and modular jacks.

*Labor Units for Undercarpet Telephone Systems:* The following procedures are generally included:

**Cable runs:** Include cable only. Additional procedures to be listed and extended are floor marking; floor preparation; top shield; tape; spray adhesive; cable folds, conduit or raceways to transition floor boxes; telephone cable to transition boxes; terminations before transition boxes; and floor preparation as described in power section.

**Transition fittings:** Include assembly of the transition block and terminating wires. Additional procedures to be listed and extended are wall outlet box with cover, and home run wires and conduit.

**Floor fittings:** Include placing the device, covers, modular jacks, and fastening in place. An additional procedure to be listed and extended is cutting carpet holes.

Be sure to include all cable folds when pricing labor.

*Takeoff Procedure for Telephone Systems:* After reviewing floor plans, identify each transition. Number or letter each cable run from that fitting.

Start at the transition fitting and survey each circuit for the components needed. List the cable type, terminations, cable length, and floor fitting type under the specific circuit number.

Use the floor plan layout scale to get the cable footage. Add some extra length (next higher increment of 5') to cable quantities.

Combine the list of components in each circuit and figure the total quantity in the last column. Calculate the necessary 5% allowance for such items as wire, tape, bottom shield and spray adhesive.

Adjust the final figures whenever possible to accommodate standard packaging. Check that items such as transition fittings, floor boxes and floor fittings — that are to be used for both power and telephone — have been priced as combination fittings, thereby avoiding duplication.

Make sure to include the marking of floors and drilling of fasteners if the fittings specified are not the adhesive type.

## Undercarpet Data Systems

An undercarpet data system is designed to interconnect remote data processing terminals to the main computer by means of undercarpet wiring.

*Units for Measure for Undercarpet Data Systems*: As above, for telephone systems.

*Material Units for Undercarpet Data Systems*: The following items are generally included:

> **Cable:** Includes the cable only. Additional items to be listed and extended are tape and adhesive.
>
> **Transition fittings:** Include 1 base plate, 1 cover, and 1 transition block.
>
> **Floor fittings:** Include 1 frame/base kit, 2 covers, and modular jacks. Additional items to be listed and extended are connectors.

*Labor Units for Undercarpet Data Systems*: The following procedures are generally included:

> As telephone cable, but without shields.

Additional procedures to be listed and extended:

> Conduit or raceways to transition or floor boxes
> Data cable to transition boxes
> Terminations before transition boxes
> Floor preparation as described in the power section
> Notching cable sides to make turns

*Takeoff Procedure for Undercarpet Data Systems*: Start at the transition fittings and take off quantities in the same manner as the telephone system, keeping in mind that data cable does not require top or bottom shields. The data cable is simply cross-taped on the cable run to the floor fitting.

Data cable can be purchased in either bulk form or in precut lengths with connectors. If it is obtained in bulk, coaxial connector material and related labor must be priced.

Data cable cannot be folded and must be notched at 1" intervals. A count of all turns must be added to the labor portion of the estimate. Notching requires:

| | |
|---|---|
| 90 degree turn | 8 notches per side |
| 180 degree turn | 16 notches per side |

Floor boxes, transition boxes and fittings are the same as described in the power and telephone procedures.

Since undercarpet systems require special hand tools, be sure to include this cost in proportion to the number of crews involved in the installation.

## Chapter 14
# BOXES AND WIRING DEVICES

A box is used in electrical wiring – at each junction point, outlet, or switch. Boxes provide access to electric connections and serve as a mounting for fixtures or switches. They may also be used as pull points for wire in long runs or conduit. A wiring device is a switch or receptacle that controls, but does not consume, electricity. This chapter has five sections to cover various types of boxes, cabinets, wiring devices, and the appropriate fasteners and hangers. Descriptions, units for measure, material and labor requirements, and a takeoff procedure are given for each.

## Pull Boxes & Cabinets

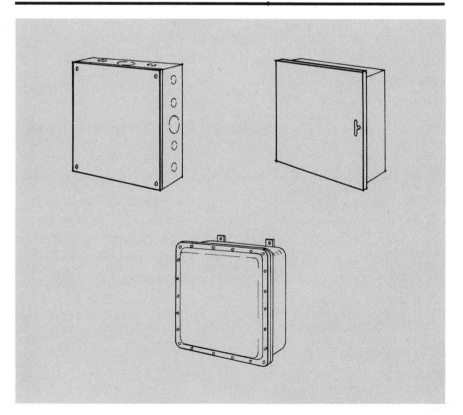

Pull boxes are inserted in a long run of conduit to facilitate the pulling of wire. They are also used where conduit changes direction or wires divide into different directions. Cabinets are used where wire terminates. Both come in various NEMA types to match the requirements of a given environment.

Boxes are usually made of galvanized steel, epoxy-painted steel, stainless steel, or aluminum. Less common materials include malleable iron, cast iron, cast aluminum, and high-density plastic.

## Units for Measure:
Boxes and cabinets are counted in terms of each (EA.) and are tabulated by size and type.

## Material Units:
The following items are generally included per unit:

    Boxes
    Covers
    Fasteners

These additional items are listed and extended:

    Support racks
    Wire terminations

## Labor Units:
Different types of installations require different types of labor units:

**Wall mount (indoor or outdoor) installations:** Include unloading and uncrating, hauling of enclosures from the loading dock, measuring and marking, and mounting and leveling boxes. Note: A plywood backboard is not included.

**Ceiling mounting:** Includes unloading and uncrating, hauling boxes from the loading dock, measuring and marking, drilling four lead anchor-type fasteners using a hammer drill, and installing and leveling boxes using rolling staging.

**Freestanding cabinets:** Includes unloading and uncrating, hauling cabinets from the loading dock, marking of floor, drilling four lead anchor-type fasteners using a hammer drill, and leveling and shimming.

**Telephone cabinets:** Include unloading and uncrating; hauling cabinets; measuring and marking; and mounting and leveling, using four lead anchor-type fasteners.

These additional items must be listed and extended:

    Punching or drilling holes for conduit
    Wire and terminations

## Takeoff Procedure:
Verify the box requirements per the specifications. List cabinets and pull boxes by NEMA type and size.

| Example: | Quantity | Type | Size |
|---|---|---|---|
| | 3 | NEMA 1 | 6"W x 6"H x 4"D |
| | 2 | NEMA 3R | 6"W x 6"H x 4"D |

Itemize and detail any special support or hanger requirements.

# Outlet Boxes

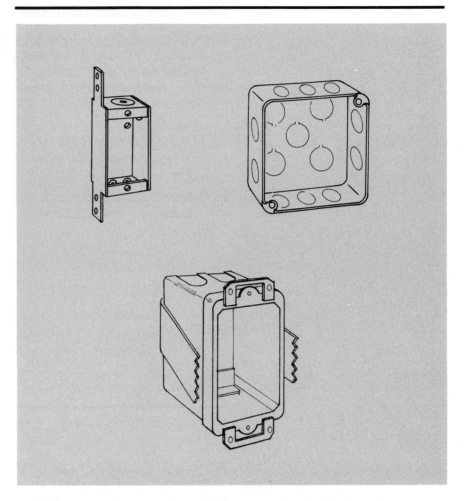

Outlet boxes made of steel or plastic are used to hold wiring devices, such as switches and receptacles. They are also used as a mount for lighting fixtures. Some outlet boxes are ready for flush mounting; others require a plaster frame. Some have plain knockouts for pipe up to 1¼", while others have built-in brackets for Romex or BX wire. The capacity of outlet boxes ranges from one to six devices. They may have brackets for direct stud mounting or plaster ears for mounting in existing walls.

**Units for Measure**: Outlet boxes are counted as each (EA.). Related accessories, such as plaster rings and covers are also counted individually.

**Material Units**: The following items are generally included per unit:

> Box and fastener

**Labor Units**: The following procedures are generally included per outlet box:

> Marking the box location
> Mounting the box
> Material handling

*Takeoff Procedure:* Outlet boxes should be included on the same takeoff sheet as branch wiring or devices to better explain what is included in each circuit. Each unit price in this section is a stand-alone item and contains no other component unless specified. To estimate a duplex outlet, for example, the following components may be listed:

> Outlet box
> Plaster ring
> Duplex receptacle from Div. 16.2-30
> Device cover

*Cost Modifications:* For large concentrations of plastic boxes in the same area, the following percentages can be deducted from man-hours totals:

| | | | |
|---|---|---|---|
| 1 | to | 25 | – 0% |
| 26 | | 50 | –15% |
| 51 | | 75 | –20% |
| 76 | | 100 | –25% |
| Over 100 | | | –30% |

Note: It is important to understand that these percentages are not used on the total job quantities, but only in areas where concentrations reach the levels specified.

# Wiring Devices and Low-Voltage Switching

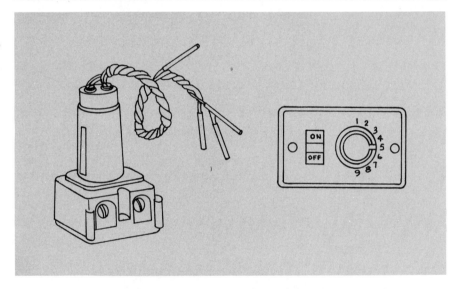

A device is, by NEC definition, "a unit of an electrical system which is intended to carry but not to utilize electric energy." Wiring devices include receptacles, wall switches, pilot lights and wall plates. Low-voltage switching includes the above plus relays, transformers, rectifiers and controls.

Receptacles are a convenient means of connecting portable equipment and appliances to power. Receptacles are available in voltage ratings ranging from 125 to 600, and in amperages from 10 to 400. A wide variety of configurations is used to avoid having a plug or cap with a certain voltage and ampere rating inserted into a receptacle of another rating. Switches, used to turn off power to lights and receptacles are also rated for different voltages and amperages. Low-voltage switching operates at

24V through a transformer and uses relays to control 120V lighting and above. Operating at a low voltage allows the use of multiconductor #18 wire — a size much smaller than regular wire. A master control can be used to operate all lighting from one location.

**Units for Measure:** Wiring devices are counted as each (EA.). It is important to list and quantify all associated items with each device, such as plates, boxes, and plaster rings.

**Material Units:** The following items are generally included per unit:

Receptacles
Wall switches
Pilot lights

Low-voltage switching should include a count of the above plus these items:

Relays
Transformers
Rectifiers
Controls

**Labor Units:** The following procedures are generally included per unit:

Stripping of wire insulation
Attaching wire to device using terminals on the device itself
Mounting of device in box

These additional items are listed and extended:

Conduit
Wire
Boxes
Cover plates
Wire splices

**Takeoff Procedure:** Wiring devices should be priced on a takeoff form that includes associated boxes, covers, conduit and wire. List the device as indicated on the drawings and note the legend symbol beside it. Then list all associated items necessary to complete the installation, i.e., box, cover plate, etc. Count each device, marking the drawing as you go. Total and transfer to a cost analysis sheet for pricing and extension. When practical, get vendor quotes for the material portions.

**Cost Modifications:** For large concentrations of devices in the same area, deduct the following percentages from man-hours:

| | | | |
|---|---|---|---|
| 1 | to | 10 | — 0% |
| 11 | | 25 | —20% |
| 25 | | 50 | —25% |
| 51 | | 100 | —35% |
| Over 100 | | | —35% |

# Fasteners

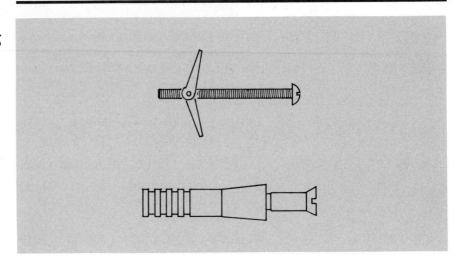

Fasteners include various types of anchors, bolts, screws, nails, rivets, and studs. These are included to provide a reasonable guide to costs. However, conditions vary so greatly that one cost cannot be applied to all situations. Thought must be given to the particulars of each installation and costs adjusted accordingly.

*Units for Measure:*  Fasteners are counted as units (EA.).

*Material Units:*  A material unit generally is an assembly of the bolt, nut, washer, and anchor (as required).

*Labor Units:*  The following procedures are generally included per unit:

> Drilling an appropriate mounting hole for the bolt (no holes are drilled in the device being mounted)
> Inserting and setting the anchor (when appropriate)
> Threading on and tightening the fastener

These additional items are listed and extended:

> Torquing and testing

*Takeoff Procedure:*  As devices, boxes, panels, etc. are taken off, be cognizant of the assumptions made for mounting. If these allowances are inadequate, add fasteners accordingly. Also, when developing new installation costs, include fastener costs.

This takeoff is performed on the same sheet as that for the device being mounted. Transfer the quantities to a cost analysis sheet and extend.

*Cost Modifications:*  Most installation costs include anchors, with several installed at one time in one area. Labor costs must be added for unusually difficult situations, or reduced for a large quantity installed at one time in the same area. These judgements must be left to the discretion of the estimator.

A fairly accurate count is important at this time so that quantities of specified or specialty fasteners can be ordered. (It is costly to waste field labor for lack of enough bolts!)

# Hangers

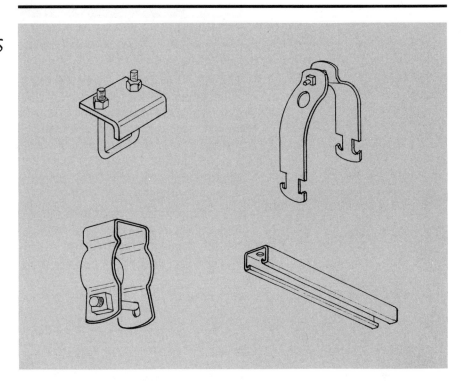

For our purposes, a hanger is a means of suspending or supporting electrical raceway installations. As mentioned in previous chapters, hangers are often included in the linear foot cost of a raceway at predetermined intervals. Variations from these standard assumptions become cost considerations for the estimator.

Hangers can be grouped into two distinct categories — those that are shop-manufactured, and those that are field-fabricated. Manufactured hangers are designed to support by attaching to an existing building structure without modification. Examples of this are one- or two-hole straps, beam clamps, or riser clamps. A field-fabricated hanger is the manual combination of two or more components to provide support. A trapeze-type hanger is a typical example. To assemble a trapeze-type hanger, the following components are required:

> Channel steel ("Unistrut")
> Threaded rod
> Nuts, bolts and washers
> Support steel
> Beam clamps or fasteners
> Conduit straps

Each of these items involves a material and labor cost. These must be combined for an accurate cost assessment. To disregard the cost of hangers (or to assume that the linear foot cost of raceways will be accurate enough) can be a costly approach. This is particularly true on projects such as process piping where whole networks of conduit racking systems are used.

## Units for Measure:
Unique hangers must first be estimated as mini-systems — arriving at a cost for material and labor for each specific configuration. The hangers are then counted as each (EA.).

## Material Units:
The material price for hangers generally includes only the hanger itself (with the particular components required to assemble it).

*Labor Units*: The assembly of each component and installation of the completed unit are generally the only items included in the man-hours.

*Takeoff Procedure*: Check the prints and specs for details on hanger arrangements. If the arrangement is similar to those priced into the linear foot quantities of raceway, then only a count for material procurement may be necessary. This count is the L.F. total for each size of raceway, divided by the specified distance between supports. If the hangers and supports are field fabricated, price them out as subassemblies in order to get the total cost of each fabricated hanger. Then divide the L.F. total of raceway by the specified distance between supports.

If the hangers are supporting groups of conduits, then count each hanger individually. Set up the work sheet by the type and size of hanger. Take off the quantities, making sure to include risers and clamps if required. Transfer the quantities to a cost analysis sheet and extend.

# Chapter 15
# STARTERS, BOARDS, AND SWITCHES

Starters are electric controllers that accelerate a motor from rest to running speed; they are also used to step the motor. Boards serve as a mounting for electric components and/or controls. Switches are devices used to open, close, or change the connection of an electric circuit. This chapter contains 18 sections and includes components ranging from panelboards and circuit breakers to motor control centers and meter sockets. Each of these components is first defined, then supplied with the appropriate units for measure. Following are material and labor requirements and the procedures involved in the takeoff.

## Circuit Breakers

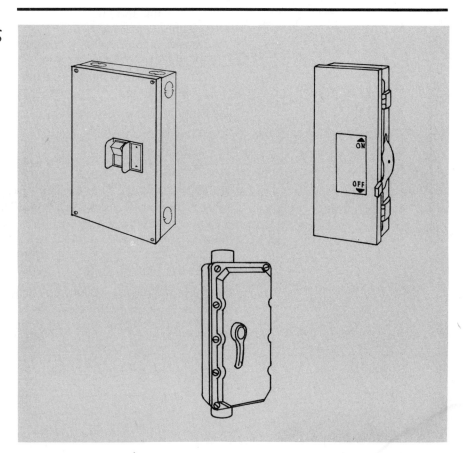

Circuit breakers are used in general distribution and in branch circuits to protect the wires and equipment downstream from current overload. Circuit breakers are rated in amperes and are capable of interrupting their rated current at their rated voltage. There are several types of circuit breakers: magnetic trip only, molded case, current-limiting (which includes three coordinated current-limiting fuses), and electronic trip. An important parameter in selecting a breaker is its fault current interrupting capacity. Several different grades of breakers are available in most sizes, for different uses and conditions. The common Ampere Interruption Capacity ratings are 10,000 AIC; 22,000 AIC; 42,000 AIC; and 65,000 AIC. Their application is determined by the amount of energy available at the breaker's line side during a short-circuit fault.

Breakers are available in several voltages from 120 to 600, in one, two, or three poles. Enclosed breakers are available up to 2000 Amps. Breakers come in frame sizes of 100, 225, 400, 1000, 1200 and 2000 Amps. Several different ampere ratings use the same frame size.

Example: 400-Ampere frame size has ratings of 125, 150, 175, 200, 225, 250, 300, 350, and 400 Amps, all of which have the same price. When estimating circuit breakers, it is important to use correct NEMA designations pertaining to areas of use.

## Units for Measure: Circuit breakers are characterized in terms of each (EA.) breaker.

## Material Units: The cost for each circuit breaker generally includes 4 lead anchor-type wall fasteners.

## Labor Units: The following procedures are generally included per unit:

Unloading and uncrating
Hauling equipment
Measuring and marking the circuit breaker location
Installing the circuit breaker
Connecting and phase marking wire on primary and secondary
    terminals

These additional procedures are listed and extended:

Primary or secondary conduit
Backboards
Wire

## Takeoff Procedure: First, set up the work sheet by circuit breaker type, voltage, number of poles, NEMA classification, and ampere rating. Count the circuit breakers, listing each one under the proper designation on the quantity sheet as you count. In this way, there is no need to hunt for like circuit breakers to count them before proceeding with the next type. Then transfer the total quantities to a cost analysis sheet and extend.

## Cost Modifications: If several circuit breakers are installed in a common location by the same electrician or crew, an adjustment can be made. Deduct the following percentages from the labor units:

| | |
|---|---|
| 0-5 | — 0% |
| 5-10 | — 5% |
| 10-20 | —10% |
| 21-30 | —15% |
| Over 30 | —18% |

Example: Six 30-Amp. 600V circuit breakers are to be installed from common 12" x 12" wireway. Adjust the man-hours in this way.

| Quantity | | Man-Hours | | Adjustment | | Adjusted Man-Hours |
|---|---|---|---|---|---|---|
| 6 | x | 2.5 | x | .95 | = | 14.25 M.H. |

## Control Stations

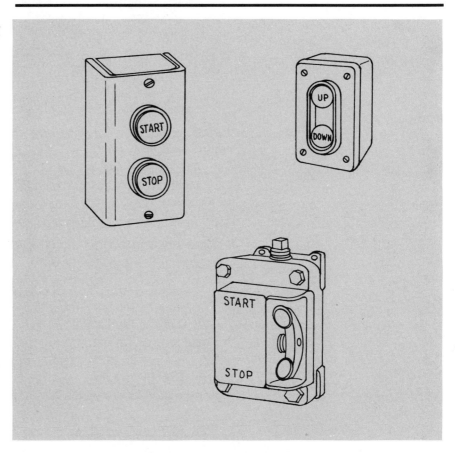

(Also see "Control Switches", listed separately in this section.)

Control stations are enclosures for mounting control switches and pilot lights.

Standard-duty control stations usually come completely assembled as an enclosure with control switches and legends. Standard-duty control stations have a capacity for up to 3 units and use either push buttons or selector switches. Pilot lights are also available with these stations.

Control stations are assembled in a wide variety of configurations and NEMA classifications. Standard-duty stations with NEMA enclosures have 1 to 3 positions (holes), while NEMA 4, 7, and 9 have 1 or 2 positions.

Other control stations (without switches) are heavy-duty oiltight NEMA 4 and 13 with up to 30 positions; or watertight, corrosion-resistant NEMA 4, 4X, or B for up to 25 units.

NEMA 1 enclosures come with prepunched knock-outs. However, other NEMA enclosures should be specified for hole sizes and arrangement at the time of purchase.

Oiltight and watertight enclosures may be ordered with controls premounted. This saves considerable man-hours in the field.

**Units for Measure:**  Each station (enclosure) is counted as a unit (EA.).

**Material Units:**  The following items are generally included per control station:

>   Enclosure
>   Switch assembly (for some stations)
>   Wire marker for each contact
>   Legend tag for station

These additional items are listed and extended:

>   Branch wire
>   Branch conduit
>   Switches, pilot lights, and legends for each station

**Labor Units:**  The following procedures are generally included per control station:

>   Installing enclosure
>   Mounting switches if required
>   Testing

These additional procedures are listed and extended:

>   Installing branch wire to enclosure
>   Terminations of control wires
>   Marking of control wires
>   Installing conduit to enclosure
>   Cutting or punching mounting hole (see "Cutting and Drilling" in Chapter 12: Raceways)

**Takeoff Procedure:**  Set up the work sheet by type. Take care to note how many devices are in each enclosure and in each type of enclosure. In most cases, this work can be taken off most conveniently at the same time as motors and motor controls are taken off.

## Fuses

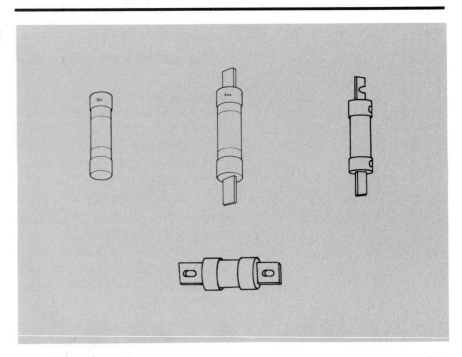

Fuses are used to interrupt a circuit in the event of overload. Fuses have a fusible link through which the current must flow. If too much current flows, the resulting heat melts the link. This stops the flow of current. There are many different types of fuses, such as plug, cartridge, and bolt-on. Fuses are designed with fast-acting or time-delay links. Both types will have the ability to react quickly to short-circuit currents.

Some plug fuses have an Edison screw base, which is the same as a medium base lamp. Others have a type "S" base, which requires an adapter screwed into the medium base socket. The purpose of these adapters is to discourage the use of fuses larger than the size appropriate to the existing wiring. Plug fuses are made up to 30 amps at 125 volts. Cartridge fuses are available in many types, including renewable and non-renewable, dual element, time delaying, and current limiting. Most are rated 250-volt or 600-volt and ampacities up to 600 amps. Bolt-on (cartridge) fuses are available to 6000 amps. Cartridge fuses are categorized by class designations such as Class H, K, RK5, J RK1, or L. These designations indicate the interrupting rating (i.e., 10,000 AMP; 100,000 AMP; or 200,000 AMP) and the performance characteristics of the fuse.

### Units for Measure: Fuses are counted as each (EA.). Categorize fuses by voltage rating, current rating, and class.

### Material Units: The fuse itself is generally the only item included.

### Labor Units: The following procedures are generally included per unit:

    Replacing the fuse if necessary
    Installing new fuses
    Testing with a continuity tester

### Typical Job Conditions: The fuseholder is to be in an easily accessible location in an unlocked enclosure.

### Takeoff Procedure: Be careful to identify the particular type of fuse that is to be used. List by Amp., voltage, and type or class. Be sure to count one fuse for each line (or phase) being protected. List on the quantity sheet and mark the drawing.

# Load Centers

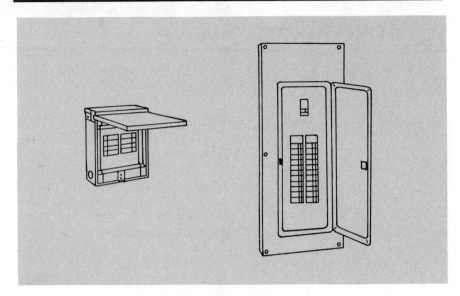

Load centers are a specialized type of panelboard used principally for residential applications. These panels are designed for lighter sustained loads than those used in industrial and commercial applications.

Load centers are available from 100-amp to a maximum 200-amp rating in several configurations: i.e., 120/240 V, 1 phase, 3 wire; 120/208 V, 3 phase, 4 wire; 240 V, 1 phase, 2 wire; 120/240 V, 3 phase, 4 wire; and 240 V, 3 phase, 3 wire. Load centers generally use plug-in circuit breakers. The single-pole breakers range from 15-50 amps; double-pole breakers from 15-100 amps; and three-pole breakers from 15-100 amps. Two- and three-wire switched neutral and ground fault interrupting styles are also available.

Lighting and appliance load centers are limited to a maximum of 42 overcurrent devices in each cabinet.

*Units for Measure*: Load centers are counted as each (EA.) unit by type and size. While standard breakers may be included, care must be taken to ensure that all special service breakers are noted and priced as additional items.

*Material Units*: The following items are generally included with load centers:

> Breakers
> Box
> Cover
> Trim

*Labor Units*: The following procedures are generally included per load center:

> Receiving
> Measuring and marking walls
> Drilling 4 lead anchor-type fasteners using a hammer drill
> Mounting and leveling panel
> Preparation and termination of feeder cable to lugs or main breaker
> Testing and load balancing
> Marking panel directory

These additional procedures are listed and extended:

> Modifications to enclosure (extra breakers)
> Structural supports
> Additional lugs
> Plywood backboards
> Painting or lettering

Note: Knockouts are included in the price of terminating conduit runs and need not be added to the load center costs.

*Takeoff Procedure*: When pricing load centers, list panels by size and type. List breakers in a separate column of the quantity sheet, and define by phase (poles) and ampere rating.

*Cost Modifications*: If a load center does not include a full panel of single-pole breakers, use the following method to adjust material and labor costs. For example, in an 18-circuit panel, only 16 single-pole breakers are needed.

> 18 breakers included
> 16 breakers needed
> _____
> 2 breakers need to be deleted

To adjust for the breakers that are not required, find the unit cost of a 20A single-pole breaker:

|  | **Typical Costs** |
|---|---|
| Single pole breaker = | $ 5.80 for material |
|  | $14.95 for labor |
|  | $20.75 total material and labor |

Multiply the material price of the breaker by .50.

> $5.80 x .50  =  $2.90

Multiply the labor price by .60.

> $14.95 x .60  =  $8.97

The cost to delete per breaker will be $2.90 for material and $8.97 for labor. Multiply these figures by the number of breakers not required, in this case, two.

> $2.90 x 2  =  $ 5.80 material
> $8.97 x 2  =  $17.94 labor
> _____
> $23.74 to be deducted from the cost of the panel

# Meter Centers and Sockets

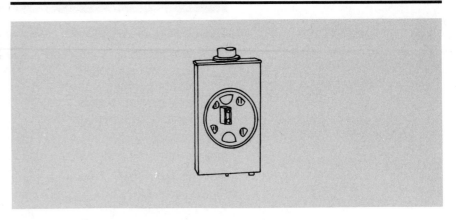

Meter Sockets are enclosures designed to receive the plug-in utility watt meters that monitor a customer's power usage. For multiple tenants in commercial or residential buildings, Meter Centers may be used to monitor and distribute a single service entrance cable to two or more different users. This is accomplished with multiple sockets.

A Meter Center may have a main breaker, a fusible disconnect, or a remote protection device. If it feeds two or more branch services, a Meter Center will usually have individual meter breakers. Meter Centers are available in a variety of configurations for one to six meters and can be bussed together for larger groupings. Bus capacities normally range from 125 amps to 800 amps, though Meter Centers with a capacity to 1200 amps may be special ordered.

The two styles of enclosures commonly used are surface-mounted and semiflush; NEMA 1 (general, indoor) and NEMA 3R (outdoor, rainproof) are available with provisions for top or bottom feed of the main service cable.

Meters are owned and installed by the utility company. Therefore, no cost allowance needs to be made for material or labor.

*Units for Measure*: Meter Sockets are counted as units (EA.). A Meter Center is a unit (EA.) for multiple sockets.

*Material Units*: The following items are generally included per unit:

> Meter Socket or a Meter Center
> Wall anchor fasteners

These additional items must also be listed and extended:

> Main circuit breaker or fused disconnect
> Tenant (or Meter) circuit breakers
> Conduit
> Termination lugs

*Labor Units*: The following procedures are generally included per Meter Socket or Meter Center:

> Receiving
> Handling
> Mounting and alignment
> Legend identifications

These additional procedures must be listed and extended:

Conduit
Cables
Terminations
Breakers
Testing

## Takeoff Procedure:

For Meter Sockets:

List all sizes and types of the sockets required. Count each Meter Socket and list it under the appropriate heading. Total each column, transfer these figures to a cost analysis sheet, and extend.

For Meter Centers:

List each Meter Center, noting its bus capacity, number of Meter Sockets, and type of enclosure. Take off the size of the main breaker or fusible disconnect switch. List the size of each tenant breaker. Total each column, transfer these figures to a cost analysis sheet, and extend.

# Motor Control Center

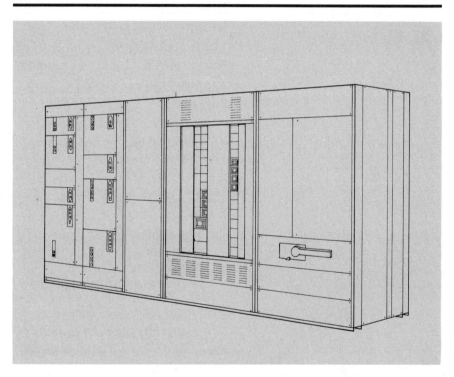

Many commercial buildings and process plants maximize efficiency by grouping electrical controls into centralized locations. Motor control centers (MCC's) serve this need by providing a structure for mounting a variety of motor starters, auxiliary controls, and feeder tap units. A motor control center is a collection of motor control equipment and bus bars assembled in a series of steel-clad enclosures. The enclosure, including the bus, is called a ''structure'' and will take any combination of starters up to 72" high. Some structures allow for mounting starters on both the front and back sides.

A starter combination MCP, FVNR with control transformer size 1 is 12" high; thus, six size 1 starters could fit into one structure. Size 2 is 18" high, so four size 2, or two size 2 with three size 1 would be acceptable. Size 3 is 24", 4 is 30", 5 is 48", and 6 is 72". Assorted combinations of these (and other) controls can be assembled. Also, many structures may be "bussed" together to expand the MCC. The enclosures are commonly available in NEMA classes:

NEMA 1:  General purpose, indoor
NEMA 2:  Dripproof
NEMA 3R: Rainproof
NEMA 12: Dusttight and driptight, indoor

There are two class designations for MCC wiring. In Class I, all internal control wiring is done in the field. In Class II, the control circuits are factory-wired to terminal strips. There are also three NEMA-type designations: Type A, B, or C. The type indicates how the control and power cables are terminated (i.e., wired to terminal blocks or direct to the devices).

Internal bus bars are usually aluminum, but copper is often specified. We recommend that you contact manufacturers for complete material prices. Generally, an MCC is ordered as a complete assembly, including starters, breakers, doors, etc. The MCC can, however, be modified with field-installed components.

## Units for Measure:  Each section is counted as one unit (EA.). All components are listed and totalled.

## Material Units:  The following items are generally included in the takeoff:

Motor control center complete as required
Thermal heaters as required
Fuses as required
Wire markers

## Labor Units:  The following procedures are generally included in the man-hours:

Receiving
Handling
Setting equipment
Leveling and shimming
Anchoring
Cable and wire terminations (power only, not control cables)
Cable identification (power)

These additional procedures are listed and extended:

Rigging
Equipment pad
Steel channels embedded or grouted in concrete
Conduit and wire
Testing

## Takeoff Procedure:  When pricing the structure, be careful to check the specs for NEMA Class, bus size, copper or aluminum bus, and NEMA type. List each MCC by the engineer's designation, and every section in each MCC. List (per MCC) all components which will be ordered in the MCC. (Do not add labor when components are factory-installed.)

# Motor Control Center Components

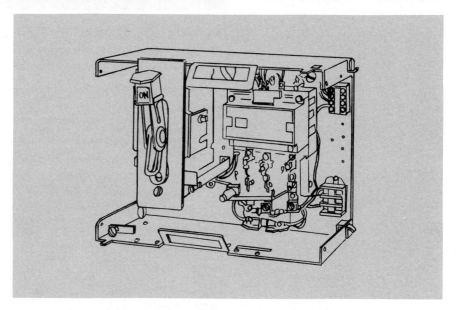

Motor control center (MCC) components are modular devices that plug into an MCC structure's bus. They are generally called out, ordered, and shipped with the MCC as a complete assembly. Occasionally, components are purchased individually to be installed into an existing MCC. When ordering components to fit an existing installation, it is necessary to order the same make and model of components. The products of different manufacturers may not be interchangeable.

Some of the commonly used components are combination starters with either circuit breakers or fused disconnects; reversing and two-speed motor starters; feeder circuit breakers; main circuit breakers; lighting panelboards; and transformers.

The NEC wiring class and type must also be noted. For example, with type A components, all field control wiring must be terminated directly to the device. This unit is less expensive to purchase but more costly to install. A type B component has all intra-wiring brought out to terminal blocks (only the power cable is terminated directly to the starter). This approach simplifies field installation and maintenance repairs. Type C components are similar to type B except that terminal blocks are also provided for the load terminals in starters size 3 and under. Installation labor is about the same for types B and C, but maintenance and service are easier to perform.

While some general material costs are available from reference books, by far the most reliable pricing source will be the specified manufacturer or his representative.

*Units for Measure*: These components are listed as each unit (EA.) with careful attention to the use and to specifications.

*Material Units*: The following items are generally included per MCC component:

> Starter or other components as required
> Thermal heater as required
> Fuses as required
> Wire markers

These additional items are listed and extended:

   Motor control center structure
   Any wire or raceways to the structure

*Labor Units:* The following procedures are generally included per MCC component:

   Receiving
   Equipment handling
   Installing components in structure
   Connecting power wires to components
   Wire identification
   Installing 3 thermal heaters where required

These additional procedures are listed and extended:

   Testing
   Control cables

*Typical Job Conditions:* Productivity (labor) is based on the MCC being in place, ready to receive components.

*Takeoff Procedure:* List all components on the quantity sheet, being careful to categorize each component by location, size, type, class, and service. Total like items, transfer these figures to a cost analysis sheet, and extend.

## Motor Connections

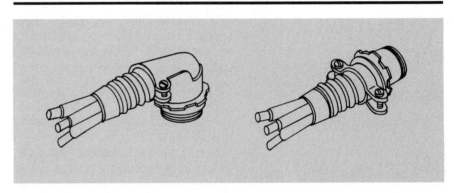

A motor connection is the means by which the power leads are terminated to a motor. From the estimator's perspective, the motor terminations must also account for connecting the flexible conduit and fittings from the rigid conduit to the motor terminal box.

Flexible conduit to motors serves four functions. First, it can absorb the normal motor vibrations that would, over time, fatigue and crack rigid conduit connections. Second, since the vibration does not loosen flexible conduit as it would rigid parts, the ground path continuity is not disrupted. Third, most motor installations require that the motor be movable so that its shaft can be aligned with the shaft of the driven device. Finally, there is a distinct labor savings in mating a flexible conduit to a motor box over field-bending rigid conduit to fit.

Many motors have a winding space heater. That cable is often routed and counted as a separate connection.

*Units for Measure*: Each cable to a motor is counted as one unit (EA.).

*Material Units*: This is actually a mini-system. The size of each component is pre-determined by the standard size of cable and raceway used by each horsepower motor. This approach saves the estimator from having to dig through a conduit plan to find the size of cable and conduit servicing the motor.

The following items are generally included for motor connectors:

> Flex connectors
> Flexible metallic conduit
> Wire markers
> Termination for cable conductors

*Labor Units*: The following procedures are generally included in the man-hours for motor connectors:

> Cutting flexible metal conduit
> Installing two flex connectors
> Routing wire through flex
> Terminating leads to motor
> Identifying leads

These additional procedures are listed and extended:

> Mounting or placing motor (only when done by electricians)
> Disconnect switch
> Motor starter
> Motor controls
> Conduit and wire
> Checking motor rotation

*Takeoff Procedure*: When taking off motor connections, it is advisable to list connections on the same work sheet as motors and/or motor starters. Set up columns by the horsepower rating of the motors. Count the number of motors to be connected and transfer the results to the quantity sheet. Be sure to mark all motors that are counted on the print. Transfer quantities from the takeoff sheet to a cost analysis sheet and extend.

*Cost Modifications*: If several motors are to be connected in a concentrated area by the same electrical crew, the following percentages can be deducted from labor only:

| | | | |
|---|---|---|---|
| 1 | to | 10 | – 0% |
| 11 | | 25 | –20% |
| 26 | | 50 | –30% |

# Motors

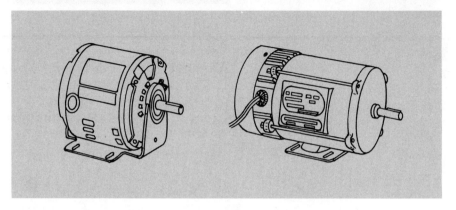

Motors are machines for converting electrical energy into mechanical energy. They are used in appliances, elevators, and all kinds of machinery. Motors are available in a multitude of sizes, voltages, types, and enclosures. Motor sizes vary from fractional HP, single phase to as high as 2000 HP, 3 phase. Voltages typically range from 120 V to 4160 V AC. The most common are fractional 110 volt to 200 HP 600 volt. The most common housings are dripproof and totally enclosed.

A range of DC motors operating at voltages from 12 V DC to 250 V DC are used for some special applications (such as elevators). Their use is, however, quite rare. For estimating purposes, use cost data from the next size larger AC motor.

Motors are not usually placed by the electrician, but rather by the trade furnishing the equipment they power.

*Units for Measure*: The unit for measure in this case is each motor (EA.).

*Material Units*: Included in the material units is the motor only.

These additional items are listed and extended:

> Pulleys
> Sheaves
> Couplings

*Labor Units*: The following procedures are generally included per motor:

> Receiving
> Handling
> Setting the motor in place
> Bolting or fastening
> Preliminary alignment

These additional procedures are listed and extended:

> Connecting the motor
> Testing for rotation
> Final alignment (usually by millwrights)
> Heater connections (temporary power) for motors in storage

*Takeoff Procedure*: Carefully list all sizes of motors by horsepower, number of phases, and voltage. Motors should be listed at the same time as starters, disconnects, controls, and motor connections. Before pricing motors, check to see who furnishes them.

# Motor Starters and Controls

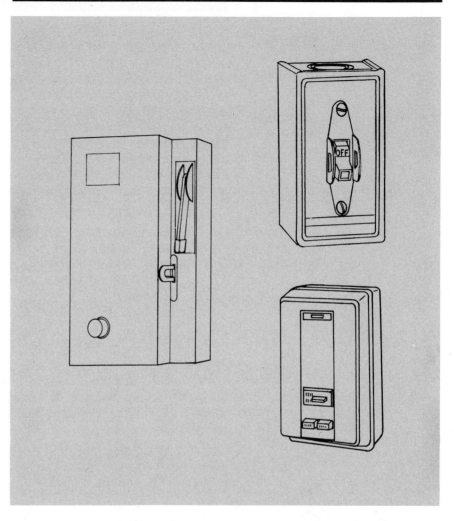

Motor Starters and across-the-line starters are manufactured in a variety of sizes and types. The most common types of across-the-line starters are the following:

Motor-starting switch with no overload protection.
Single throw switch with overload protection.
Magnetic switch with thermal overload protection.

A *motor-starting switch* is simply a tumbler rotary, lever, or drum switch. It does not provide any protection against overload or inadequate voltage. It is used for nonreversing duty in small size motors up to 2 HP., and for reversing duty up to 10 HP.

A *single throw, across-the-line switch* can obtain thermal overload protection by means of thermal cutouts, time-lag fuses, or by a thermal overload release device. In switches with a thermal overload release device, the overload trips the holding catch of the switch and allows the switch to open. These starters do not provide protection against inadequate voltage.

*Magnetic across-the-line starting switches* (motor contactors) are magnetically operated and controlled from a push-button station — either integral or remote. They are manufactured with thermal-relay protection against overload and inadequate voltage. Pressing the start button closes the circuit to the operating coil of the magnetic switch. When the stop button is pressed, the operating coil circuit is opened, allowing the switch to spring open; the motor is stopped. Motor contactors are made in a variety of types, including both non-reversing and reversing duty, and for dual-speed motors. The reversing types consist of two contactors electrically and mechanically interlocked so as to prevent the closing of both switches at the same time.

Variations of magnetic across-the-line starting switches are as follows:

> Combination, with motor circuit-breaker protection
> Combination, with fused switch
> Magnetic (FVR) full voltage reversing with control circuit transformer
> Magnetic (FVR) full voltage dual-speed with control circuit transformer
> Combination (FVR), fused, with control transformer and push buttons

It is important to identify not only the types of starters, but also the type of enclosure for each starter. Most estimating standards price motor starters with enclosures included in order to show the costs of the different types. For example, the material cost of a 10 H.P. size 1 starter with a NEMA 1 enclosure is $125.00. The same starter with a NEMA 7 enclosure is $450.00.

Starter sizes are determined by the voltage and horsepower rating of the motors being controlled. The following chart can be used if the horsepower of the motor is known, but the starter size is not specified.

## STARTER SIZES

| Motor Voltage (60 Hz) | Motor Horsepower (max) | Starter Size |
|---|---|---|
| Single Phase | | |
| 115 volt | 1/3 | 00 |
| | 1 | 0 |
| | 2 | 1 |
| | 3 | 2 |
| 230 volt | 1 | 00 |
| | 2 | 0 |
| | 3 | 1 |
| | 5 | 1 |
| | 7-1/2 | 2 |
| Three Phase | | |
| 200/230 volt | 1-1/2 | 00 |
| | 3 | 0 |
| | 7-1/2 | 1 |
| | 10/15 | 2 |
| | 25/30 | 3 |
| | 40/50 | 4 |
| | 75/100 | 5 |
| | 150/200 | 6 |
| | 300 | 7 |
| | 450 | 8 |
| | 800 | 9 |
| 460/575 volt | 2 | 00 |
| | 5 | 0 |
| | 10 | 1 |
| | 25 | 2 |
| | 50 | 3 |
| | 100 | 4 |
| | 200 | 5 |
| | 400 | 6 |
| | 600 | 7 |
| | 900 | 8 |
| | 1600 | 9 |

*Units for Measure*: Starters and contactors are priced as each (EA.). Starters may or may not designate a mini-system (i.e., with circuit protection and/or controls included).

Overload heaters are priced per line phase (EA.).

*Material Units*: The following items are generally included in the material cost for each starter:

Starter and enclosure
Four lead anchor-type fasteners
Six wire markers
Three thermal overload heaters
Three dual-element cartridge-type fuses (for a combination-type starter)

Note: When pricing motor controls, be sure to consider the control system to be utilized. If the controls are factory-installed and located in the enclosure itself, then two options must be considered in pricing. (1) Is the integral control already included in the price of the starter? (2) If the control is not included, then its price must be added to the cost of the starter.

These additional items are listed and extended:

> Raceway to or from the starter
> Wire
> Control switches (unless included in material price line)
> Control transformer
> Plywood backboard (if required)

*Labor Units*: These procedures are generally included in the man hours for starters:

> Unloading, uncrating, and handling of starters
> Measuring and marking
> 4 lead anchor-type fasteners
> Termination of power (line) and load side cables
> Wire identification
> Installing 3 thermal overload heaters

This additional procedure must be listed and extended:

> Testing

*Typical Job Conditions*: Productivity is based on new construction and an installation height of 5' above grade.

*Takeoff Procedure*: Note: Make sure that the starters are not furnished by others. Check the specifications. (Very often the starters are furnished as part of a mechanical or HVAC package.)

List the starters by size, voltage, NEMA enclosure, and type. Count each starter and list it in the proper grouping, rather than searching for all of one type and size before proceeding to another. Most starters are listed on a legend schedule which will indicate if integral controls are to be included. If the control is remotely located, the starter can be priced alone. If it is not, then an integral control system must be added to that starter's price. Transfer quantities to a cost analysis sheet and extend.

Be aware of what is included in the material cost of each starter. If a total material list is required for purchasing, then list each component on the same sheet, extend against the quantity of each size starter. Do not assign a material or labor cost, because these are included in the cost units.

Note: The selection and purchasing of overload heaters for starters must be based on the actual name plate amps of the motors installed. It is not prudent to purchase overload heaters from general charts or tables.

*Cost Modifications*: When several magnetic starters are installed in a common location by the same electrician or crew, the following percentages can be deducted from the labor units:

| | |
|---|---|
| 0-5 | − 0% |
| 5-10 | − 5% |
| 11-20 | −10% |
| 21-30 | −15% |
| Over 30 | −18% |

# Contactors

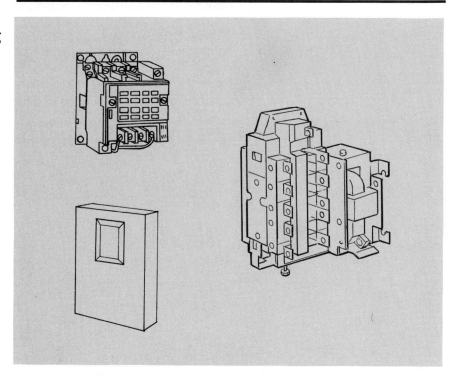

A magnetic contactor is basically a relay with heavy-duty contacts. An electromagnetic coil is energized to close the contacts and hold them closed. Magnetic contactors are used to switch heating loads, capacitors, transformers, or electric motors. They do not have overload protection and are available from 1 to 5 poles (i.e., separate circuits). Contactors are controlled by a wall switch or push-button station. Magnetic contactors are similar to magnetic motor starters, with the exception that they do not contain overload protection devices.

Lighting contactors are a combination of a magnetic contactor with either fuse clips or a circuit breaker. They are used to control lighting and electric resistance heating loads.

*Units for Measure*: Count each contactor as one unit (EA.).

*Material Units*: The following items are generally included per unit:

> Contactor
> Enclosure
> Fasteners

These additional items must be listed and extended:

> Conduit
> Wire
> Control switches
> Tags or labels

*Labor Units*: The following procedures are generally included per unit:

> Receiving
> Handling
> Mounting

These additional procedures must be listed and extended:

> Control and power terminations
> Control switches

*Takeoff Procedure*: List the contactors by size, voltage, and amp ratings. Note the enclosure types required. Enter each contactor (as it is encountered) onto the takeoff sheets, rather than trying to search out all the devices of one category before proceeding to the next. Transfer the quantity totals to a cost analysis sheet and extend.

## Relays

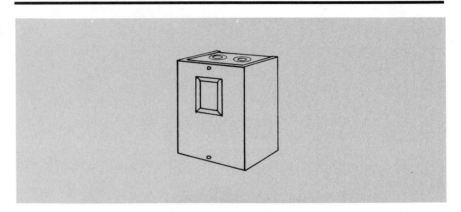

A relay is a control device that takes one input to a coil and operates a number of isolated circuit contacts in a control scheme. An electromagnetic coil is used to operate an armature, which holds the contacts. The relay's contacts may be set up in the following ways: open at rest (normally open), and closing when energized; or closed at rest (normally closed), and opening when energized. A relay may have from 1 to 12 or more poles.

A wide variety of relay configurations is available. Some common variations are time-delay relays (either delay on or delay off); latching relays (mechanically held needing two coils: one for on, another for off); and AC or DC relays.

Coils for relays can be specified for either AC or DC circuits and for voltages from 6 to 480.

Although relays can be purchased in an enclosure such as a NEMA 1 general-purpose indoor, they are usually bought individually for assembly into a larger enclosure.

*Units for Measure*: Relays are counted as units (EA.).

*Material Units*: Generally, only the relay itself is included.

These additional items are listed and extended:

      Enclosure
      Fasteners

*Labor Units*: The following procedures are generally included per unit:

      Installing the relay in an enclosure
      Connecting the wires (coil only)

These additional procedures are listed and extended:

      Mounting the enclosure
      Fastening to the wall
      Conduit
      Wire
      Control terminations

*Takeoff Procedure*: Relays should be taken off on the same sheet as the item they control. Price this item by carefully checking the catalogue number. There are many types of relays, and prices vary considerably.

Carefully note the coil voltage, contact voltage and amp rating, number of poles, and type of operation. List each relay on the quantity takeoff and total. Transfer the totals to a cost analysis sheet and extend.

## Panelboards

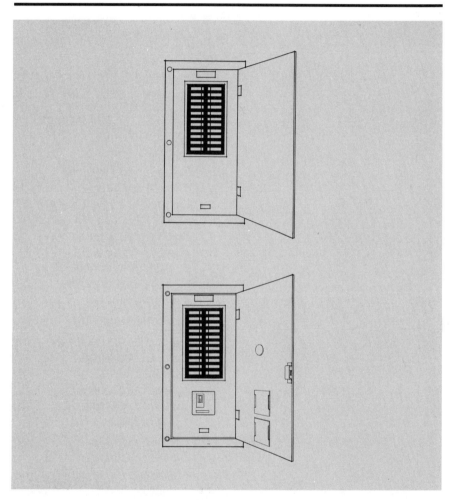

Panelboards are used to group circuit switching and protective devices into one enclosure. Lighting and appliance panelboards for residential applications are referred to as "load centers". (See the section on "Load Centers" earlier in this chapter.)

Panelboards consist of an assembly of bus bars and circuit protective devices housed in a metal box enclosure. The box provides space for wiring and includes a trim plate with a cover.

Although panelboards are available as fused models, styles using molded case circuit breakers are far more common. Panelboards are available in a variety of configurations, including AC styles up to 600 V and DC styles up to 250 V. The AC styles can be single- or three-phase. Models for AC up to 240 volts, with mains up to 600 amps and branch circuits up to 100 amps are commonly available for lighting and power distribution. Panelboards for motor starters and power distribution may have mains up to 1200 A and branch breakers up to 600 A.

Panelboard enclosures are also available to suit a variety of NEMA classifications, including NEMA 1, 3, and 12.

Due to the wide variety of options and configurations available, it is best to consult a vendor or manufacturer's representative for pricing information.

**Units for Measure:** List by size (ampacity) and type, voltage, and fault current capacity. List each panelboard (EA.). Then record the breakers with each board and price each (EA.). (Note: Some standard board and breaker assemblies are available as pre-assembled units.)

**Material Units:** The following items are generally included per panelboard:

> Panelboard
> Fasteners

**Labor Units:** The following procedures are generally included for panelboard installation:

> Receiving
> Measuring and marking
> Drilling 4 lead anchor-type fasteners using a hammer drill
> Mounting and leveling panel
> Preparation and termination of feeder cable to lugs or main breaker
> Branch circuit identification
> Lacing using tie wraps
> Testing and load balancing
> Marking panel directory
> Cover and trim

These additional procedures are listed and extended:

> Modification to enclosure (additional breakers)
> Structural supports
> Additional lugs
> Plywood backboards
> Painting or lettering
> Adding breakers
> Field assembly
> Terminating branch wiring

Note: Knockouts are included in the price of terminating conduit runs and need not be added to the panelboard costs.

Material and Labor prices generally include breakers.

**Takeoff Procedure:** When pricing panelboards, list the panels by size, type and voltage. List breakers in a separate column of the quantity sheet, and define by phase (poles) and ampere rating.

**Cost Modifications:** If you do not plan to include a full panel of single pole breakers, use the method as described in "Load Centers".

# Panelboard Circuit Breakers

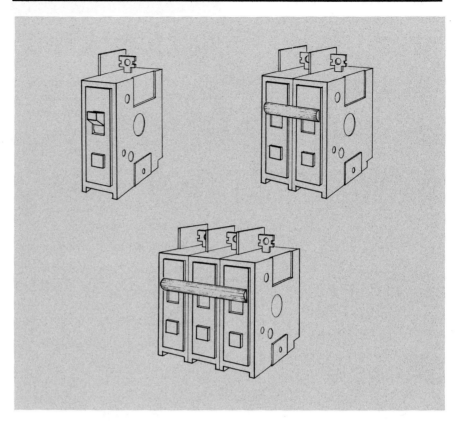

A panelboard circuit breaker is a molded-case plastic breaker (with one, two, or three poles). Two styles are available: plug-on and bolted (to the bus bars). There are models for either DC service up to 250 volts or for AC service up to 600 volts. Main circuit breakers come in many sizes ranging from 100 amps up to 1200 amps and branch circuit breakers from 15 amps to 800 amps.

A major consideration in selecting and pricing circuit breakers is interrupting capacity (I.C.). This is a breaker's ability to stop the flow of electricity when a high-current fault exists (which is dependent upon the amount of energy available from the source). Breakers are made with I.C. ratings as low as 5,000 amps and range up to 200,000 AIC. A typical lighting and branch power panel might have a 150-amp main rated at 65,000 AIC with branch breakers from 15 to 30 amps rated at 10,000 AIC.

Most circuit breakers have two modes of tripping. These are time (thermal) trip and instant (magnetic) trip. Some breakers also have a third mode — ground fault (shunt coil). The first type of trip occurs in the case of mild overloads and works by heating a thermal trip device. The time required for this function is related to the amount of the overload current. For example, a 200% overload may trip the breaker in 100 seconds, while a 600% overload can trip in two seconds.

Breakers also operate on high-fault overloads via a magnetic trip device. This function is referred to as an "instantaneous" trip. The time required for this function is from ½ to 1 cycle or .0083 to .0166 seconds. The magnetic trip function generally operates at between 10 and 20 times (1000 to 2000%) the rated current and above. The wide variety of breakers available includes many that are designed to differ from these characteristic tripping curves.

Ground fault tripping breakers are the third type. Most breakers do not have this kind of trip function. Nevertheless, it is a requirement for specific applications that involve potential personal hazards. In ground fault tripping breakers, a sensitive current coil is used to assure that the amperage going out of the breaker closely equals the current returning through the neutral wire. When these two currents are unbalanced by as little as 5 MA, the coil activates a shunt-tripping device. This process occurs as quickly as 8 milliseconds.

*Units for Measure*:  List by size, type, AIC rating, and manufacturer, and price per each (EA.) breaker.

*Material Units*:  The following items are generally included for panelboard circuit breakers:

> The breaker itself and the lugs

These additional items must be listed and extended:

> Panelboard
> Wiring
> Tags and markers

*Labor Units*:  The following procedures are generally included:

> Removing the trim from the panelboard
> Installing the circuit breaker
> Connecting the wires to the circuit breaker
> Installing the trim

These additional procedures must be listed and extended:

> Wire and conduit connections to the panelboard
> Stripping jackets from the wires

*Takeoff Procedure*:  List the breakers by amps, number of poles, voltage and plug-in or bolt-on. Note on the takeoff the brand and model of panelboard in which the breaker is to be installed. Use the catalogue number if it is available. (Use a cost book for pricing, but the catalogue number when purchasing.)

Note:  Some manufacturers' breakers are interchangeable, but many are not.

## Safety Switches

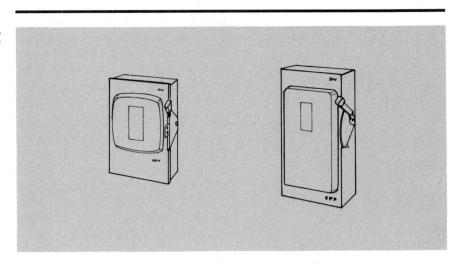

Safety switches are intended for use in general distribution and in branch circuits. They provide an assured means of manually disconnecting a load from its source. The salient feature of a safety switch is that its operating handle is capable of being padlocked in the "OFF" position. This feature protects those working on the equipment from the possibility that someone might inadvertently energize the circuit. For some devices, such as fans, not only is a safety switch required by code, but the switch must be installed within sight of the fan.

Safety switches are designated in three different categories. General-duty safety switches are rated from 30 to 600 amps at 240 V AC. They are available in 2-, 3-, and 4-pole styles, fusible or non-fusible.

Heavy-duty safety switches are rated from 30 to 1200 amps and 250 volts AC or DC, or 600 volts AC. They are available in 2-, 3-, or 4-pole and in either fusible or non-fusible models.

Motor circuit switches rated at 30 to 200 amps are intended specifically for applications where the switch must be within sight of the motor. They are non-fusible and have either 3 poles or 6 poles (for dual-speed motors).

The safety switches described above are furnished in a number of different NEMA class enclosures. It is important to the estimate that the appropriate enclosure be priced. For example, a general-duty switch in a NEMA 1 enclosure may cost around $220, while the same switch in a NEMA 4X enclosure could cost about $1200.

Units for Measure: Safety switches are priced each (EA.) by size, type, poles, voltage, and enclosure ratings.

Material Units: The following items are generally included per unit:

> One safety switch
> Four lead anchor-type fasteners
> Fuses (if fused type)

These additional items must be listed and extended:

> Cable and conduit terminations
> Tags and labels
> Brackets or supports (if needed)

Labor Units: The man-hours for each safety switch generally include these items:

> Receiving
> Material handling
> Measuring and marking the switch location
> Installing the switch on a masonry wall
> Connecting and phase marking wire on primary and secondary
>    terminals

These additional procedures are listed and extended:

> Primary or secondary conduit
> Backboards, brackets, or supports for enclosure
> Wire

*Takeoff Procedure:* Set up the work sheet by duty type, voltage, number of poles, NEMA classification, and ampere rating. Count the switches, transferring each into its proper designation on the quantity sheet as you count. In this way, you can avoid hunting for like switches before proceeding to the next type. Finally, transfer the quantities to a cost analysis sheet and extend.

*Cost Modifications:* If several disconnects are being installed in a common location by the same electrician or crew, the following percentages can be deducted from the labor units:

| | | | |
|---|---|---|---|
| 0 | to | 5 | – 0% |
| 5 | | 10 | – 5% |
| Over 10 | | | –10% |

# Switchboards

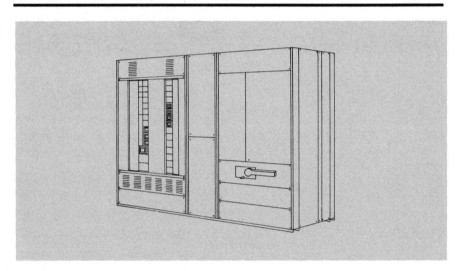

Switchboards are used in buildings that have larger load requirements than can be serviced by a single load center panelboard and its disconnect device. A switchboard is a modular assembly of functional sections or compartments. These include the service section, auxiliary section, metering compartment, and distribution sections.

The service section contains the main (incoming) breaker, which may be rated from 200 to 4,000 amps. An auxiliary section is a blank compartment which may be used to facilitate cable pulling. A transition section is provided where needed to interface with transformers, special distribution sections, or an MCC. A metering compartment may contain current transformers (CT's), potential transformers (PT's), and relays, as well as meters for amps, volts and watts. The distribution sections may contain any combination of fused disconnect switches, branch circuit breakers, or motor starters. The branch circuit breakers may be rated from 30 to 1200 amps.

Bus bars are available in ratings ranging from 200 through 4,000 amps at 250 V DC or 600 V AC. Bus bars are generally braced to withstand 50,000-amp fault currents, but optional bracing is available to increase that rating – up to as much as 200,000 amps. Bus bars are aluminum, but copper bars are an available option.

Three options are available for wiring. These are Type A — without terminal blocks; Type B — with terminal blocks for control wiring at the side of each unit; and Type C — with master terminal blocks at the top or bottom of each switchboard section.

## Units for Measure:
Switchboards must be itemized per section (EA.); the subcomponents of each distribution section must be listed and detailed per unit (EA.). Note any modifications or extras (such as the copper bus option) and quantify accordingly (per section, for example).

The material and labor units for each section type generally assume that the entire unit is specified in detail during the procurement phase and that the equipment will be factory assembled with all requisite components in place. Some sections may be shipped separately and bolted together in the field.

## Material Units:
The material cost units for switchboards include:

> The equipment as shipped from the manufacturer
> Bolts, bus ties, and insulators to connect shipping segments
> Anchor bolts and fasteners
> Minor components (such as relays) which may be shipped for field insertion — does not include wiring or connections to such devices by field personnel

These additional items are listed and extended:

> Equipment pads
> Steel channels
> Special knockouts
> External controls
> Conduit
> Wire and cables

## Labor Units:
The following procedures are generally included for the installation of switchboard components:

> Receiving
> Equipment handling to location
> Assembly of shipping sections
> Anchors and bolting in place
> Leveling, shimming, and alignment
> Interconnection of bus bars and wiring
> Marking panel legend

The following activities are listed and extended as additional labor items:

> Conduit
> Wiring and terminations
> Wire tagging and identifications
> Testing
> Adding breakers, switches, and meters to existing switchboards

## Takeoff Procedure:
It is recommended that switchboards and their associated components be quoted from equipment manufacturers as a factory-assembled unit. The following items should be included:

> Switchboards
> Feeder sections
> Instruments
> Distribution sections and subcomponents
> Options and modifications

The takeoff procedure must identify the equipment by section and by the subcomponents within each section. Parameters such as amperage, voltage, type, bus material, and breaker sizes, should be detailed. List the sections by amps, volts, and type. Identify the section's components by type and size. List any options or modifications per section and extend.

## Substations

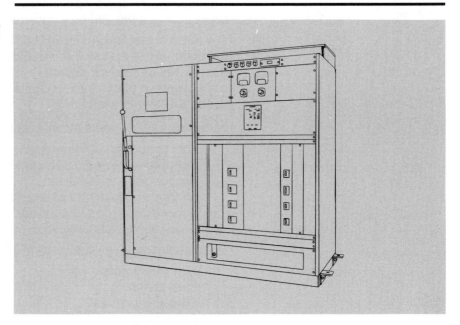

A substation, or secondary unit substation, is an assembly consisting of a high-voltage incoming line section, a step down, three-phase power transformer, and a secondary low-voltage distribution section. Substations have two principle applications, both of which are industrial. The first is as an incoming service center; the second is as a power center with dense, high-power requirements.

There are many advantages in locating the power transformer and its distribution section close to the load. First, the cable and or bus lengths can be reduced. There is also better voltage control, reduced power loss, and a minimum installed cost.

Typical transformer ratings are from 112 KVA through 2500 KVA. The most common primary voltages are 4.16 KV and 13.8 KV, but the available range is from 2.4 KV through 34.5 KV. The most common secondary voltage is 480/277 V, but 208, 240 and 600 volts are also used.

Depending upon the application, the high-voltage line section may be a terminal compartment for cables to connect to the transformer, or it may contain a disconnect switch (fused or non-fused).

The distribution section is an attached switchboard which usually uses draw-out circuit breakers for power distribution. The bus ratings typically run from 800 to 4,000 amps. The draw-out breakers can be operated manually or electrically. Distribution sections with molded case breakers are also available. In most cases, one compartment of the distribution section is devoted to meters, instruments, and trip relays.

*Units for Measure*: Each substation must be taken off individually and listed in terms of its component sections and their subcomponents (EA.).

As is the case with most major electrical assemblies, substations are normally specified for purchase as a factory-assembled unit. The various sections, however, are usually split for shipping and are then set and bolted together in place at the site. Draw-out breakers are shipped individually to be slid in after the equipment is set.

*Material Units*: The following items are generally included in the material cost unit:

> The transformers
> Incoming line section and switch
> Distribution section with breakers
> Anchor bolts
> Bus bars and bolts for interconnections

These additional items are taken off and listed:

> Equipment pads and rails
> Rigging
> Conduit
> Cable
> Terminations

*Labor Units*: The following procedures are generally included per unit:

> Receiving
> Equipment handling
> Setting, shimming, and leveling
> Anchoring in place

These additional procedures must be listed and extended:

> Cables and terminations
> Potheads
> Testing

*Takeoff Procedure*: List each substation individually by size, voltage, and type of transformer. Then list each component section with its subcomponents. Transfer to a cost analysis sheet and extend.

## Control Switches

Control switches provide manual input to the control circuits of relays or magnetic starters. Control switches are used for equipment such as valves, fans, pumps, conveyors, and air conditioners. Control switches are assembled or "built-up" using the following three types of components: 1) the legend plate, 2) the operator mechanism, and 3) the contact blocks. Each of these three is available in a multitude of styles and types which can be combined in nearly endless variations to suit the particular application.

*Legend plates* are used to describe the functions controlled by a particular control switch. They are ordered either by stock descriptions (stop, start, jog, up, down, etc.) or with unique inscriptions. The legend plate is usually included in the material and labor pricing of the operator.

*Operators* are the mechanical devices that position the contacts. A whole range of operators is available. Some common types are listed below:

Selector switch – maintained position
Selector switch – spring return
Push button – spring return
Push button – with illuminated button

Also available is a variety of indicating lights, styled to match the operator. These lights indicate the operating status of the equipment.

*Contact blocks* provide control and are mounted behind the operator in modular (stackable) fashion. Contacts are described at their "rest" position as either "normally open (N.O.)" or "normally closed (N.C.)". A third contact arrangement is a block which has one N.O. pole and one N.C. pole. The majority of switches use some combination of these three basic contact blocks. An operator can move the contact to the "held" position in order to effect control.

Common operator and contact arrangements are assembled and "in stock", while others must be ordered as components and field assembled. Attention must be paid to the voltage and amp ratings of the contacts when ordering.

Control switches may be installed into distribution equipment such as motor control centers for local control. Control switches may also be installed into control station boxes for remote control functions.

## Units for Measure: Control switches are counted individually (EA.) and listed according to type, contact arrangement, and legend function. Each pilot light is counted and listed according to color and voltage.

## Material Units: The following items are generally included in the material cost of each control switch:

Legend plate
Retaining rings
Operator
Contact block
Wire marker for each contact

These additional items are listed and extended:

Enclosure
Wire
Branch conduit

## Labor Units: The following procedures are generally included in the man-hours for control switches:

Installing control switches in a pre-punched enclosure
Marking control wires
Testing switch

These additional procedures are listed and extended:

Installing wire to control switch
Termination of control wires
Cutting or punching mounting hole
Installing enclosure
Branch conduit to enclosure

## Takeoff Procedure: The biggest misconception on pricing control switches is that the estimator needs to understand all the functions being controlled by the switches. Most control prints contain a detailed legend of control components, including types and catalogue numbers. Thus, the estimator need not involve himself with the operation of the equipment.

Set up the work sheet by the type of control switch. Quantities are best taken off from a schematic control drawing. Be sure to identify those installations that involve manually punching or cutting mounting holes. Transfer the quantities to the work sheet and extend.

## Cost Modifications: Most control station enclosures are prepunched. However, if control switches are to be installed in a custom-made enclosure where the holes must be field determined and punched, the labor cost must be increased. Refer to "Cutting and Drilling" in the "Raceways" chapter, or to other standards being used; then choose the hole size required.

For a large concentration of control switches in the same enclosure, the following percentages can be deducted from man-hours:

| | | | |
|---|---|---|---|
| 1 | to | 10 | – 0% |
| 11 | | 25 | –20% |
| 26 | | 50 | –25% |
| 51 | | 100 | –30% |
| Over 100 | | | –35% |

# Chapter 16
# TRANSFORMERS AND BUS DUCT

Transformers are devices with two or more coupled windings, with or without a magnetic core. They introduce mutual coupling between circuits and are used to convert a power supply from one voltage to another. Bus duct, or "busway", is a prefabricated unit that contains and protects one or more busses. In this chapter, there are four sections on transformers, isolating panels, and two types of bus duct. Units for measure, material and labor requirements, and a procedure for takeoff follow each description.

## Transformers

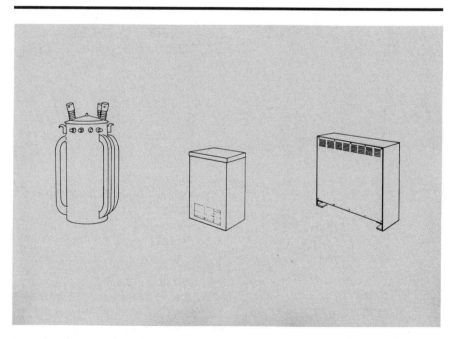

Transformers are used in Alternating Current (AC) systems to convert from one voltage to another. They cannot be used in Direct Current (DC) systems. The total energy entering a transformer is equal (except for minor losses) to the energy leaving the transformer. Thus, when the voltage is changed (stepped down, for instance), the current changes in inverse proportion (in this case, it increases). A transformer's capacity can, therefore, be designated in terms of the product of the voltage (V) and the amperage (A), i.e., as volt-amps (VA) at either side. Larger transformers are often designated in thousand volt-amps (KVA).

Transformers are used for four basic applications: 1) instrument transformers, 2) control transformers, 3) isolating transformers, and 4) power transformers. Their capacities range from fractional VA to thousands of KVA. The first two are single-phase types only, but the second two may be single- or three-phase designs.

Power transformers may be either dry-type air-cooled or liquid-cooled. Both can be furnished for indoor or outdoor use (dry-type in a NEMA 3R enclosure). Voltage ratings are always described from the primary (incoming) to the secondary (output).

*Units for Measure*: Each transformer is counted as a unit (EA.).

*Material Units*: The following items are generally included in the material unit price:

> Transformer
> Anchor bolts or screws
> Identification tag

These additional items must be listed and extended:

> Transformer pad
> Grounding
> Conduit
> Wire and cables
> Terminations plugs

*Labor Units*: The following procedures are generally included in the man-hours for transformers:

> Receiving
> Handling
> Crane (if needed)
> Setting or hanging in place
> Anchor bolts

These additional procedures must be listed and extended:

> Supports, brackets, etc.
> Conduit and connections
> Cables and terminations
> Grounding
> Testing

*Typical Job Conditions*: The labor units assume reasonable access and distance from the loading dock at about 100 feet. Adjustment factors will need to be applied for work in congested areas or for material handling that involves especially long distances.

*Takeoff Procedure*: List the transformers by VA rating and service voltages on the material takeoff sheet, and total. Transfer the totals to a cost analysis sheet. Price and extend.

# Isolating Panels

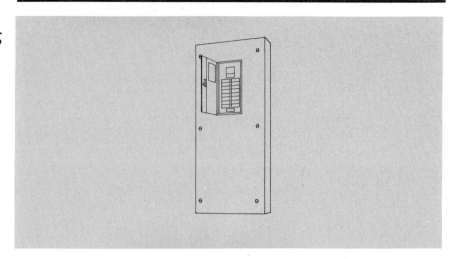

Isolating panels are used in hospitals, where they serve as added protection – to patients and sensitive monitoring equipment – from the effects of ground potentials. This kind of protection is particularly important in a hospital environment because of the numerous electrical instruments and appliances with wet systems. An operating room panel typically consists of a back box with stainless steel trim. It contains an isolating transformer, a primary circuit breaker, eight 20-amp 2-pole secondary circuit breakers, and a line isolation monitor.

The critical care area panels are the same as operating room panels, with the addition of 8 power receptacles and 6 grounding jacks.

The X-ray panels are also similar to operating room panels, but they have 1 60-amp, 2-pole breaker instead of 8 20-amp, 2-pole breakers. Up to 8 outlets can be controlled from this panel.

**Units for Measure**:  Each panel is a unit (EA.) including breakers, transformers, etc.

**Material Units**:  A complete panel with anchors is generally included in the material unit price.

**Labor Units**:  The following procedures should be included in the man-hours per unit:

> Receiving
> Handling
> Setting in place with wall-mounting hardware

These additional procedures are listed and extended:

> Making the connections to primary and secondary
> Testing
> Structural support

**Takeoff Procedure**:  List the panels individually on the price sheet.

# Bus Duct

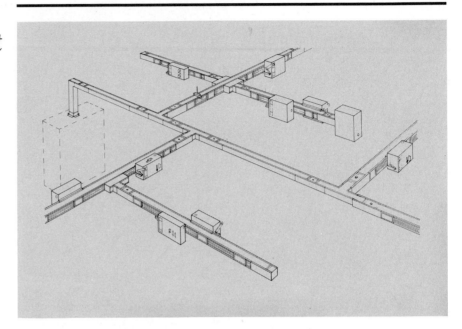

Bus duct provides a flexible distribution system for power in industrial and commercial buildings. The busway itself consists of copper or aluminum bars mounted in a sheet metal enclosure. There are several types available. The most common are feeder-type (indoor), plug-in type (indoor), and weatherproof feeder-type.

In some styles of bus ducts, branch taps can be readily changed to conform to new locations of motors and equipment. This system is widely used in industrial plants where equipment is continually changed to meet new manufacturing conditions.

Feeder-type bus duct is used primarily in industrial buildings to connect the service entrance to the main switchboard, or for high-capacity feeders to distribution centers, including feeder risers. Feeder types have no access (or taps) between the service starting point, or main, and the point of use.

Plug-in bus duct is used for indoor systems where flexibility is needed for power over a wide area. It is available in ratings from 225 amps to 3000 amps in a 3-pole, 4-wire arrangement. Sections are available in 1' through 5', and 10' lengths. Provisions for plug-in taps are made through outlets located on both sides of the duct.

Weatherproof feeder-type bus duct performs the same function as feeder type busway, but it is enclosed in a weatherproof casing for use outdoors or in damp, indoor areas.

*Plug-in Units:* After the bus or feeder duct has been installed, plug-in, modular units are used to tap or branch off from the main bars of the bus duct. There are various types of modules designed for specific branch circuit needs. Some of the most common are listed below:

> Fusible 30-amp to 400-amp capacity
> Fusible with contactor 30-amp to 200-amp capacity
> Circuit breaker type 30-amp to 1600-amp capacity
> Circuit breaker with contactor 30-amp to 225-amp capacity

**Bus Duct Fittings:** Standard components are available to provide elbows, offsets, "T" connections, and crossovers. Other fittings include reducers for changing the size of the busway, end closers, and expansion fittings. Cable tap boxes are available to provide cable or conduit tap-offs or feeders. They are made in two types:

> End cable tap boxes for attachment at the ends of a plug-in busway run
>
> Center cable tap boxes for installation at any point in a busway run

**Units for Measure:** Straight sections of bus duct are measured in linear feet (L.F.) and any part sections (less than 10') as each unit (EA.). Fittings are counted and listed as units (EA.).

**Material Units:** The following items are generally included in the L.F. cost of bus duct straight sections:

> Bus duct (based on 10' length)
> Hanger every 5 L.F.
> 2' of threaded rod per hanger
> 1 beam type clamp per hanger

These additional items must be listed and extended:

> Bus duct fittings
> Support brackets
> Auxiliary steel
> Wall penetrations
> Scaffold rentals

**Labor Units:** The following procedures are generally included in the man-hours for straight sections of bus duct:

> Receiving
> Handling
> Measuring and marking
> Set up of rolling staging
> Installing hangers (beam clamp type)
> Hanging and bolting sections
> Aligning and leveling

These additional procedures are listed and extended:

> Bus duct fittings
> Modifications to structure for hanger supports (auxiliary steel)
> Threaded rod in excess of 2' per hanger
> Welding (if required)
> Penetration through walls
> Installing fittings, 90's, T's, crosses, etc.
> Testing

**Takeoff Procedure:** Set up the quantity sheet by type, ampere rating and voltage. Measure from the source and work out to the point of use or end, depending on the type of bus duct.

Caution: When pricing the material, do not assume that all bus duct sections are 10-foot lengths. While most estimating standards base their L.F. price on the cost of 10' sections, those that are less than 10' are much more expensive per linear foot. Care should be taken to identify variable lengths. Keep these odd lengths separate from the linear foot totals and price them using manufacturers' quotations.

Go back over the print; count the fittings; and identify them by type, ampere rating, and configuration. Count accessories, such as wall flanges, weather stops, enclosures, Switchboard stubs, and spring-type hangers. Transfer the quantities' totals to a cost analysis sheet and extend.

**Cost Modifications:** A large quantity of bus duct might be installed at varying heights. The estimate can be adjusted for these conditions.

Man-hours are generally based on elevations up to 15'. Add the following percentages for higher runs:

| | | | |
|---|---|---|---|
| 16' | to | 20' high | + 10% |
| 21' | | 25' | + 20% |
| 26' | | 30' | + 30% |
| 31' | | 35' | + 40% |
| 36' | | 40' | + 50% |
| Over 40' | | | + 60% |

Add these percentages to the man-hours only for the installation of those quantities that exceed the specified height levels.

## Bus Duct – 100 Amp and Less

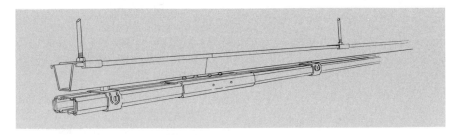

100-amp plug-in busway is similar to larger busway (as described previously in this chapter). It differs in that it has plated round aluminum or copper bars instead of flat bars. 100-amp bus duct is typically used for branch power feeders to panel boards or motors, or for small, distributed loads.

30- and 60-amp plug-in busway has an open bottom into which the terminal plug is inserted. It is available in 4' or 10' lengths and in two- and three-wire configurations. The bus can be cut with a hacksaw to length, but it must then be capped. A strength beam is available to aid in mounting and to add rigidity. This kind of busway is designed to distribute power for all types of lighting, but it can also be used for light-duty power.

Like 30- and 60-amp busway, 20-amp plug-in busway also has an open bottom and has similar applications. It is also used for small portable tools and for drop light cords and receptacles.

**Units for Measure:** Take off the linear feet (L.F.) of straight sections and the required fittings (EA.).

**Material Units:** The following items are generally included with bus duct for 100 amps and less:

Straight sections of bus
Connectors and bolts
1 hanger every 5'
Beam clamp and 2' threaded rod for each hanger

These additional items must be listed and extended:

- Fittings
- Scaffolding
- Brackets
- Auxiliary steel

## Labor Units: The following procedures are generally included:

- Receiving
- Handling
- Installing hangers
- Installing bus duct
- Leveling and alignment
- Rolling scaffold

These additional procedures must be listed and extended:

- Wall penetrations
- Supports and auxiliary steel
- Welding (if needed)
- Fittings

## Takeoff Procedure: See "BUS DUCT" for takeoff methods, labor assumptions, and cost modifications.

# Chapter 17
# POWER SYSTEMS AND CAPACITORS

Power systems are stand-by or emergency generator sets for providing power to essential services during a loss of normal power. Capacitors are used for power factor correction to increase efficiency, thus reducing electrical utility bills. Definitions are provided for each of the components in this chapter. Units for measure, material and labor requirements, and a takeoff procedure are also provided.

## Capacitors

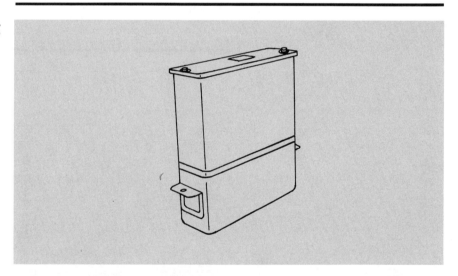

Power factor is a measurement of the ratio between true power and apparent power, i.e., P.F. = W / VA. The ideal P.F. is 1.0. In industrial and commercial buildings, reactive loads, such as motors, can cause phase shifts between the voltage (V) and the current (A) which, in turn, create undesirable power factors. By installing a capacitive load in the system, the P.F. can be adjusted toward 1.0.

Industrial capacitors are metal-clad units which include one (single-phase) or three (three-phase) capacitors and discharge resistors. Industrial capacitors may be connected to the loads of each individual motor or to the main incoming lines. Capacitors are sized in terms of KVAC at a specific voltage and according to the number of phases.

**Units for Measure**: Capacitors are counted as each unit (EA.).

**Material Units**: The following items are generally included per unit:

Capacitor
Fasteners

These additional items must be listed and extended:

Cable
Terminations

**Labor Units**: The following procedures are generally included per unit:

Receiving
Handling
Measuring and marking the capacitor location
Mounting

These additional procedures must be listed and extended:

Wire
Connecting and marking wire terminals
Conduit
Testing

**Takeoff Procedure**: Take off capacitors at the same time as motors and controls. Check to see if they are included in the price of switchgear. Be sure to specify voltage, phase(s) and KVAC size. List each capacitor according to its size and type; add the total quantities. Transfer the quantities to a cost analysis sheet and extend.

## Generator Set

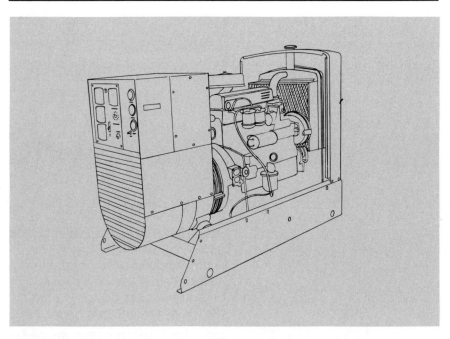

Generator sets are used as an emergency or stand-by power source in the event of the loss of normal power. They are rated in terms of their capacity in kilowatts (KW). Generators are available in single-phase or three-phase and may be specified at 120 through 6600 volts. (Typical units are used at 120 V single-phase or 480 V three-phase.)

The motors may be fueled by gasoline, propane (or natural) gas, and diesel fuel. Gasoline engines have a low initial cost and start well in cold weather. Gaseous fuels require less engine maintenance and are generally more convenient to supply. Diesel units consume far less fuel than gasoline and gaseous units, are generally available in larger capacities, and the motors require less maintenance. Diesel fuel also has a higher flash point than gasoline or gaseous fuels and, thus, is safer to handle.

In general, gas generator sets are available from 10 to 170 KW and diesel-powered sets range from 15 to 1000 KW. (Even larger diesel units are available as custom-ordered units.) Although some small models may be started with a recoil rope starter, most units are fitted with electric starters and batteries. Control options include local or remote start/stop functions, engine heaters, exhaust mufflers (larger units), and annunciators.

*Units for Measure*:  Generators are counted as units (EA.).

*Material Units*:  The following items are generally included per unit:

> An engine generator with a control panel
> A battery
> A charger
> A muffler (small units)
> A day tank for fuel (small units)

These additional items must be listed and extended:

> Control and accessory items
> Conduit
> Cables
> Transfer switch
> Weather-tight enclosure

*Labor Units*:  The following procedures are generally included per unit:

> Receiving
> Equipment handling
> Setting, leveling, and shimming
> Anchoring

These additional items are listed and extended:

> A muffler and exhaust piping (large units)
> Fuel and fuel piping
> A large fuel tank
> A concrete pad
> Steel channels other than normally supplied by the manufacturer
> Rigging
> Connecting of wire and cable
> Testing

*Takeoff Procedure*:  Check the generator specs and get quotes from manufacturers. The generator is usually priced as a lump sum for all parts listed under "Materials". Check to see who furnishes the material listed above as "These additional items. . .". It might be necessary to get the prices for these items from other trades (such as for cooling water). List each generator on a cost analysis sheet and extend. Note all pertinent options and accessories. Also note any special cranes or equipment needed for handling. Determine if any accessories will need to be field assembled.

Note: Some manufacturers will require that their service representative be present for initial start-up in order to validate the warranty. This will usually be an extra-cost item, but it can be a very sound investment.

## Automatic and Manual Transfer Switches

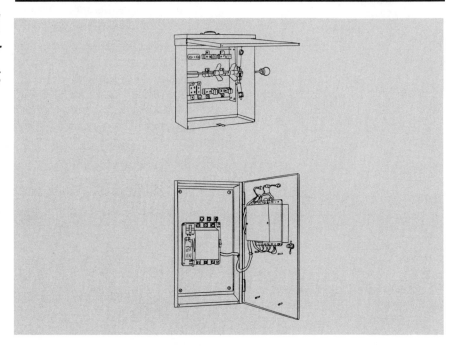

Transfer switches are used to change from the normal source of power to an alternate source, such as a generator set or a battery-powered inverter, for example. There are two basic types of transfer switches — manual or automatic.

A manual transfer switch is a lever- or handle-operated double-throw device with a pole for each line wire. This type of switch is available from 30 to 600 amps.

Automatic transfer switches are designed to monitor the normal line source and to switch the load electrically to the backup source should the normal source fail. Automatic transfer switches are rated from 30 to 2000 amps.

A number of extra-cost options and accessories are available for automatic transfer switches. A few of the options are auto-start relays to signal a generator when normal power fails, sensing relays to prevent transfer until the backup source is at full voltage, status-indicating lights, timed relay transfer to restore the load to normal power after it has been energized for several minutes, time-delay engine-stop relay to keep the generator running unloaded for a few minutes in order to cool the windings, and test and exercise control switches to permit the generator set to be started periodically and run.

*Units for Measure:* Switches are counted as units (EA.).

*Material Units:* The following items are generally included per unit:

Transfer switch, complete in enclosure
Fasteners

These additional items are listed and extended:

Optional control functions
Conduit
Cable
Terminations

*Labor Units:* The following procedures are generally included per unit:

Receiving
Material handling
Measuring and marking
Drilling four lead-type anchors, using a hammer drill
Mounting and leveling

These additional procedures are listed and extended:

Modifications in the enclosure
Structural supports
Preparation and terminating the cable to the lugs
Circuit identification
Testing
Plywood backboard
Painting or lettering
Conduit

*Takeoff Procedure:* When taking off transfer switches, identify by type (auto or manual), voltage, amperage, and number of poles. List each on a cost analysis sheet, along with any required options or accessories, and extend.

Note: It is best to get a vendor's quote for pricing, particularly if control options are specified. Also, check the generator package, as the switches can often be purchased from the same manufacturer. If not, the control interface may not be compatible.

# Chapter 18
# LIGHTING

By far the most elemental portion of electrical construction is lighting. The first commercial use of electricity was for lighting. Today, virtually every building, industrial plant, house, bridge, and roadway sign makes extensive use of lighting fixtures. In building construction, lighting is still the largest single electrical cost center. This fact is evidenced in the sample estimate, Part III of this book.

This chapter contains descriptions of seven different types of electrical lighting, fixtures, and fixture whips. In addition to the description, each component section also includes guidelines on units for measure, material and labor requirements, and a step-by-step takeoff procedure. Where applicable, cost modification factors — for economy of scale or difficult working conditions — are provided.

## Interior Lighting Fixtures

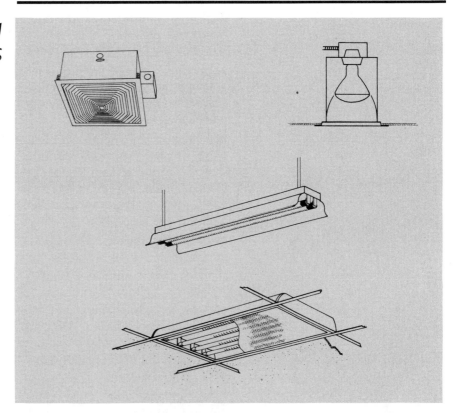

Fixture styles for interior building lighting can be either surface-mounted or recessed in a wall or ceiling. Other options are pendant or hanging fixtures. There are, of course, numerous variations of all of these styles.

Not only is lighting a major cost center for installation, but it is also a significant factor in the operating costs of a facility. In the past 15 years, much attention has been given to the development of more efficient lamps. Efficiency in fixtures is measured in the number of lumens per watt. The color spectrum of light from various lamps is also a prime factor in judging efficiency. That is to say, if a bulb could be made that was, for example, 80% efficient at converting watts to lumens, but the light energy was entirely in the ultraviolet region (invisible to human eyes), its useful output would be zero.

Lamps for interior lighting fixtures can be incandescent, fluorescent, or high-intensity discharge (HID). HID lamps include mercury vapor, metal halide, or high-pressure sodium. (Low-pressure sodium lamps are used principally for outdoor applications because of their strong yellow color.)

*Incandescent* bulbs are the least efficient of the above, but their low initial cost keeps them attractive. Also, their "warm" (high red content) color spectrum is pleasing for many applications. One significant advantage of incandescent bulbs is that they light up instantly.

*Fluorescent* bulbs require the use of ballast coils, which makes initial fixture costs somewhat higher than the cost of incandescent fixtures. Fluorescent bulbs are, however, more efficient, and their color spectrum far more closely imitates natural daylight. A few different fluorescent lamp constructions are available that offer trade-offs among efficiency, color spectrum, and bulb cost. Fluorescents find their widest application in office buildings where people spend long periods of time reading and/or working with documents. The output for 48" fluorescents is approximately 60 lumens per watt. Fluorescent fixtures are available with a rapid-start requiring only a couple of seconds.

HID lamps generally are more efficient than fluorescent but do not generate the same broad light spectrums. HID lamps require special ballasts and require several minutes to warm up before full output is reached.

*Mercury vapor* lamps put out about 65 lumens/watt and have a green-blue color. Because they have little or no red component in their spectrum, they distort the actual color of objects. They require 3 to 7 minutes after starting to reach their full brilliance. Since the bulb must cool before it can restrike, it takes about the same amount of time to light up again after a power disruption.

*Metal halide* lamps are 1.5 to 2 times more efficient than mercury vapor lamps. Their color is a blue-white, and they render color slightly better than MV lamps. They take longer to warm up or restart − about 15 minutes.

*High-pressure sodium* lamps have an efficiency of about 110 lumens/watt. They have a golden-white color and distort colors considerably. About 15 minutes are required for a high-pressure sodium light to reach full luminance, but it can restart quickly after momentary interruption − in only 1½ to 2 minutes.

## Units for Measure: Lighting fixtures are counted on a unit basis (EA.). The number of lamps (for purchasing purposes) is determined per fixture and listed as each lamp (EA.).

**Material Units:** The items included in the material unit for fixtures depend somewhat upon the style of fixture. In general, the unit will include the following:

Fixture
Lamps
Wire nuts
Fasteners or hangers as required (chain, S-hooks, bolts, screws, etc.)
Canopies as required by the type of fixture

These additional items must be listed and extended:

Boxes
Plaster rings
Conduit or raceways
Wire
Fixture whips
Switches

**Labor Units:** The following procedures are generally included in the labor units for fixtures:

Receiving
Fixture handling
Layout
Installing (hanging or mounting) the fixture
Terminating wires at the fixture
Installing lamps and assembling the fixture
Testing

These additional procedures must be listed and extended:

Conduit
Wire
Fixture whips
Special support brackets or channels
Switches

**Takeoff Procedure:** When taking off interior lighting fixtures, it is advisable to set up the quantity work sheet to conform to the lighting schedule as it appears on the print. Include both the alphanumeric code and the symbol on your work sheet.

Take off one particular section or floor of the building at a time, counting the number of each type of fixture before going on to the next type. It is also advantageous to include on the same work sheet the conduit, wire, and fittings associated with a particular lighting system. This method makes material purchases more specific, helping to define when and how much to order.

After totalling all quantities, add 5% to the conduit takeoff and 10% to wire for waste. Transfer the quantities to a cost analysis sheet and extend.

**Cost Modifications:** Productivity is based on new construction in an unobstructed floor location, using rolling staging to 15' high. Add the following percentages to labor only for elevated installations:

| | | | |
|---|---|---|---|
| 15' | to | 20' high | +10% |
| 21' | | 25' | +20% |
| 26' | | 30' | +30% |
| Over 30' | | | +40% |

For large concentrations of lighting fixtures in the same area, deduct the following percentages from the labor units:

| | | | |
|---|---|---|---|
| 25 | to | 50 fixtures | −15% |
| 51 | | 75 | −20% |
| Over 75 | | | −25% |

## Exit and Emergency Lighting

Exit lights are available with mounting arrangements for walls or ceilings and with or without an arrow. Explosion-proof enclosures can be obtained for these lights. Exit signs have either incandescent or fluorescent lamps. Six-inch "EXIT" letters are standard.

Battery-operated emergency lights are available with either battery-mounted or remote heads. Several different types of batteries and voltages are available. Emergency lights are usually surface-mounted. (A special lamp pack is available for use in fluorescent fixtures; it mounts either in the ballast channel or on top of the fixture.)

### Units for Measure:  Exit light fixtures are counted individually (EA.).

Emergency lights are counted as each unit (EA.). (If the battery is used with remote heads, count each head (EA.) and battery unit (EA.).

### Material Units:

**Exit lights:** generally include the Exit sign with lamps, and anchors and wire nuts as required.

**Emergency lights:** generally include the battery unit and remote head complete with lamps and mounting brackets, anchors, and plug and cord or wire nuts.

### Labor Units:  The following procedures are generally included per unit:

**Exit Lights:**

Receiving
Material handling
Layout
Installing the fixture and lamps
Testing

Additional items to be listed and extended:

Conduit
Wire

Boxes
Connectors

**Emergency Lights:**

Receiving
Handling
Layout and installing mounting bracket
Connection in outlet box
Battery connection and testing

Additional items to be listed and extended:

Receptacle or boxes
Wire
Conduit

*Takeoff Procedure:* Generally, emergency lighting fixtures are taken off with interior lighting. (Refer to "Interior Lighting" in this chapter for guidelines.)

Note: Take care to list and count battery packs and/or remote heads if they are used.

# Exterior Fixtures

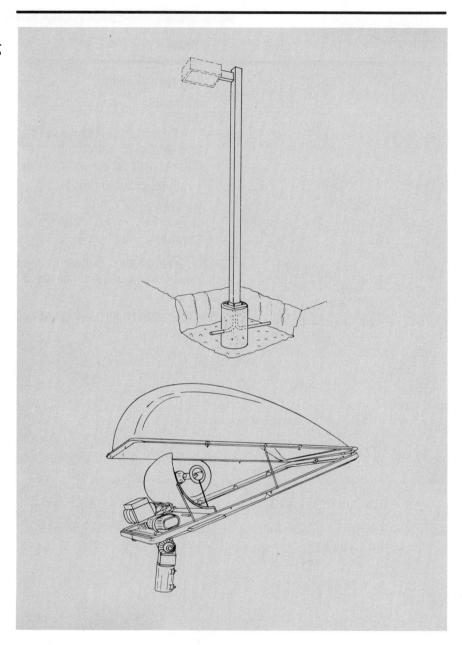

Exterior lighting serves both utilitarian and decorative purposes. The same assortment of lamps used for interiors (discussed previously in this chapter) is also available for exterior use. Low-pressure sodium (LPS) is an additional option for exterior fixtures.

*Low-pressure sodium* lamps generate an almost monochromatic, yellow light. They are, however, highly efficient, putting out over 170 lumens per watt. They require 7 to 15 minutes to reach full light output but will restart immediately after interruption of the power source.

*Fluorescent* lamps have special concerns when used outdoors. They are sensitive to low temperatures and require more starting energy in cold weather.

Ballasts are available for each of the following temperature ranges: above 50 degrees F for indoor applications, above zero degrees F for outdoor temperature applications, and above -20 degrees F for outdoor applications. Also, the surface temperature of fluorescent lamps has a direct impact on light output (it drops by 35% at zero degrees F). Again, there are fluorescent lamps specially designed to compensate for this factor.

Most exterior fixtures are either wall- or pole-mounted.

*Units of Measure*:  Each fixture is counted as a unit (EA.).

*Material Units*:  The following items are generally included per unit:

> Fixture
> Lamp
> Wire nuts
> Fasteners or brackets as required

These additional items must be listed and extended:

> Conduit
> Wire
> Pole
> Pole arms
> Photocell control

*Labor Units*:  The following procedures are generally included per unit:

> Receiving
> Fixture handling
> Layout
> Mounting or installing (up to 15′ high)
> Lamp and fixture assembling
> Wire connections
> Testing and aiming

These additional procedures must be listed and extended:

> Conduit
> Boxes and fittings
> Wire
> Pole and arm installation (pole base)
> Lightning protection
> Photocell control

*Takeoff Procedure*:  When taking off exterior lighting fixtures, it is advisable to set up the quantity worksheet to conform to the lighting schedule as it appears on the print. Include both the alphanumeric code and the symbol on your work sheet.

Take off one particular section or area of the building at a time, counting the number of each type of fixture before going on to the next type. It is also advantageous to include on the same work sheet the conduit, wire, and fittings associated with a particular lighting system. This method makes material purchases more specific, helping to define when and how much to order.

After totalling all quantities, add 5% to conduit takeoff and 10% to wire for waste. Transfer the quantities to a cost analysis sheet and extend.

## Lamps

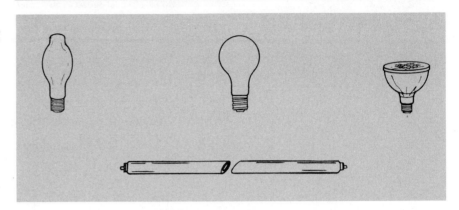

Please refer to the sections on Interior and Exterior Lighting (at the beginning of this chapter) for a detailed discussion of the various lamps that are used in most fixtures. Since fixture costs are normally calculated to include lamps, this category is most useful for listing lamps to be replaced.

*Units for Measure*:  Lamps are counted individually (EA.).

*Material Units*:  The material cost generally includes only the cost of the lamp.

*Labor Units*:  The following procedures are generally included per unit:

> Receiving
> Material handling
> Use of rolling staging
> Opening of the fixture
> Removal of the old lamp
> Installation of the new lamp and testing
> Hauling the old lamp for disposal

Note: The labor units should presume installing new lamps for entire areas — not just one or two at a time.

*Takeoff Procedure*:  List the number of each type of fixture. Then multiply this figure by the number of lamps in each fixture. List the lamps by type and wattage. Total the quantities, transfer these figures to a cost analysis sheet, and extend.

# Track Lighting

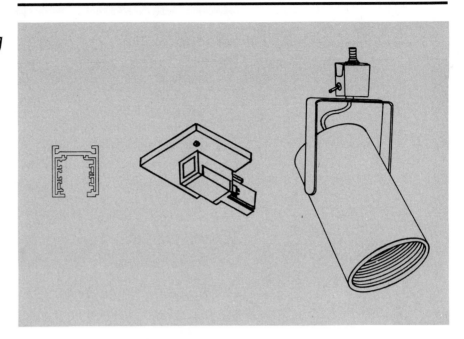

Track lighting is a versatile system designed to adapt easily as requirements change. Track lights can be swivelled and rotated to aim in any direction. This kind of flexibility is in highest demand and most often used in offices, homes, and galleries.

The basic component of a track light system is the track, which serves as both a raceway and a source of power for the lights. Tracks may be mounted on walls or ceilings, and come in standard lengths of 4', 8', and 12'. They may contain 1, 2, or 3 circuits. To connect sections of track, a joiner coupling is required; this coupling must also be 1, 2, or 3 circuits. Joiners are available in the following configurations: straight, "L"s, "T"s, and cross connectors. A feed kit is required to connect branch power sources to the track.

Track lighting fixtures are manufactured in a wide range of styles and finishes. Wattages range from 50 to 300. Fixtures can be mounted anywhere along the track simply by twisting the base of the fixture 180 degrees into the track and depressing a locking tab on the fixture itself.

Caution: Although most estimating standards give a typical price for an average fixture, prices for track lighting fixtures range from $20.00 to $300.00 per unit. Consult the specifications for the exact type of fixture before pricing. If the type of fixture is not specified, use the standard, but note that it is an allowance. Then, if a more expensive fixture is specified later, the estimator can submit a change order for the balance.

*Units for Measure:* Count the track in standard length segments (EA.). Count the fixtures as separate units (EA.).

*Material Units:* The following items are generally priced separately:

> Track (including mounting screws)
> Light fixtures

These additional items must be listed and extended:

> Feed kits
> Joiner couplings
> Switches

**Labor Units:** The following procedures are generally included in the man-hours for track lighting:

> Receiving
> Handling
> Measuring and marking
> Installing each component
> Aiming and testing

These additional procedures are listed and extended:

> Branch conduit
> Wiring to the track system
> Remote switching or dimmers

**Takeoff Procedure:** Organize your work sheet by:

> Track — according to size and number of circuits
> Feed kits by number of circuits
> Fixtures — by type, catalogue number, and wattage

Measure the track lengths, noting the different lengths of sections, and transfer this information to the takeoff sheets. Count all feed and end kits and transfer these figures to the takeoff sheet. Then count and list all fixtures. Transfer the quantity totals from the takeoff sheet to a cost analysis sheet and extend.

**Cost Modifications:** Add the following percentages to labor only for elevated installations:

| | | | |
|---|---|---|---|
| 15' | to | 20' high | +10% |
| 21' | | 25' | +20% |
| 26' | | 30' | +30% |
| Over 30' | | | +40% |

## Fixture Whips

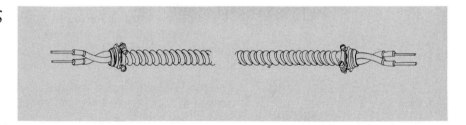

A fixture whip is a flexible connection consisting of wire in flexible metallic conduit. A fixture whip is used to connect the fixture, such as a troffer fixture for a suspended ceiling, to an outlet box when the fixture is not mounted directly to the box. Fixture whips are usually 4' to 6' long.

**Units for Measure:** Fixture whips are counted as units (EA.).

**Material Units:** The following items are generally included per unit:

> Fixture whip complete with 2 greenfield connectors and the ends of the wire already stripped
> Wire nuts for one end

**Labor Units:** The following procedures are generally included per unit:

Receiving
Material handling
Connecting to fixture and outlet box
Use of rolling stage in unobstructed area up to 15' high

**Takeoff Procedure:** Check the fixture schedule for suspended fixtures. Allow one whip per fixture (or one whip per row of fixtures) as shown on the drawings. For detailed takeoff guidelines, see "Interior Lighting Fixtures" at the beginning of this chapter.

**Cost Modifications:** Productivity is based on new construction in an unobstructed floor location, using rolling staging to 15' high. Add the following percentages to labor only for elevated installations:

| | | | |
|---|---|---|---|
| 15' | to | 20' high | +10% |
| 21' | | 25' | +20% |
| 26' | | 30' | +30% |
| Over 30' | | | +40% |

For large concentrations of lighting fixtures in the same area, deduct the following percentages from the labor units:

| | | | |
|---|---|---|---|
| 25 | to | 50 fixtures | −15% |
| 51 | | 75 | −20% |
| Over 75 | | | −25% |

# Chapter 19
# ELECTRICAL UTILITIES

Electric utilities include the site work required for the installation of electrical or telephone wires and the actual underground installation of those wires or cables. This chapter addresses electrical and telephone site work, and underground duct bank. Each of these components is described in terms of what is included and required — both for materials and labor. A takeoff procedure is also provided.

## Electric and Telephone Site Work

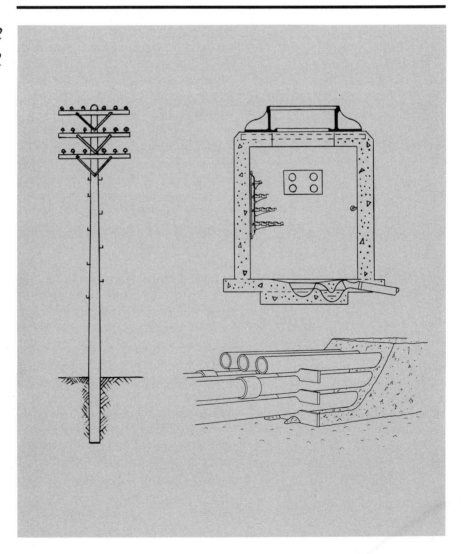

This category covers the distribution methods used to route power, control, and communications cables onto a facility's property and between its buildings and structures. There are three basic options: 1) direct burial cables, 2) underground in duct banks, and 3) overhead on poles.

*Direct burial cables* are the least versatile; they are generally used for residential applications where aesthetics are a more important factor than flexibility or an allowance for future changes. Occasionally, direct burial is used in commercial and industrial facilities to route a branch feeder to unique equipment, such as a well pump or roadway lighting. Two techniques are used to place direct buried cables. A trenching machine may dig a narrow slot 2 to 4 feet deep for a single cable or a trench 1 to 2 feet wide for multiple cables. The trench is usually backfilled with a few inches of sand — below and again above the cables — for protection. About 12 inches above the cable, concrete planks or plastic marker tape may be placed as a warning to future excavation and digging. Finally, on the finished grade, concrete, monuments, or markers may be placed every 100' to indicate the path of the buried cable and to denote changes in direction.

*Duct banks* are a group of 2 or more underground raceways (conduits) usually encased in concrete. Although more costly than direct burial systems, duct banks offer three advantages. First, the cables are far better protected from hazards and the elements. Secondly, groups of several cables can be pulled through each conduit (and many conduits may be installed in each duct bank). Finally, new cables may be pulled (or failed cables replaced) quickly and economically to meet future needs.

For duct banks that cover long distances, access must be provided to the run for pulling cables. When only a few conduits are installed, a hand-hole meets this requirement. For multiple conduit banks and for large conduits, a manhole is installed. Hand-holes and manholes may also serve to change the direction of a run or to split up a run. Both are usually made of precast concrete, and manholes are sized large enough for both men and pulling equipment. Pulling eyes (steel inserts) are normally cast into manholes opposite the cable entry points. This is done to facilitate the rigging process for cable pulling. Hangers may also be cast into the boxes to carry or support the installed cables.

Duct banks are generally buried, allowing 2' to 4' from grade to the top row of conduits; the conduits are separated by plastic spacers and held fast with tie-wires. When the conduit is encased in concrete, it is important to prevent the conduit from "floating" in the pour. A minimum of 2" of concrete is required on all sides.

The most common duct bank conduit material was formerly asbestos cement. In recent years, however, asbestos has been discontinued entirely in favor of plastic (PVC) and fiber duct. Galvanized steel may also be used, especially when power cables and instrumentation will be pulled into separate conduits in close proximity to each other. Plastic conduit comes in 20' lengths and is bent in the field with a heater. Couplings and fittings are glued on. Unless the run is very short (under 100'), the installation of duct conduits of less than 2" in diameter is very rare.

*Poles and overhead routing* represent the most conventional method of distributing power and communication cables. Many cables are built and rated for aerial service. Some cables include "strength members" to carry the tensions of the stretched cables. Still other types of cable, such as service drops and telephone lines, will be supported by a steel messenger wire. Note: Code requirements may dictate certain minimum heights for suspended cable. Be sure that these codes are understood before pricing aerial installations.

The hole for a line pole is usually made with an auger machine. After the pole is set, crushed limestone or similar fill is placed and compacted around the pole. In some areas, poles will need lightning rods and grounding. Don't forget to look for these requirements on the drawings.

## Units for Measure: Several quantity designations may apply to electrical and telephone site work.

> For direct burial trenching, the work is measured per linear foot (L.F.).
> For duct bank conduits, the work is measured per linear foot (L.F.).
> For sand backfill or concrete encasement, measurements are made in cubic yards (C.Y.).
> Manholes and hand-holes are counted as each (EA.).
> Fittings, spacers, and elbows are also counted individually (EA.).
> Poles and crossbars are counted as each (EA.).
> Cable is measured in units of 100 linear feet (C.L.F.).

## Material Units: The following summaries list the material items usually included in the material unit for each component:

**Trenching**: includes no materials.

> Additional items to be listed and extended are sand or concrete as required.

**Underground conduits:**

Conduit
Spacers
End fittings
Tie-wire
Additional items to be listed and extended are elbows and special adaptors.

**Manholes and hand-holes:**

Precast unit with cover and inserts (if required)

**Poles:**

Hole
Compacted steel fill
Additional items to be listed and extended are cross arms, guy wire and anchors, and lightning protection.

**Cable:**

Cable only
Additional items to be listed and extended are messenger wire (if not "built in"), terminations, and testing.

## Labor Units: The following is a summary of the activities usually included in the labor unit for each component:

**Trenching:**

Trench excavation
Borrowing material for backfill
Compacting

### Underground conduits:

Placing and fitting conduits, spacers, and wire ties. (Note: Spacers and fittings are counted for ordering only, not for pricing.)

Additional procedures to be listed and extended are special fittings and elbows.

### Manholes and hand-holes:

Hole excavation
Setting the precast unit in place
Backfill and grading
Placing covers

### Poles:

Auger excavation
Placing the pole
Backfill with limestone or suitable material
Compacting the fill
Rough grading
Additional procedures to be listed and extended are cross arms, guy wire and anchor, and lightning protection.

### Cable:

Cable only
Additional procedures to be listed and extended are messenger wire, terminations, and testing.

*Takeoff Procedure:* First, list all components on the takeoff sheets. Measure or determine linear distances between points and note these on the drawings. Next, determine the amount of earth work, i.e., trenching or hole excavation, as well as backfill, materials, etc. Count the number of underground conduits in each run by size, and list the total lengths. Count each manhole or post. Then determine the pulling length required for each cable run, including the indoor portions to their destinations. List each run for each cable type separately. This information may be useful later when ordering precut cables. For more information, see the section on shielded cable in the "Conductors and Grounding" chapter. Total the above; add 5% to the conduit lengths and 10% to the cable lengths for waste. Finally, transfer the total quantities to a cost analysis sheet and extend.

# Chapter 20
# SPECIAL SYSTEMS

Special systems include unique control or monitoring equipment used for specialized applications in residential, commercial, and industrial construction. Included in this chapter are sections ranging from electric heating and residential wiring, to television, public address, and sound systems. Doorbell and hospital paging systems are also covered. Each of these items is first described, then given appropriate units for measure, along with material and labor requirements and instructions for the takeoff procedure.

## Clock Systems

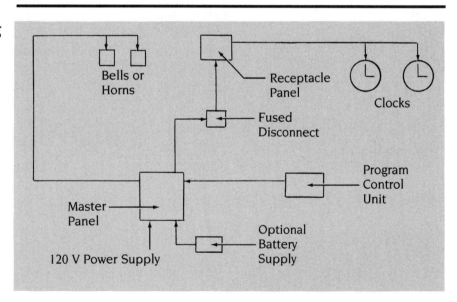

A variety of clock systems is available. The most common types have a master clock, a tone generator and control, an antenna, a lightning arrester, and slave clocks. The system picks up radio signals every hour from the National Bureau of Standards to keep the master clock set at the correct time. In turn, high-frequency tone signals are sent out over the regular receptacle wiring to keep the system's clocks on time. In this type of system, the clocks are plugged into the regular receptacle.

Other systems use different methods of resetting clocks. Some have two-motor clocks, with a control that speeds up the clock with the second motor. These systems require separate conduit and an independent wire system. Program signaling is also controlled through the master clock and controller.

Another option, a combination clock and speaker, consists of a back box, trim, and grille for the speaker and clock. The clock is included in the price of the box; the speaker is not.

**Units for Measure:** Each component is listed (EA.). [For convenience, some estimating standards list a complete system cost based upon the number of rooms. In that case, each system is listed (EA.).]

**Material Units:** A complete component with fasteners, box, and connections (wire nuts) is generally included in the material price.

These additional items must be listed and extended:

> Cable and wire
> Conduit

**Labor Units:** The following procedures are generally included per unit:

> Receiving
> Material handling
> Mounting as required
> Terminations
> Testing

These additional items are listed and extended:

> Conduit
> Wire

**Takeoff Procedure:** List the items on the takeoff sheet. Transfer this information to a cost analysis sheet and extend.

Note: Because of the variety of systems and manufacturers, it is recommended that a vendor price quote be obtained for material pricing.

# Detection Systems

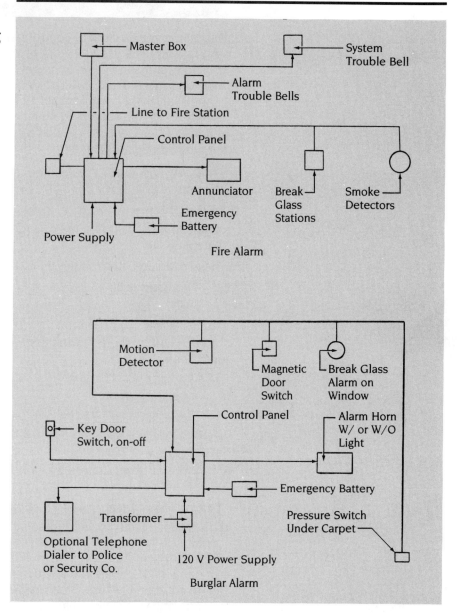

Master Box

System Trouble Bell

Alarm Trouble Bells

Line to Fire Station

Control Panel

Annunciator

Emergency Battery

Break Glass Stations

Smoke Detectors

Power Supply

Fire Alarm

Motion Detector

Magnetic Door Switch

Break Glass Alarm on Window

Control Panel

Alarm Horn W/ or W/O Light

Key Door Switch, on-off

Emergency Battery

Transformer

Pressure Switch Under Carpet

Optional Telephone Dialer to Police or Security Co.

120 V Power Supply

Burglar Alarm

Fire alarm systems and burglar alarm (intrusion detection) systems are similar in both their principals of operation and their installation techniques. Nevertheless, they are two separate and distinct installation components. Rarely do they share any hardware or wiring.

*Burglar alarm systems* consist of control panels, indicator panels, various types of alarm devices, and switches. The control panel is usually line powered with a battery backup supply. Some systems have a direct connection to the police or protection company, while others have auto-dial telephone capabilities. Most, however, simply have local control monitors and an annunciator. The sensing devices are various pressure switches, magnetic door switches, glass break sensors, infrared beams, infrared sensors, microwave detectors, and ultrasonic motion detectors. The alarms are sirens, horns, and/or flashing lights.

Fire alarm systems consist of control panels, annunciator panels, battery with rack and charger, various sensing devices, such as smoke and heat detectors, and alarm horn and light signals. Some fire alarm systems are very sophisticated and include speakers, telephone lines, door closer controls, and other components. Some fire alarm systems are connected directly to the fire station. Requirements for fire alarm systems are generally more closely regulated by codes and by local authorities than are intrusion alarm systems.

*Units for Measure*: The components of a system are listed and counted as each (EA.).

*Material Units*: For all of the above listed components, the component itself and its mounting hardware are generally included in the material price.

These additional items must be listed and extended:

> Wire connections
> Wire
> Conduits

*Labor Units*: The following items are generally included in the man-hours per unit:

> Receiving
> Material handling
> Installation and testing

These additional items are listed and extended:

> Conduit
> Wire or boxes, unless the box is part of the component
> Terminations

*Takeoff Procedure*: First, list each component on the takeoff sheet. Then count each type of component, listing all of one type before continuing to the next. Measure conduit lengths with a rotometer; mark the lengths on the drawing as well as on the takeoff sheets. Using the conduit lengths, figure the lengths of each cable type and list this information on the takeoff sheets. Next, total all components and add 5% to conduit and 10% to cable for waste. Transfer the total quantities to a cost analysis sheet and extend.

## Doctors' In-Out Register

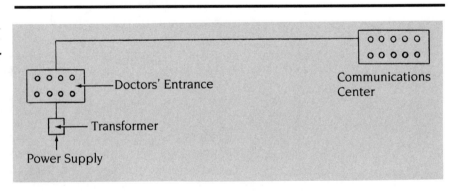

This is a system used in hospitals to facilitate communication between staff members. A recording register is located at the entrance. The doctors push their buttons, sending a signal to a register at the hospital control center. The attendant is then able to page or signal the doctor. Pocket paging devices are a material cost only.

*Units for Measure*:  Each system is counted as one unit (EA.). Pocket paging devices, if needed, are counted as each (EA.).

*Material Units*:  The following items are generally included in the material unit price:

> Push button board
> Console monitor
> Fasteners and brackets

These additional items must be listed and extended:

> Wire
> Conduit
> Terminations
> Pocket paging devices

*Labor Units*:  Man-hours for a Doctors' In and Out Register generally include the following procedures:

> Receiving
> Material handling
> Mounting
> Testing

These additional procedures must be listed and extended:

> Wire
> Conduit
> Terminations

*Takeoff Procedure*:  List the components on the takeoff sheet. Total and add 5% to conduit and 10% to wire quantities for waste. Transfer these figures to the cost analysis sheets and extend.

Note: Whenever practical, obtain material cost quotes from manufacturers' representatives or vendors. This is also a good source of information regarding special installation requirements or instructions.

# Doorbell Systems

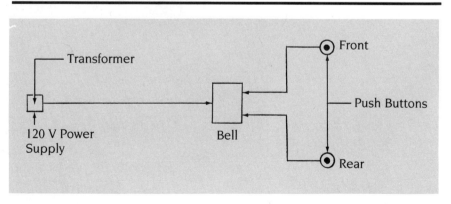

This system includes a low-voltage transformer (120 V / 24 V), push button(s), and a signal device. The signal device can be a bell, buzzer, or chimes. Because the wiring is low-voltage, conduit is rarely used.

Units for Measure: Each component is listed as a unit (EA.).

Material Units: The following items are generally included in the material cost per unit:

Doorbell
Buttons
Transformer
Wire terminations

These additional items must be listed and extended:

Wire
Box for transformer

Labor Units: Man-hours generally include the following procedures:

Installation
Terminations
Testing

These additional procedures are listed and extended:

Wire
Outlet boxes

Takeoff Procedure: List the components on the takeoff sheet (usually taken off with branch wiring). Total and add 10% to wire. Transfer the total quantities to the cost analysis sheets and extend.

Note: The cost of the chime units can vary widely. Be sure to understand the specified requirements when pricing this item.

## Electric Heating

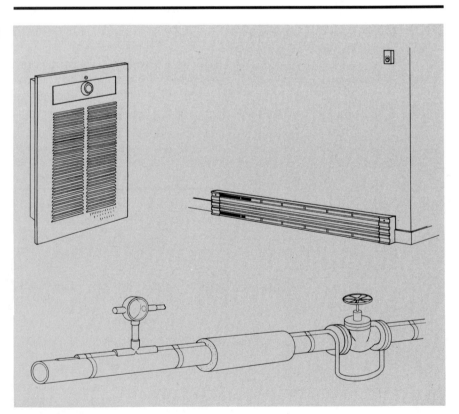

Electric heating is an item that covers a wide range of equipment and applications. A few subcategories are baseboard or wall unit heaters for residential or commercial buildings, heat trace systems for freeze protection of water lines or for maintaining liquid temperature in chemical and oil pipes, radiant ceiling heat panels, and unit space heaters. Clearly, a number of different estimating problems are posed by an electrical component that covers such a broad range. Nevertheless, two specific applications can serve to guide the estimator through the majority of cases. These are space heaters and piping trace cables.

Early in the estimating process, NEC and local building code regulations must be considered as a factor in electric heating. For example, receptacles cannot be located above electric baseboard heaters because of danger to hanging cords. Minimum wall insulation values must also be met when electric heating is installed in residences. It is important to become familiar with these and other requirements before estimating various forms of electric heat.

*Electric space heat* may be provided with a baseboard arrangement, wall-mounted units, or ceiling-hung heaters. The baseboard units are usually convection types. They are classed in terms of watts per foot, i.e., watt density. Low watt density models produce approximately 175 W/L.F., while high watt density units run 257 W/L.F. (This number suggests the operating temperature of the element.) Baseboard heaters are available in sizes that increase in 2' increments — from 2' to 12' long. They may be rated for operation on 120, 240, or 277 volts.

Unit space heaters generally operate with forced air (fans) blowing across an electric coil. These units are rated in terms of their heat output in kilowatts (KW). Sizes run from 10 KW to 50 KW. Although 208 to 240-volt single-phase units are used, 240 and 480-volt three-phase units are far more common.

There are two types of thermostat controls for heater circuits. The first is full (line) voltage thermostat switches. The second is low-voltage, relay-equipped thermostats. Line thermostats are not suited for more than 240 volts, nor are they used for large heaters at lower voltages.

If no other information is given about heating capacity requirements, there is a general rule of thumb that provides 10 watts per square foot of floor area.

*Heat trace* is a parallel-resistance heating cable. The purpose of heat trace is to protect a pipe from freezing. This is done by laying a continuous heater wire against a pipe and wrapping both with insulation. Thermostats (with sensing elements against the opposite side of the pipe or at least 90 degrees away) are set to maintain about 40 degrees F Some liquids, such as heavy fuel oils, require higher temperatures to remain in a fluid state. The heat tracing used on these systems is designed for these higher temperatures. Heat trace systems typically operate from 100 to 300 degrees F. Mineral insulated heat trace cable is commonly used in high-temperature systems.

Heat trace cable is rated in watts per foot. If extra heat is needed, the cable can be wrapped in a spiral around the pipe, thereby increasing the number of watts per linear foot of pipe. The number of turns or wraps per foot is called the "pitch". Tables are available that relate pitch and pipe diameter to total cable footage. Valves, flanges, and other fittings receive extra wrapping to compensate for added heat losses. Special low-temperature heat cable can be used for plastic PVC pipes.

When estimating for heat trace systems, it is important to understand the intended use of the cable. Freeze-protection cable is simply attached to the pipe with a polyester tape band every 12". Another method is to attach the cable to the pipe with a continuous coverage of 2"-wide aluminum tape. This increases heat transfer and temperature distribution around the pipe, allowing higher watt densities without spiral wrapping. Polyester tape bands must still be used every 12". Cable covered with a factory-extruded heat transfer cement maximizes heat transfer. This covering allows temperature maintenance up to 300 degrees F.

*Units for Measure*: Heaters are counted as units (EA.). Heat trace cables are measured in linear feet (L.F.).

*Material Units*: The following items are included for heat component units:

### Baseboard heaters:

The heater
Mounting screws
Additional items to be listed and extended are terminations, wire, and thermostats.

### Unit heaters:

The heater (with thermostat)
Brackets or hangers
Wall or ceiling fasteners as required (up to 15' high)
Additional items to be listed and extended are terminations, wire, conduit, and thermostats (some models).

### Heat trace:

Heat cable
Attaching tape
Additional items to be listed and extended are a heat cable connection kit, end kit, thermostat, branch circuit wire, and conduit.

*Labor Units*: The following procedures are generally included in the man-hours for heating units:

### Heaters:

Receiving
Material handling
Installation or mounting of heaters in place
Additional items to be listed and extended are branch wiring, conduit, terminations, and installation of thermostats (if not integral to the unit)

### Heat trace:

Receiving
Material handling
Installing heat cable by wrapping with tape (up to 15' high)
Additional items to be listed and extended are installing end termination kits, terminal kits, thermostats, branch wiring, conduit, and terminations.

*Takeoff Procedure*: Set up the work sheet by individual components (with heat trace cable, subdivide into volts and watts per linear foot). Add accessories, such as metallic raceway and support clips. Transfer all quantities to the cost analysis sheet and extend.

Notes:
1. If the installation uses transfer cement, keep in mind that 1 gallon applies approximately 60 linear feet of raceway.
2. If the pipe system contains valves, be sure to add enough cable to wrap them. This procedure is described below, in the "Cost Modifications" section.

*Cost Modifications for Heat Trace*:  For each type of valve, add the following quantities to the linear foot totals of heat trace cable:

| Butterfly | | Flange | | Screwed or Welded | |
|---|---|---|---|---|---|
| 1/2″ | = 0′ | 1/2″ | = 1′ | 1/2″ | = .5′ |
| 3/4 | = 0 | 3/4 | = 1.5 | 3/4 | = .75 |
| 1 | = 1 | 1 | = 2 | 1 | = 1 |
| 1.5 | = 1.5 | 1.5 | = 2.5 | 1.5 | = 1.5 |
| 2 | = 2 | 2 | = 2.5 | 2 | = 2 |
| 2.5 | = 2.5 | 2.5 | = 3 | 2.5 | = 2.5 |
| 3 | = 2.5 | 3 | = 3.5 | 3 | = 2.5 |
| 4 | = 3 | 4 | = 4 | 4 | = 4 |
| 6 | = 3.5 | 6 | = 8 | 6 | = 7 |
| 8 | = 4 | 8 | = 11 | 8 | = 9.5 |
| 10 | = 4 | 10 | = 14 | 10 | = 12.5 |
| 12 | = 5 | 12 | = 16.5 | 12 | = 15 |
| 14 | = 5.5 | 14 | = 19.5 | 14 | = 18 |
| 16 | = 6 | 16 | = 23 | 16 | = 21 |
| 18 | = 6.5 | 18 | = 27 | 18 | = 25.5 |
| 20 | = 7 | 20 | = 30 | 20 | = 28.5 |
| 24 | = 8 | 24 | = 36 | 24 | = 34 |
| 30 | = 10 | 30 | = 42 | 30 | = 40 |

Caution: Pipe insulation is installed after the heat tracing is in place. This is usually performed by the mechanical contractor, but it can sometimes be specified as the responsibility of the electrical contractor.

# Lightning Protection

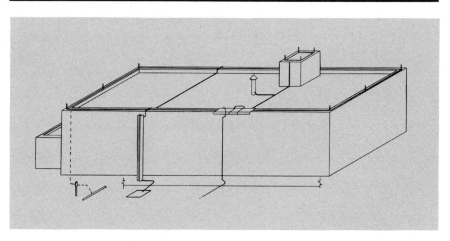

Lightning protection for the rooftops of buildings is achieved by a series of lightning rods or air terminals joined together by either copper or aluminum cable. The cable size is determined by the height of the building. The lightning cable system is connected through a download to a ground rod that is a minimum of 2′ below grade and 1-1/2′ to 3′ out from the foundation wall. In order to effectively estimate lightning grounding systems, we need to incorporate many of the components described in the grounding section of this book. By code, all metal bodies on the roof that are located within six feet of a lightning conductor must be bonded to the lightning protection system. As a result, the lightning protection system is listed with the equipment grounding in our estimating procedure.

One of the basic components of a lightning grounding system is the air terminal. These are manufactured in either copper or aluminum. The most common sizes used are 3/8" diameter copper at 10" high for roofs under 75' high. For roofs over 75', a 1/2" diameter by 12" high terminal is most common. For aluminum air terminals, a 1/2" diameter by 12" high is used for roofs under 75', and a 5/8" diameter by 12" high is used for heights over 75'. Air terminals are usually mounted in the perimeter parapet wall and include a masonry cable anchor base. A wide range of configurations is available from manufacturers for mounting air terminals to different roof surfaces. Most do not vary significantly in price.

Cable for main conductor use is calculated in pounds per 1,000'; it is available on 500' spools. Conductors are manufactured in copper and aluminum. The industry standard for copper is 204 lbs. per 1000' when used on a roof of less than 75' high, and 375 lbs. per 1000' for structures over 75' high. The main conductor for aluminum cable is 104 lbs. per 1000' for structures under 75' high, and 190 lbs. per 1000' for structures over 75' high. Connections to main grounding conductor cable are accomplished in one of three ways: clamping, heat or exothermic welding, or brazing.

Units for Measure: Air terminals are counted as units (EA). Cable is taken off by the linear foot (L.F.). Terminations are counted as units (EA.).

Material Units: The following items are generally included in the material price of

### Air terminals:

The terminal or rod
One mounting base
Four mounting lead-type anchors and screws

### Cable:

Base copper or aluminum cable
One cable clamp for every three linear feet of cable
One lead-type anchor and screw every three linear feet of cable

Included in the material cost of *ground connections* is only the connector itself.

Labor Units: Man-hours generally include the following procedures:

### Air Terminals:

Mounting on concrete surface
Installing four anchor-type fasteners

### Cable:

Receiving main conductor cable on 500' reels
Hauling to central roof location
Setting up cable reel
Drawing cable
Fastening cable every 3 linear feet using a lead-type anchor

### Connections:

Setting up joint configurations using hardware or molds
Attaching and terminating the connection

*Takeoff Procedure*:  Set up the quantity takeoff sheet listing air terminals by size, type, and configuration. List the wire size in pounds per 1000 L.F. and by type of material. List the connections by type, wire-to-wire size, and configuration. Start at one corner of the building and work clockwise around the perimeter of the building, counting the air terminals. Check each terminal with a marker as you count, and avoid jumping from the perimeter to air terminals located in other sections of the roof. After the perimeter count, take off all the other air terminals.

Transfer the count of all quantities to the takeoff sheet. Measure the main ground cable, again following the perimeter route. (Do not, at this point, attempt to incorporate equipment ground cables.) Transfer the quantities (in linear feet) to the work sheet. Then, count the number of drops to the grounding electrode, and multiply this figure by the distance between the connection and the ground rod. Measure the equipment cable runs and terminations to the equipment casing and the main grounding conductor. Transfer the quantities to a cost analysis sheet and extend.

*Cost Modifications*:  Man-hours are based on the installation of up to 10 air terminals on a common roof. Deduct the following percentages from both air terminals and connections for quantities of:

| | | | |
|---|---|---|---|
| 1 | to | 10 | 0% |
| 11 | | 25 | 20% |
| Over 25 | | | 25% |

# Nurses' Call Systems

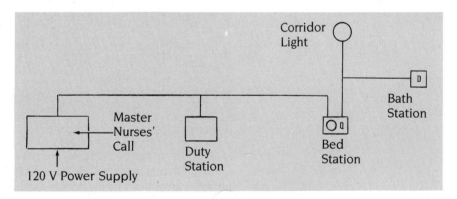

These systems, used in hospitals a.   nursing homes, enable the patient to signal the nurse's station, and may also turn on a door light in the corridor outside the room. The central component is a master control station, which includes an amplifier, microphone, and speaker. In some systems, patients can speak to the nurses through a bedside call station, or a pillow speaker. It is best to call the manufacturer for a price on the complete system.

*Units for Measure*:  The master control unit, bedside station, and corridor light are all counted as units (EA.).

**Material Units:** The following items are generally included per unit:

Master control unit
Bedside stations
Corridor light for each room

**Labor Units:** Generally included in the man-hours are the following procedures:

Receiving
Material handling
Installation
Testing of devices and master control station

These additional items are listed and extended:

Conduit
Wire
Outlet boxes (unless the box is part of the component)
Terminations

**Takeoff Procedure:** List the components on the takeoff sheet (usually taken off with branch wiring). Total and add 10% to wire. Transfer the total quantities to a cost analysis sheet and extend.

## Television Systems

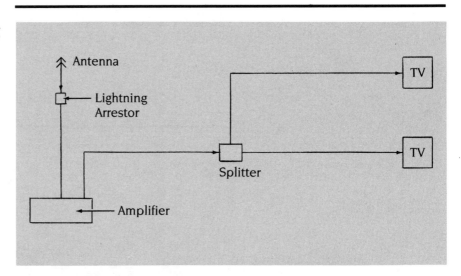

Master T.V. antenna systems are used in schools, dormitories and apartment buildings. Each system consists of the antenna, lightning arrester, amplifier, splitters, and outlets. The signal is received by the antenna and increased by the amplifier. It then goes through the main cable to the splitter. Here, the signal is split among several branch circuit cables.

A closed-circuit surveillance system consists of a T.V. camera and monitor. These systems are used indoors or outdoors for security surveillance. Some applications require pan, tilt, and zoom (PTZ) mechanisms for remote control of the camera.

Educational T.V. studio equipment is comprised of 3 cameras and a monitor.

Notes: When estimating material costs for T.V. camera systems and monitors, it is always prudent to obtain manufacturers' quotes for equipment prices. Also, some installations will require special accessories. Often, the sales engineers are a good source of information on these requirements.

Some general precautions are listed below:

- Various models of cameras are sold with a wide choice of lens options. Many are priced above "standard" lens costs. It is common to have one type and model camera specified for a site, with each camera having a different lens. As a result, each camera could have a different price.
- Camera models may also have several different video tube options. Those designed for "low light" (used outdoors at night) are more expensive than "standard" tubes.
- For long video cable runs, or runs near power cables and equipment, signal accessories may need to be added to the system. These accessories may include: hum clampers to remove 60 HZ, AC noise; equalizers to balance the signal's frequency spectrum after a long cable run; and line amplifiers to boost signal strength at the midpoint of the run.
- As the camera-to-monitor distance increases, the type of coaxial cable used becomes critical. High-efficiency cables (low signal loss) are, naturally, more expensive to buy and to install.
- Monitors are generally specified in terms of resolution in lines. A high-resolution monitor, say 750 lines, will be more expensive than one rated at 550 lines.
- Specifications may require the use of expensive test equipment during check-out and start-up. The rental cost or purchase price must be included.
- The presence of the manufacturer's service representative or engineer is often required during testing. This expense will not be included in the material cost and must be added to the estimate.

## Units for Measure: Each component is listed (EA.).

## Material Units: The following items are generally included per unit:

Components, complete with fasteners and brackets
Antenna, complete with mounting straps or guys

These additional items must be listed and extended:

Cable (both 120 V power and coaxial signal)
Conduit
Terminations

## Labor Units: The following procedures are generally included per unit:

Receiving
Material handling
Installation and testing of components

These additional procedures must be listed and extended:

Conduit
Wire or cable
Outlet boxes

*Takeoff Procedure:* First, list each type of component on the takeoff sheet. Count the like components and enter each on the takeoff sheet. Then measure each type of wire and enter on the takeoff sheet. Next, count coaxial cable connections carefully (watch for long runs that may require splices). Total the items and add 5% to conduit and 10% to cable for waste. Finally, transfer the total quantities to a cost analysis sheet and extend.

# Residential Wiring

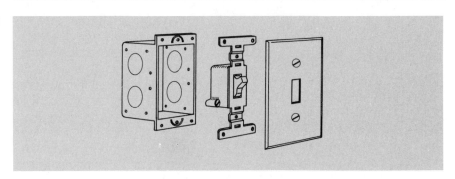

For estimating purposes we will organize residential wiring into certain defined systems. These systems, when combined, comprise all the electrical components needed for a complete residential electrical installation. Because of the relatively low cost of residential electrical wiring, and the competitive nature of the residential business, most estimating standards combine the expense of assorted materials and list the cost per outlet or per switch. This approach involves little risk, provided the estimator knows in advance what is included in each of these mini-systems. Why use a mini-system approach? If an estimator spends the time to measure every L.F. of #14-2 Romex at $.05 per L.F., he would still arrive at a cost very close to that derived from a mini-system. Furthermore, he may have spent six times as long to prepare an estimate that might not be any more accurate. The residential industry does not afford that degree of precision, and most residential prints do not include wiring details.

Residential wiring can be broken down into the following categories:

> Service
> Branch circuit wiring
> Appliances
> Heating and A/C
> Intercom and/or doorbell systems
> Light fixtures
> Special needs

For purposes of residential estimating, we will define branch circuits as circuits containing any combination of receptacles, switches and lighting outlets. Using a mini-system approach, the estimator needs only count the number of these items.

Appliance circuits are defined as specific needs, or direct-connected outlets. Examples are range circuits, water heater circuits, exhaust fans, disposal wiring, and dryer circuits. These circuits are run to, but do not include, the appliance itself. There is little variation in the voltage and amperage requirements of appliances. Thus, many estimating standards apply the same mini-system approach to both appliance circuits and branch circuit devices or outlets. The wire for appliance circuits must be measured and priced individually.

Residential intercom systems consist of a master station (most commonly located in the kitchen area), up to 8 remote speakers, and associated low-voltage wiring. The transformer steps voltage down from 110 V to 24 V. Remote speakers are interconnected with #18 gauge nonmetallic sheathed cable.

Heating and air conditioning circuits are direct, or "home runs". Most estimating standards do not include the equipment, which is instead supplied and installed by the mechanical contractor; the circuit and tie-in are done by the electrical contractor.

The raceways used in residential wiring vary according to specifications as well as state, local, and national electrical codes. Because of the wide variations in cost, it is important to note the type of raceway material being used. Some of the more common types are listed below:

> Nonmetallic sheathed cable (Romex)
> BX cable
> EMT conduit

Aluminum or steel conduit may also be used.

## Units for Measure: Residential wiring devices and outlets are counted on a "mini-system" basis (EA.).

## Material Units:

### Service:

The following items are generally included in the material price for a 100-amp service using nonmetallic cable:

> 15' of S.E. type cable
> 1 service head
> 1 weatherproof S.E. connector
> 2 standard S.E. non-watertight connectors
> 4 S.E. two-hole clips
> 1 8' x 3/8" ground rod
> 1 ground clamp
> 5' bare CU ground wire
> 1 residential 100-amp 20-circuit panel with main breaker
> 20 single-pole 15- and 20-amp breakers

Note: For a 200-amp service, the same quantities apply, but they are sized for the larger installations.

### Branch circuits:

Branch circuit wiring is based on an average of 20' between devices, including an allowance for the home run. The following items are generally included in the material unit price:

> 20' of 14-2 Romex
> 1 device (residential grade)
> 1 single gang plastic box
> 1 plastic device cover
> 4 staples or clips, and nails

Note: Although the above listing is for a Romex installation, the quantities and component logic also apply to BX, EMT, and conduit installations. Lighting circuits do not contain a fixture.

### Appliance circuits:

The following items are generally included per unit for appliance circuits:

20' of cable or raceway (the size depends upon the appliance; for example, an exhaust fan requires only #14-2 cable, while a range requires #6-3 cable)

2 connectors – one at the panel, one at the appliance

4 staples of conduit clip

Note: For range and dryer circuits, a surface-mounted receptacle is included.

### Residential intercom systems:

The following items are generally included per unit for residential intercom systems:

Master control unit
Seven remote speakers
Seven mounting rings
300' #18-4 nonmetallic sheathed cable
1 low-voltage .50 transformer

## Labor Units:

### Service:

Man-hours for residential service generally include the following procedures:

Ordering
Measuring and marking
Installing service
Tying in branch wiring to circuit breakers
Wire identification and marking panel schedule

### Branch circuits:

Man-hours for each branch outlet or switch generally include:

Ordering, receiving and storing
Marking box locations
Mounting boxes
Drilling wood studs
Pulling cable or installing raceway
Making up the gang box
Installing the device (switch, receptacle)
Installing the finish cover for the device
Testing for power

### Appliance Circuits:

Measuring and marking
Drilling wood studs
Running cable or conduit
Connecting the appliance

Additional procedures to be listed and extended for appliance circuits:

Thermostat or controls
Home run conduit or wiring (over 20')

**Residential intercom systems:**

    Measuring and marking
    Mounting retaining rings
    Drilling wood studs
    Running low-voltage wire
    Tie-in of speakers and low-voltage transformer
    110-volt feed to transformer

*Takeoff Procedure*: Set up the work sheet and define the units by type and print graphic.

For example:    115V Receptacle
                S.P. Switch

Count all units of one type before proceeding to the next. Make sure to mark all components with a colored pencil. As mentioned in the appliance section, home runs should be measured back to the panel. Transfer all quantities to a cost analysis sheet and extend.

The fixtures should be counted last. Manufacturers' prices should be used, since fixtures are the single most expensive component and their cost has such a significant impact on the estimate. Because of this effect, many contractors issue a fixture allowance for each house or customer, with the electrical contractor responsible for installation.

Check the size and type of the service to be used. Transfer the lighting and service quantities to a cost analysis sheet and extend.

Note: If using the mini-system approach, the subcomponents of each system can be identified and listed separately, then extended against the unit count. This gives the estimator a quantity count of all subcomponents that need to be purchased.

# Chapter 21
# FIELD
# FABRICATION

Field fabrication involves the construction of stands or supports for electrical equipment, conduits, or cable trays. Field fabrications are performed when there is no stock item available to conform to existing field conditions. This chapter describes both field fabrication and the appropriate estimating procedure for this aspect of electrical work.

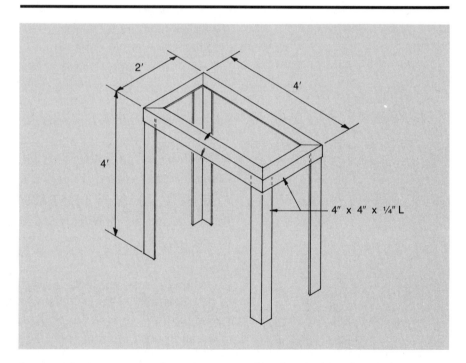

In many instances, stands and supports for electrical equipment, conduits, or cable trays, are not stock items. They must be shop or field fabricated to conform to existing field conditions.

Steel supports are fabricated of angles designated "L", channels "C", wide flange beams "W", American Standard beams "S", plates, rods, or bars. Steel members are usually priced per pound, so it is necessary to convert each component into weight before pricing. Beams, both "W" and "S", and channels "C" are designated by weight. A "W 8 x 31" designates a nominal 8" deep wide flange beam 31 pounds per foot; "S 5 x 14.75" refers to a 5" deep American Standard beam (formerly called an "I beam") 14.75 pounds per foot; and "C 7 x 9.8" indicates an 8" deep channel section 9.8 pounds per foot. (For any given size, the weight per foot is a function of the member's thickness and, therefore, is also indicative of its load capability.)

Angles are rolled with equal and unequal legs in various thicknesses and are designated "L 3 x 3 x 1/4" or "5 x 3 x 7/16".

The weight of angles, bars, rods, or plates is not usually shown on drawings. To convert linear footage to weight requires a handbook with the various members listed per size and weight. If a handbook is not readily available, one linear foot of 1" x 1" steel or 12 cu. in. of steel weighs 3.4 pounds. Therefore, one foot of a 4" x 4" x 1/4" L weighs approximately 8" x 1/4" = 2 cu. in./ft. x 3.4 pounds = 6.8 pounds per foot. The actual weight from a handbook lists a 4" x 4" x 1/4" L as 6.6 pounds per foot, which includes the deductions for the radius at the toe of the legs. A plate 12" x 1" x 12" would weigh 12 x 3.4 pounds = 40.8 pounds.

Large quantities of steel may be purchased from the mill cut to specified lengths; smaller quantities can be obtained from a warehouse at a higher price and in standard available lengths. If small quantities of shapes are to be purchased from a warehouse, allowance for waste must be included in the estimate.

The following example demonstrates the estimating process for a field-fabricated transformer stand:

| | | |
|---|---|---|
| Top: | 2 @ 4' + 2 @ 2' | = 12.00 L.F. |
| Legs: | 4 @ 3' 11-3/4" | = 15.92 L.F. |
| | | 27.92 L.F. |

Angle is available from a warehouse in 5' increments. Buy 1 @ 20'-0" and 1 @ 15'-0".

| | |
|---|---|
| Material: 35 L.F. x 6.6 LB./FT. = 231 LB. @ $.35 = | $80.85 |
| Fabricate: Cut to length 12 each x 1/4 HR. = 3 HR. @ $22.40 = | $67.20 |
| Weld: 4.5'/4 L.F./HR. = 1.125 HRS. @ $22.40 = | $25.20 |
| Torch Gas and Air: 3 HRS. @ $16.20 = | $48.60 |
| Welder (machine): 1.125 HRS. @ $15.80 = | $17.80 |
| Total: | $239.65 |

The estimate does not include overhead and profit. Painting is assumed to be performed by the painting contractor.

Estimates for most field fabricated support structures may be broken down and priced utilizing cost manuals, such as Means' *Building Construction Cost Data* .

# Part III
# SAMPLE ESTIMATE

## Chapter 22

# HOW TO USE MEANS ELECTRICAL COST DATA

Parts I and II of this book are directed to the principles of estimating — from an overview of the entire process to a detailed discussion of techniques for measuring and pricing individual components. Part III, the Sample Estimate, demonstrates these principles and techniques in practice. The prices used throughout the sample estimate are from *Means Electrical Cost Data*. This chapter describes the components and uses of this annual cost source, along with the methods by which these costs are obtained and organized.

## Format and Data

Users of *Means Electrical Cost Data* are chiefly interested in obtaining quick, reasonable, average prices for electrical construction estimates. This is the primary purpose of the annual book — to eliminate guesswork when pricing individual items of an installation. Many persons use the cost data, whether for bid verification, quotations or budgets, without being fully aware of how the prices are obtained and derived. Without this understanding, the resource is not being used to fullest advantage. In addition to the basic cost data, the book also contains a wealth of information to aid the estimator, contractor, designer, and the owner to better plan and manage electrical construction projects. For example, productivity data is provided in order to assist with scheduling; national labor rates are analyzed; and tables and charts for location and time adjustments are included and help the estimator tailor the nationally averaged prices to a specific project. The costs in *Means Electrical Cost Data* consist of over ten thousand unit price line items as well as thousands of electrical systems, or assembly price line items. This information can also provide an invaluable checklist to the construction professional to assure that all required items are included in a project.

The major portion of *Means Electrical Cost Data* is Section A — Unit Prices. This data is organized according to the UCI format (See the list of index divisions below). This index was developed by representatives of many parties concerned with the building construction industry. It has been accepted by the American Institute of Architects (AIA), the Associated

General Contractors of America, Inc. (AGC), and the Construction Specifications Institute, Inc. (CSI). In *Means Electrical Cost Data*, relevant parts of other divisions are included along with Division 16 — Electrical. Appropriate items from Divisions 1, 2, 3, 5, 6, 10, 11, 13, and 15 all appear prior to the Division 16 entries.

## Uniform Construction Index (UCI) Divisions:

Division 1  — General Requirements
Division 2  — Site Work
Division 3  — Concrete
Division 4  — Masonry
Division 5  — Metals
Division 6  — Wood & Plastics
Division 7  — Moisture-Thermal Control
Division 8  — Doors, Windows & Glass
Division 9  — Finishes
Division 10 — Specialties
Division 11 — Equipment
Division 12 — Furnishings
Division 13 — Special Construction
Division 14 — Conveying Systems
Division 15 — Mechanical
**Division 16 — Electrical**

In addition to the sixteen UCI divisions of the Unit Price section, Division 17, Square Foot and Cubic Foot Costs, presents consolidated data from over 10,500 actual reported construction projects and provides information based on total project costs as well as costs for some major segments.

Section B, Systems Cost Tables, contains thousands of costs for electrical and appropriate mechanical systems. Components of the systems are fully detailed and accompanied by illustrations.

Section C contains technical tables and reference information. It also provides estimating procedures for some types of work and explanations of cost development which support and supplement the unit price and systems cost data. Also included in Section C are the City Cost Indexes, representing the compilation of construction data for 162 major U.S. and Canadian cities. Cost adjustment factors are given for each city, by trade, relative to the national average.

The prices presented in *Means Electrical Cost Data* are national averages. Material and equipment costs are developed through annual contact with manufacturers, dealers, distributors and contractors throughout the United States. Means' staff of engineers is constantly updating prices and keeping abreast of changes and fluctuations within the industry. Labor wage rates are the national average of each trade as determined from union agreements from thirty major U.S. cities.

Following is a list of factors and assumptions on which the costs presented in *Means Electrical Cost Data* have been based:

> *Quality* — The costs are based on methods, materials and workmanship in accordance with U.S. Government standards and represent good, sound construction practice.

> *Overtime* — The costs as presented, include *no* allowance for overtime. If overtime or premium time is anticipated, labor costs must be adjusted accordingly.

*Productivity* — The unit man-hour figures are based on an eight hour workday, during daylight hours. The chart in Figures 11.7 shows that as the number of hours worked per day (over eight) increases, and as the days per week (over five) increase, production efficiency decreases. (See Chapter 11)

*Size of Project* — Costs in *Electrical Cost Data* are based on commercial and industrial buildings which cost $400,000 and up. This equates to an electrical construction package of about $35,000 to $40,000 and up. Large residential projects (i.e., apartment or condominium complexes) are also included.

*Local Factors* — Specific local conditions are not included in national averages. Weather conditions, season of the year, local union restrictions, and unusual building code requirements can all have a significant impact on construction costs. The availability of a skilled labor force, sufficient materials and adequate energy and utilities are assumed. The lack of any of these will affect the project's costs. These factors vary in impact and are not dependent solely upon location. They must be reviewed for each project in every area.

In presenting prices in *Means Electrical Cost Data*, certain rounding rules are employed to make the numbers easier to use without significantly affecting accuracy. The rules are used consistently and are as follow:

| From | To | Rounded to nearest |
|---|---|---|
| $ 0.01 | $ 5.00 | 0.01 |
| 5.01 | 20.00 | 0.05 |
| 20.01 | 100.00 | 1.00 |
| 100.01 | 1000.00 | 5.00 |
| 1,000.01 | 10,000.00 | 25.00 |
| 10,000.01 | 50,000.00 | 100.00 |
| 50,000.01 | up | 500.00 |

## Section A — Unit Price

The Unit Price section of *Means Electrical Cost Data* contains a great deal of information in addition to the unit cost for each construction component. Figure 22.1 is a typical page, showing costs for transformers. Note that prices are included for several types of transformers, each in a wide range of capacity ratings. In addition, suggested crews of workers are indicated. These crews are the basis of the installed costs and man-hour units. The information and cost data is broken down and itemized in this way to provide for the most detailed pricing possible. Both the unit price and the systems sections include detailed illustrations for rapid location of components.

Within each individual line item, there is a description of the construction component, information regarding the typical crews as anticipated to perform the work, and productivity shown in man-hours (labor units). Costs are presented both as "bare", or unburdened, and with mark-ups for overhead and profit. Figure 22.2 is a graphic representation of how to read a Unit Price page as presented in *Means Electrical Cost Data*.

**Transformers & Bus Duct**

| | | CREW | MAN-HOURS | UNIT | MAT. | INST. | TOTAL | TOTAL INCL O&P |
|---|---|---|---|---|---|---|---|---|
| 01-001 | OIL FILLED TRANSFORMER Pad mounted, Primary delta or Y, | | | | | | | |
| 005 | 5 KV or 15 KV, with taps, 277/480 secondary, 3 phase | | | | | | | |
| 010 | 150 KVA | R-3 | 30.770 | Ea. | 5,225 | 825 | 6,050 | 6,900 |
| 011 | 225 KVA | | 36.360 | | 6,350 | 975 | 7,325 | 8,325 |
| 020 | 300 KVA | | 44.440 | | 7,825 | 1,200 | 9,025 | 10,300 |
| 030 | 500 KVA | | 50.000 | | 9,750 | 1,350 | 11,100 | 12,600 |
| 040 | 750 KVA | | 52.630 | | 12,700 | 1,400 | 14,100 | 15,900 |
| 050 | 1000 KVA | | 76.920 | | 14,900 | 2,075 | 16,975 | 19,200 |
| 060 | 1500 KVA | | 86.960 | | 19,000 | 2,325 | 21,325 | 24,100 |
| 070 | 2000 KVA | | 100.000 | | 23,600 | 2,675 | 26,275 | 29,700 |
| 071 | 2500 KVA | | 105.000 | | 27,100 | 2,825 | 29,925 | 33,700 |
| 072 | 3000 KVA | | 118.000 | | 28,600 | 3,150 | 31,750 | 35,800 |
| 080 | 3750 KVA | ▼ | 125.000 | ▼ | 32,400 | 3,350 | 35,750 | 40,300 |
| 05-001 | TRANSFORMER, SILICON FILLED, Pad mounted | | | | | | | |
| 002 | 5 KV or 15 KV primary 277/480 volt secondary, 3 phase | | | | | | | |
| 005 | 225 KVA | R-3 | 36.360 | Ea. | 14,000 | 975 | 14,975 | 16,700 |
| 010 | 300 KVA | | 44.440 | | 14,900 | 1,200 | 16,100 | 18,000 |
| 020 | 500 KVA | | 50.000 | | 17,500 | 1,350 | 18,850 | 21,100 |
| 025 | 750 KVA | | 52.630 | | 21,700 | 1,400 | 23,100 | 25,800 |
| 030 | 1000 KVA | | 76.920 | | 25,700 | 2,075 | 27,775 | 31,100 |
| 035 | 1500 KVA | | 86.960 | | 33,300 | 2,325 | 35,625 | 39,900 |
| 040 | 2000 KVA | | 100.000 | | 40,600 | 2,675 | 43,275 | 48,400 |
| 045 | 2500 KVA | ▼ | 105.000 | ▼ | 46,800 | 2,825 | 49,625 | 55,500 |
| 10-001 | DRY TYPE TRANSFORMER | | | | | | | |
| 005 | Single phase, 240/480 volt primary 120/240 volt secondary | | | | | | | |
| 010 | 1 KVA | 1 Elec | 4.000 | Ea. | 100 | 90 | 190 | 240 |
| 030 | 2 KVA | | 5.000 | | 145 | 110 | 255 | 320 |
| 050 | 3 KVA | | 5.710 | | 185 | 130 | 315 | 390 |
| 070 | 5 KVA | | 6.670 | | 255 | 150 | 405 | 495 |
| 090 | 7.5 KVA | | 7.270 | | 360 | 165 | 525 | 630 |
| 110 | 10 KVA | ▼ | 10.000 | | 440 | 225 | 665 | 805 |
| 130 | 15 KVA | 2 Elec | 13.330 | | 595 | 300 | 895 | 1,075 |
| 150 | 25 KVA | | 16.000 | | 775 | 360 | 1,135 | 1,375 |
| 170 | 37.5 KVA | | 20.000 | | 1,025 | 450 | 1,475 | 1,775 |
| 190 | 50 KVA | | 22.860 | | 1,250 | 510 | 1,760 | 2,100 |
| 210 | 75 KVA | ▼ | 24.620 | | 1,700 | 550 | 2,250 | 2,675 |
| 211 | 100 KVA | R-3 | 22.220 | | 2,175 | 595 | 2,770 | 3,225 |
| 212 | 167 KVA | " | 25.000 | | 3,225 | 670 | 3,895 | 4,475 |
| 219 | 480V primary 120/240V secondary, nonvent., 15 KVA | 2 Elec | 13.330 | | 820 | 300 | 1,120 | 1,325 |
| 220 | 25 KVA | | 17.780 | | 1,200 | 400 | 1,600 | 1,900 |
| 221 | 37 KVA | | 21.330 | | 1,800 | 480 | 2,280 | 2,675 |
| 222 | 50 KVA | | 24.620 | | 2,000 | 550 | 2,550 | 3,000 |
| 223 | 75 KVA | | 26.670 | | 2,800 | 595 | 3,395 | 3,950 |
| 224 | 100 KVA | | 32.000 | | 3,450 | 715 | 4,165 | 4,825 |
| 225 | Low operating temperature(80° C), 25 KVA | | 16.000 | | 1,100 | 360 | 1,460 | 1,725 |
| 226 | 37 KVA | | 20.000 | | 1,375 | 450 | 1,825 | 2,150 |
| 227 | 50 KVA | | 22.860 | | 1,825 | 510 | 2,335 | 2,750 |
| 228 | 75 KVA | | 24.620 | | 2,450 | 550 | 3,000 | 3,500 |
| 229 | 100 KVA | ▼ | 29.090 | ▼ | 2,750 | 650 | 3,400 | 3,950 |
| 230 | 3 phase, 240/480 volt primary, 120/208 volt secondary | | | | | | | |
| 231 | 3 KVA | 1 Elec | 8.000 | Ea. | 330 | 180 | 510 | 620 |
| 270 | 6 KVA | | 10.000 | | 380 | 225 | 605 | 740 |
| 290 | 9 KVA | ▼ | 11.430 | | 505 | 255 | 760 | 925 |
| 310 | 15 KVA | 2 Elec | 14.550 | | 760 | 325 | 1,085 | 1,300 |
| 330 | 30 KVA | | 17.780 | | 1,025 | 400 | 1,425 | 1,700 |
| 350 | 45 KVA | | 20.000 | | 1,225 | 450 | 1,675 | 2,000 |
| 370 | 75 KVA | ▼ | 22.860 | | 1,850 | 510 | 2,360 | 2,775 |
| 390 | 112.5 KVA | R-3 | 22.220 | ▼ | 2,450 | 595 | 3,045 | 3,525 |
| 410 | 150 KVA | " | 23.530 | ▼ | 3,200 | 630 | 3,830 | 4,400 |

C9.3 -215

*Figure* 22.1

EXAMPLES OF HOW FIGURES ARE DEVELOPED
on page 156 under OIL FILLED TRANSFORMER, 150 KVA is shown as follows:

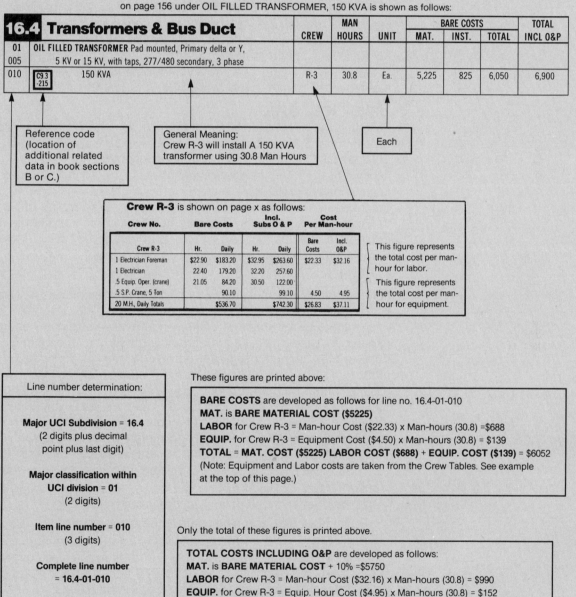

| 16.4 Transformers & Bus Duct | CREW | MAN HOURS | UNIT | BARE COSTS MAT. | INST. | TOTAL | TOTAL INCL O&P |
|---|---|---|---|---|---|---|---|
| 01 005 | OIL FILLED TRANSFORMER Pad mounted, Primary delta or Y, 5 KV or 15 KV, with taps, 277/480 secondary, 3 phase | | | | | | |
| 010 C9.3 -215 | 150 KVA | R-3 | 30.8 | Ea. | 5,225 | 825 | 6,050 | 6,900 |

Reference code
(location of
additional related
data in book sections
B or C.)

General Meaning:
Crew R-3 will install A 150 KVA
transformer using 30.8 Man Hours

Each

**Crew R-3** is shown on page x as follows:

| Crew No. | Bare Costs | | Incl. Subs O & P | | Cost Per Man-hour | |
|---|---|---|---|---|---|---|
| Crew R-3 | Hr. | Daily | Hr. | Daily | Bare Costs | Incl. O&P |
| 1 Electrician Foreman | $22.90 | $183.20 | $32.95 | $263.60 | $22.33 | $32.16 |
| 1 Electrician | 22.40 | 179.20 | 32.20 | 257.60 | | |
| .5 Equip. Oper. (crane) | 21.05 | 84.20 | 30.50 | 122.00 | | |
| .5 S.P. Crane, 5 Ton | | 90.10 | | 99.10 | 4.50 | 4.95 |
| 20 M.H., Daily Totals | | $536.70 | | $742.30 | $26.83 | $37.11 |

This figure represents
the total cost per man-
hour for labor.

This figure represents
the total cost per man-
hour for equipment.

**Line number determination:**

**Major UCI Subdivision = 16.4**
(2 digits plus decimal
point plus last digit)

**Major classification within
UCI division = 01**
(2 digits)

**Item line number = 010**
(3 digits)

**Complete line number
= 16.4-01-010**

These figures are printed above:

**BARE COSTS** are developed as follows for line no. 16.4-01-010
**MAT.** is **BARE MATERIAL COST ($5225)**
**LABOR** for Crew R-3 = Man-hour Cost ($22.33) x Man-hours (30.8) =$688
**EQUIP.** for Crew R-3 = Equipment Cost ($4.50) x Man-hours (30.8) = $139
**TOTAL = MAT. COST ($5225) LABOR COST ($688) + EQUIP. COST ($139) = $6052**
(Note: Equipment and Labor costs are taken from the Crew Tables. See example
at the top of this page.)

Only the total of these figures is printed above.

**TOTAL COSTS INCLUDING O&P** are developed as follows:
**MAT.** is **BARE MATERIAL COST** + 10% =$5750
**LABOR** for Crew R-3 = Man-hour Cost ($32.16) x Man-hours (30.8) = $990
**EQUIP.** for Crew R-3 = Equip. Hour Cost ($4.95) x Man-hours (30.8) = $152
**TOTAL = MAT. COST** $5750 + **LABOR COST** $990 + **EQUIP. COST** $152 = $6892
(Rounded to $6900)

*Figure 22.2*

## Line Numbers

Every construction item in the Means unit price cost data books has a unique line number. This line number acts as an address so that each item can be quickly located and/or referenced. The numbering system is based on the UCI classification by division. In Figure 22.2, note the bold number in reverse type, "16.4". This number represents the major UCI subdivision, in this case "Transformers & Bus Duct", of the major UCI Division 16 – Electrical. All 16 UCI divisions are organized in this manner. Within each subdivision, the data is broken down into major classifications. These major classifications are organized alphabetically and are designated by bold type for both numbers and descriptions. Each item, or line, is further defined by an individual number. As shown in Figure 22.2, the full line number for each item consists of: a major UCI subdivision number – a major classification number – an item line number. Each full line number describes a unique construction element. For example, in Figure 22.1, the line number for an oil-filled transformer, pad-mounted, primary delta or Y, 150 KVA is 16.4-01-010.

## Line Description

Each line has a text description of the item for which costs are listed. The description may be self-contained and all inclusive as is line 16.4-10-230 in Figure 22.1. If indented, the complete description for a line is dependent upon the information provided above. All indented items are delineations (by size, color, material, etc.) or breakdowns of previously described items. Note that it may be necessary to refer back one, two, three, or more successively indented headings in order to get the full description of a line item. An index is provided in the back of *Means Electrical Cost Data* to aid in locating particular items.

## Crew

For each construction element (each line item), a minimum typical crew is designated as appropriate to perform the work. The crew may include one or more trades, foremen, craftsmen and helpers, and any equipment required for proper installation of the described item. If an individual trade installs the item using only hand tools, the smallest efficient number of tradesmen will be indicated ("1 Elec", "2 Elec", etc.). Most electrical items are based upon a one-electrician crew. Abbreviations for trades are shown in Figure 22.3. If more than one trade is required to install the item and/or if powered equipment is needed, a crew number will be designated (B-5, D-3, etc.). A complete listing of crews is presented in the Foreword pages of *Means Electrical Cost Data* (see Figure 22.4). On these pages, each crew is broken down into the following components:

1. Number and type of workers designated.
2. Number, size, and type of any equipment required.
3. Hourly labor costs listed two ways: "bare" – base rate including fringe benefits; and billing rate – including installing contractor's overhead and profit. (See Figure 22.3 from the inside back cover of *Means Electrical Cost Data* for labor rate information).
4. Daily equipment costs, based on the weekly equipment rental cost divided by 5, plus the hourly operating cost, times 8 hours. This cost is listed two ways: as a bare cost and with a 10% percent markup to cover handling and management costs.
5. Labor and equipment are broken down further into: cost per man-hour for labor, and cost per man-hour for the equipment.

Labor rates used in this edition are listed below for January 1, 1986. The base rates are the averages of the 30 largest cities in the U.S. The rates have been rounded out to the nearest 5¢ and the base rates include fringe benefits but do not include insurance or taxes. The billing rate figures include taxes, insurance and the Subcontractor's Overhead and Profit.

| Abbr. | Trade | Base Rate Incl. Fringes | | Workers' Comp. Ins. | Average Fixed Overhead | Subs Overhead | Subs Profit | Subs Total Overhead & Profit | | Rate with Subs O & P | |
|---|---|---|---|---|---|---|---|---|---|---|---|
| | | Hourly | Daily | | | | | % | Amount | Hourly | Daily |
| Skwk | Skilled Workers Average (35 trades) | $20.50 | $164.00 | 9.3% | 13.8% | 12.8% | 10% | 45.9% | $ 9.40 | $29.90 | $239.20 |
| | Helpers Average (5 trades) | 15.55 | 124.40 | 9.8 | | 13.0 | | 46.6 | 7.25 | 22.80 | 182.40 |
| | Foremen Average, Inside (50¢ over trade) | 21.00 | 168.00 | 9.3 | | 12.8 | | 45.9 | 9.65 | 30.65 | 245.20 |
| | Foremen Average, Outside ($2.00 over trade) | 22.50 | 180.00 | 9.3 | | 12.8 | | 45.9 | 10.35 | 32.85 | 262.80 |
| Clab | Common Building Laborers | 15.90 | 127.20 | 10.1 | | 11.0 | | 44.9 | 7.15 | 23.05 | 184.40 |
| Asbe | Asbestos Workers | 22.75 | 182.00 | 7.7 | | 16.0 | | 47.5 | 10.80 | 33.55 | 268.40 |
| Boil | Boilermakers | 22.75 | 182.00 | 6.6 | | 16.0 | | 46.4 | 10.55 | 33.30 | 266.40 |
| Bric | Bricklayers | 20.50 | 164.00 | 7.6 | | 11.0 | | 42.4 | 8.70 | 29.20 | 233.60 |
| Brhe | Bricklayer Helpers | 16.00 | 128.00 | 7.6 | | 11.0 | | 42.4 | 6.80 | 22.80 | 182.40 |
| Carp | Carpenters | 20.00 | 160.00 | 10.1 | | 11.0 | | 44.9 | 9.00 | 29.00 | 232.00 |
| Cefi | Cement Finishers | 19.20 | 153.60 | 5.9 | | 11.0 | | 40.7 | 7.80 | 27.00 | 216.00 |
| Elec | Electricians | 22.40 | 179.20 | 4.0 | | 16.0 | | 43.8 | 9.80 | 32.20 | 257.60 |
| Elev | Elevator Constructors | 22.65 | 181.20 | 5.5 | | 16.0 | | 45.3 | 10.25 | 32.90 | 263.20 |
| Eqhv | Equipment Operators, Crane or Shovel | 21.05 | 168.40 | 7.2 | | 14.0 | | 45.0 | 9.45 | 30.50 | 244.00 |
| Eqmd | Equipment Operators, Medium Equipment | 20.60 | 164.80 | 7.2 | | 14.0 | | 45.0 | 9.25 | 29.85 | 238.80 |
| Eqlt | Equipment Operators, Light Equipment | 19.45 | 155.60 | 7.2 | | 14.0 | | 45.0 | 8.75 | 28.20 | 225.60 |
| Eqol | Equipment Operators, Oilers | 17.50 | 140.00 | 7.2 | | 14.0 | | 45.0 | 7.90 | 25.40 | 203.20 |
| Eqmm | Equipment Operators, Master Mechanics | 21.80 | 174.40 | 7.2 | | 14.0 | | 45.0 | 9.80 | 31.60 | 252.80 |
| Glaz | Glaziers | 20.15 | 161.20 | 7.9 | | 11.0 | | 42.7 | 8.60 | 28.75 | 230.00 |
| Lath | Lathers | 20.10 | 160.80 | 6.3 | | 11.0 | | 41.1 | 8.25 | 28.35 | 226.80 |
| Marb | Marble Setters | 20.10 | 160.80 | 7.6 | | 11.0 | | 42.4 | 8.50 | 28.60 | 228.80 |
| Mill | Millwrights | 20.75 | 166.00 | 6.6 | | 11.0 | | 41.4 | 8.60 | 29.35 | 234.80 |
| Mstz | Mosaic and Terrazzo Workers | 19.90 | 159.20 | 5.4 | | 11.0 | | 40.2 | 8.00 | 27.90 | 223.20 |
| Pord | Painters, Ordinary | 19.25 | 154.00 | 7.7 | | 11.0 | | 42.5 | 8.20 | 27.45 | 219.60 |
| Psst | Painters, Structural Steel | 20.00 | 160.00 | 27.0 | | 11.0 | | 61.8 | 12.35 | 32.35 | 258.80 |
| Pape | Paper Hangers | 19.50 | 156.00 | 7.7 | | 11.0 | | 42.5 | 8.30 | 27.80 | 222.40 |
| Pile | Pile Drivers | 20.10 | 160.80 | 17.0 | | 16.0 | | 56.8 | 11.40 | 31.50 | 252.00 |
| Plas | Plasterers | 19.90 | 159.20 | 7.7 | | 11.0 | | 42.5 | 8.45 | 28.35 | 226.80 |
| Plah | Plasterer Helpers | 16.50 | 132.00 | 7.7 | | 11.0 | | 42.5 | 7.00 | 23.50 | 188.00 |
| Plum | Plumbers | 22.55 | 180.40 | 4.8 | | 16.0 | | 44.6 | 10.05 | 32.60 | 260.80 |
| Rodm | Rodmen (Reinforcing) | 21.75 | 174.00 | 16.8 | | 14.0 | | 54.6 | 11.90 | 33.65 | 269.20 |
| Rofc | Roofers, Composition | 18.80 | 150.40 | 18.2 | | 11.0 | | 53.0 | 9.95 | 28.75 | 230.00 |
| Rots | Roofers, Tile & Slate | 18.95 | 151.60 | 18.2 | | 11.0 | | 53.0 | 10.05 | 29.00 | 232.00 |
| Rohe | Roofer Helpers (Composition) | 13.75 | 110.00 | 18.2 | | 11.0 | | 53.0 | 7.30 | 21.05 | 168.40 |
| Shee | Sheet Metal Workers | 22.70 | 181.60 | 6.3 | | 16.0 | | 46.1 | 10.45 | 33.15 | 265.20 |
| Spri | Sprinkler Installers | 23.25 | 186.00 | 5.5 | | 16.0 | | 45.3 | 10.55 | 33.80 | 270.40 |
| Stpi | Steamfitters or Pipefitters | 22.75 | 182.00 | 4.8 | | 16.0 | | 44.6 | 10.15 | 32.90 | 263.20 |
| Ston | Stone Masons | 20.30 | 162.40 | 7.6 | | 11.0 | | 42.4 | 8.60 | 28.90 | 231.20 |
| Sswk | Structural Steel Workers | 21.70 | 173.60 | 19.3 | | 14.0 | | 57.1 | 12.40 | 34.10 | 272.80 |
| Tilf | Tile Layers (Floor) | 19.75 | 158.00 | 5.4 | | 11.0 | | 40.2 | 7.95 | 27.70 | 221.60 |
| Tilh | Tile Layer Helpers | 15.60 | 124.80 | 5.4 | | 11.0 | | 40.2 | 6.30 | 21.90 | 175.20 |
| Trlt | Truck Drivers, Light | 16.35 | 130.80 | 8.6 | | 11.0 | | 43.4 | 7.10 | 23.45 | 187.60 |
| Trhv | Truck Drivers, Heavy | 16.60 | 132.80 | 8.6 | | 11.0 | | 43.4 | 7.20 | 23.80 | 190.40 |
| Sswl | Welders, Structural Steel | 21.70 | 173.60 | 19.3 | | 14.0 | | 57.1 | 12.40 | 34.10 | 272.80 |
| Wrck | *Wrecking | 15.90 | 127.20 | 20.7 | ↓ | 11.0 | ↓ | 55.5 | 8.80 | 24.70 | 197.60 |

*Not included in Averages.

*Figure* 22.3

# CREWS

| Crew No. | Bare Costs Hr. | Bare Costs Daily | Incl. Subs O&P Hr. | Incl. Subs O&P Daily | Cost Per Man-hour Bare Costs | Cost Per Man-hour Incl. O&P |
|---|---|---|---|---|---|---|
| **Crew B-38** | Hr. | Daily | Hr. | Daily | Bare Costs | Incl. O&P |
| 2 Building Laborers | $15.90 | $254.40 | $23.05 | $368.80 | $17.08 | $24.76 |
| 1 Equip. Oper. (light) | 19.45 | 155.60 | 28.20 | 225.60 | | |
| 1 Backhoe Loader, 48 H.P. | | 156.00 | | 171.60 | | |
| 1 Demol. Hammer, Hyd. | | 212.45 | | 233.70 | 15.35 | 16.88 |
| 24 M.H., Daily Totals | | $778.45 | | $999.70 | $32.43 | $41.64 |
| **Crew B-39** | Hr. | Daily | Hr. | Daily | Bare Costs | Incl. O&P |
| 1 Labor Foreman (outside) | $17.90 | $143.20 | $25.95 | $207.60 | $16.82 | $24.39 |
| 4 Building Laborers | 15.90 | 508.80 | 23.05 | 737.60 | | |
| 1 Equipment Oper. (light) | 19.45 | 155.60 | 28.20 | 225.60 | | |
| 1 Air Compr., 250 C.F.M. | | 111.40 | | 122.55 | | |
| Air Tools & Accessories | | 25.80 | | 28.40 | | |
| 2-50 Ft. Air Hoses, 1.5" Dia | | 13.20 | | 14.50 | 3.13 | 3.44 |
| 48 M.H., Daily Totals | | $958.00 | | $1336.25 | $19.95 | $27.83 |
| **Crew B-53** | Hr. | Daily | Hr. | Daily | Bare Costs | Incl. O&P |
| 1 Equip. Oper. (light) | $19.45 | $155.60 | $28.20 | $225.60 | $19.45 | $28.20 |
| 1 Trencher, Chain, 12 H.P. | | 135.40 | | 148.95 | 16.92 | 18.61 |
| 8 M.H., Daily Totals | | $291.00 | | $374.55 | $36.37 | $46.81 |
| **Crew B-54** | Hr. | Daily | Hr. | Daily | Bare Costs | Incl. O&P |
| 1 Equip. Oper. (light) | $19.45 | $155.60 | $28.20 | $225.60 | $19.45 | $28.20 |
| 1 Trencher, Chain, 40 H.P. | | 199.80 | | 219.80 | 24.97 | 27.47 |
| 8 M.H., Daily Totals | | $355.40 | | $445.40 | $44.42 | $55.67 |
| **Crew B-67** | Hr. | Daily | Hr. | Daily | Bare Costs | Incl. O&P |
| 1 Millwright | $20.75 | $166.00 | $29.35 | $234.80 | $20.10 | $28.77 |
| 1 Equip. Oper. (light) | 19.45 | 155.60 | 28.20 | 225.60 | | |
| 1 Forklift | | 169.40 | | 186.35 | 10.58 | 11.64 |
| 16 M.H., Daily Totals | | $491.00 | | $646.75 | $30.68 | $40.41 |
| **Crew B-68** | Hr. | Daily | Hr. | Daily | Bare Costs | Incl. O&P |
| 2 Millwrights | $20.75 | $332.00 | $29.35 | $469.60 | $20.31 | $28.96 |
| 1 Equip. Oper. (light) | 19.45 | 155.60 | 28.20 | 225.60 | | |
| 1 Forklift | | 169.40 | | 186.35 | 7.05 | 7.76 |
| 24 M.H., Daily Totals | | $657.00 | | $881.55 | $27.36 | $36.72 |
| **Crew B-89** | Hr. | Daily | Hr. | Daily | Bare Costs | Incl. O&P |
| 1 Equip. Oper. (light) | $19.45 | $155.60 | $28.20 | $225.60 | $19.45 | $28.20 |
| 1 Stake Body, 3 Ton | | 648.00 | | 712.80 | | |
| 1 Air Compressor | | 51.40 | | 56.55 | | |
| 1 Water Tank, 65 Gal. | | 18.45 | | 20.30 | | |
| 1 Generator, 10 K.W. | | 54.60 | | 60.05 | | |
| 1 Concrete Saw | | 36.60 | | 40.25 | 101.13 | 111.24 |
| 8 M.H., Daily Totals | | $964.65 | | $1115.55 | $120.58 | $139.44 |
| **Crew D-1** | Hr. | Daily | Hr. | Daily | Bare Costs | Incl. O&P |
| 1 Bricklayer | $20.50 | $164.00 | $29.20 | $233.60 | $18.25 | $26.00 |
| 1 Bricklayer Helper | 16.00 | 128.00 | 22.80 | 182.40 | | |
| 16 M.H., Daily Totals | | $292.00 | | $416.00 | $18.25 | $26.00 |
| **Crew E-4** | Hr. | Daily | Hr. | Daily | Bare Costs | Incl. O&P |
| 1 Struc. Steel Foreman | $23.70 | $189.60 | $37.25 | $298.00 | $22.20 | $34.88 |
| 3 Struc. Steel Workers | 21.70 | 520.80 | 34.10 | 818.40 | | |
| 1 Gas Welding Machine | | 56.80 | | 62.50 | 1.77 | 1.95 |
| 32 M.H., Daily Totals | | $767.20 | | $1178.90 | $23.97 | $36.83 |

| Crew No. | Bare Costs Hr. | Bare Costs Daily | Incl. Subs O&P Hr. | Incl. Subs O&P Daily | Cost Per Man-hour Bare Costs | Cost Per Man-hour Incl. O&P |
|---|---|---|---|---|---|---|
| **Crew F-2** | Hr. | Daily | Hr. | Daily | Bare Costs | Incl. O&P |
| 2 Carpenters | $20.00 | $320.00 | $29.00 | $464.00 | $20.00 | $29.00 |
| Power Tools | | 17.20 | | 18.90 | 1.07 | 1.18 |
| 16 M.H., Daily Totals | | $337.20 | | $482.90 | $21.07 | $30.18 |
| **Crew L-1** | Hr. | Daily | Hr. | Daily | Bare Costs | Incl. O&P |
| 1 Electrician | $22.40 | $179.20 | $32.20 | $257.60 | $22.47 | $32.40 |
| 1 Plumber | 22.55 | 180.40 | 32.60 | 260.80 | | |
| 16 M.H., Daily Totals | | $359.60 | | $518.40 | $22.47 | $32.40 |
| **Crew L-3** | Hr. | Daily | Hr. | Daily | Bare Costs | Incl. O&P |
| 1 Carpenter | $20.00 | $160.00 | $29.00 | $232.00 | $21.27 | $30.83 |
| .5 Electrician | 22.40 | 89.60 | 32.20 | 128.80 | | |
| .5 Sheet Metal Worker | 22.70 | 90.80 | 33.15 | 132.60 | | |
| 16 M.H., Daily Totals | | $340.40 | | $493.40 | $21.27 | $30.83 |
| **Crew L-4** | Hr. | Daily | Hr. | Daily | Bare Costs | Incl. O&P |
| 2 Skilled Workers | $20.50 | $328.00 | $29.90 | $478.40 | $18.85 | $27.53 |
| 1 Helper | 15.55 | 124.40 | 22.80 | 182.40 | | |
| 24 M.H., Daily Totals | | $452.40 | | $660.80 | $18.85 | $27.53 |
| **Crew L-6** | Hr. | Daily | Hr. | Daily | Bare Costs | Incl. O&P |
| 1 Plumber | $22.55 | $180.40 | $32.60 | $260.80 | $22.50 | $32.46 |
| .5 Electrician | 22.40 | 89.60 | 32.20 | 128.80 | | |
| 12 M.H., Daily Totals | | $270.00 | | $389.60 | $22.50 | $32.46 |
| **Crew L-7** | Hr. | Daily | Hr. | Daily | Bare Costs | Incl. O&P |
| 2 Carpenters | $20.00 | $320.00 | $29.00 | $464.00 | $19.17 | $27.75 |
| 1 Building Laborer | 15.90 | 127.20 | 23.05 | 184.40 | | |
| .5 Electrician | 22.40 | 89.60 | 32.20 | 128.80 | | |
| 28 M.H., Daily Totals | | $536.80 | | $777.20 | $19.17 | $27.75 |
| **Crew L-9** | Hr. | Daily | Hr. | Daily | Bare Costs | Incl. O&P |
| 1 Labor Foreman (inside) | $16.40 | $131.20 | $23.75 | $190.00 | $18.02 | $26.67 |
| 2 Building Laborers | 15.90 | 254.40 | 23.05 | 368.80 | | |
| 1 Struc. Steel Worker | 21.70 | 173.60 | 34.10 | 272.80 | | |
| .5 Electrician | 22.40 | 89.60 | 32.20 | 128.80 | | |
| 36 M.H., Daily Totals | | $648.80 | | $960.40 | $18.02 | $26.67 |
| **Crew Q-1** | Hr. | Daily | Hr. | Daily | Bare Costs | Incl. O&P |
| 1 Plumber | $22.55 | $180.40 | $32.60 | $260.80 | $20.29 | $29.35 |
| 1 Plumber Apprentice | 18.04 | 144.32 | 26.10 | 208.80 | | |
| 16 M.H., Daily Totals | | $324.72 | | $469.60 | $20.29 | $29.35 |
| **Crew Q-2** | Hr. | Daily | Hr. | Daily | Bare Costs | Incl. O&P |
| 2 Plumbers | $22.55 | $360.80 | $32.60 | $521.60 | $21.04 | $30.43 |
| 1 Plumber Apprentice | 18.04 | 144.32 | 26.10 | 208.80 | | |
| 24 M.H., Daily Totals | | $505.12 | | $730.40 | $21.04 | $30.43 |
| **Crew Q-3** | Hr. | Daily | Hr. | Daily | Bare Costs | Incl. O&P |
| 1 Plumber Foreman (ins) | $23.05 | $184.40 | $33.35 | $266.80 | $21.54 | $31.16 |
| 2 Plumbers | 22.55 | 360.80 | 32.60 | 521.60 | | |
| 1 Plumber Apprentice | 18.04 | 144.32 | 26.10 | 208.80 | | |
| 32 M.H., Daily Totals | | $689.52 | | $997.20 | $21.54 | $31.16 |
| **Crew Q-5** | Hr. | Daily | Hr. | Daily | Bare Costs | Incl. O&P |
| 1 Steamfitter | $22.75 | $182.00 | $32.90 | $263.20 | $20.47 | $29.60 |
| 1 Steamfitter Apprentice | 18.20 | 145.60 | 26.30 | 210.40 | | |
| 16 M.H., Daily Totals | | $327.60 | | $473.60 | $20.47 | $29.60 |

*Figure 22.4*

6. The total daily man-hours for the crew.
7. The total bare costs per day for the crew, including equipment.
8. The total daily cost of the crew is used to calculate the installing contractor's overhead and profit.

The total man-hour cost of the required crew is used to calculate the unit installation cost for each item (for both bare costs and cost including overhead and profit).

The crew designation does not mean that this is the only crew that can perform the work. Crew size and content have been developed and chosen based on practical experience and feedback from contractors. These designations represent a labor and equipment make-up commonly found in the industry. The most appropriate crew for a given task is best determined based on particular project requirements. Unit costs may vary if actual used crew sizes or content differ significantly.

Figure 22.5 is a page from Division 1.5 of *Means Electrical Cost Data*. This type of page lists the equipment costs used in the presentation and calculation of the crew costs and unit price data. Rental costs are shown as daily, weekly, and monthly rates. The Hourly Operating Cost represents the cost of fuel, lubrication and routine maintenance. Equipment costs used in the crews are calculated as follows:

| Line Number | 1.5-15-040 |
| Equipment | Diesel Engine, rotary screw, 250 C.F.M. |
| Rent per week | $275.00 |
| Hourly Operating Cost: | $7.05 |

$$\frac{Weekly\ rental}{(5\ days/week)} + (\text{Hourly Oper. Cost} \times 8\ \text{hrs/day}) = \text{Crew's Equipment Cost}$$

$$\frac{\$275}{5} + (\$7.05 \times 8) = \$111.40$$

## Unit

The unit column (see Figures 22.1 and 22.2) defines the measurement parameter for which the costs have been calculated. It is this "unit of quantity" upon which Unit Price Estimating is based. The units as used represent standard estimating and quantity takeoff procedures. However, the estimator should always check to be sure that the units taken off match those priced. For example, the estimator may measure linear feet (L.F.), but the unit prices may be per each hundred linear feet (C.L.F.). A list of standard abbreviations is included at the back of *Means Electrical Cost Data*.

## Bare Costs

The three columns listed under "Bare Costs" – "Material", "Installation" and "Total", represent the actual cost of construction items to the contractor. In other words, bare costs are those which *do not* include the overhead and profit of the installing contractor, i.e., a subcontractor or a general contracting company using its own crews.

*Material*: Material costs are based on the national average contractor purchase price delivered to the job site. Delivered costs are assumed to be within a 20 mile radius of metropolitan areas. No *sales tax* is included in the material prices because of the variations from state to state.

| 1.5 Contractor Equipment | UNIT | HOURLY OPER. COST | RENT PER DAY | RENT PER WEEK | RENT PER MONTH | CREWS EQUIPMENT COST |
|---|---|---|---|---|---|---|
| 535 Dump trailer only, rear dump, 16-1/2 C.Y. | Ea. | 4.02 | 80 | 240 | 715 | |
| 540 20 C.Y. | | 4.90 | 92 | 280 | 845 | |
| 545 Flatbed, single axle, 1-1/2 ton rating | | 3.80 | 46 | 135 | 415 | |
| 550 3 ton rating | | 4.40 | 58 | 160 | 505 | |
| 555 Off highway rear dump, 25 ton capacity | | 26 | 840* | 2,200 | 6,750 | |
| 560 35 ton capacity | | 35 | 1,150* | 3,150 | 9,475 | |
| 15-001 GENERAL EQUIPMENT RENTAL | | | | | | |
| 010 | | | | | | |
| 015 Aerial lift, scissor type, to 15' high, 1000 lb. capacity | Ea. | .16 | 79 | 255 | 765 | |
| 016 To 25' high, 2000 lb. capacity | | .19 | 120 | 375 | 1,125 | |
| 017 Telescoping boom to 40' high, 750 lb. capacity | | .22 | 245 | 720 | 2,325 | |
| 018 2000 lb. capacity | | .25 | 310 | 1,025 | 3,100 | |
| 019 To 60' high, 750 lb. capacity | | .28 | 555 | 1,275 | 3,400 | |
| 020 Air compressor, portable, gas engine, 60 C.F.M. | | 3.95 | 35 | 99 | 305 | |
| 030 160 C.F.M. | | 4.78 | 53 | 160 | 480 | |
| 040 Diesel engine, rotary screw, 250 C.F.M. | | 7.05 | 89 | 275 | 800 | |
| 050 365 C.F.M. | | 10.75 | 120 | 360 | 1,075 | |
| 060 600 C.F.M. | | 17.85 | 165 | 500 | 1,500 | |
| 070 750 C.F.M. | | 22 | 215 | 630 | 1,875 | |
| 080 For silenced models, small sizes, add | | 3% | 5% | 5% | 5% | |
| 090 Large sizes, add | | 4% | 7% | 7% | 7% | |
| 092 Air tools and accessories | | | | | | |
| 093 Breaker, pavement, 60 lb. | Ea. | .44 | 17 | 47 | 140 | |
| 094 80 lb. | | .58 | 18 | 53 | 155 | |
| 098 Dust control per drill | | .14 | 11 | 26 | 77 | |
| 099 | | | | | | |
| 100 Hose, air with couplings, 50' long, 3/4" diameter | Ea. | .10 | 4 | 13 | 27 | |
| 110 1" diameter | | .10 | 6 | 15 | 36 | |
| 120 1-1/2" diameter | | .10 | 11 | 29 | 84 | |
| 130 2" diameter | | .11 | 13 | 40 | 120 | |
| 140 2-1/2" diameter | | .12 | 14 | 42 | 125 | |
| 141 3" diameter | | .12 | 14 | 43 | 130 | |
| 153 Sheeting driver for 60 lb. breaker | | .08 | 9 | 20 | 55 | |
| 154 For 125 lb. breaker | | .08 | 9 | 22 | 66 | |
| 156 Tamper, single, 35 lb. | | .44 | 16 | 48 | 145 | |
| 157 Triple, 140 lb. | | 1.46 | 31 | 99 | 295 | |
| 158 Wrenches, impact, air powered up to 3/4" bolt | | .08 | 20 | 60 | 175 | |
| 159 Up to 1-1/4" bolt | | .09 | 27 | 79 | 225 | |
| 210 Generator, electric, gas engine, 1.5 KW to 3 KW | | .52 | 23 | 70 | 205 | |
| 220 5 KW | | .89 | 40 | 145 | 360 | |
| 230 10 KW | | 1.95 | 60 | 195 | 585 | |
| 240 25 KW | | 3.78 | 91 | 325 | 845 | |
| 250 Diesel engine, 20 KW | | 3.30 | 61 | 200 | 600 | |
| 260 50 KW | | 3.90 | 135 | 345 | 1,025 | |
| 270 100 KW | | 8.23 | 215 | 550 | 1,575 | |
| 280 250 KW | | 16.80 | 300 | 910 | 2,725 | |
| 290 Heaters, space, oil or electric, 50 MBH | | 1.03 | 15 | 45 | 135 | |
| 300 100 MBH | | 1.29 | 22 | 63 | 180 | |
| 310 300 MBH | | 1.77 | 33.68 | 100 | 300 | |
| 315 500 MBH | | 2.24 | 49 | 145 | 430 | |
| 320 Hose, water, suction with coupling, 20' long, 2" diameter | | | 7 | 17 | 46 | |
| 321 3" diameter | | | 9 | 23 | 66 | |
| 322 4" diameter | | | 12 | 36 | 110 | |
| 323 6" diameter | | | 28 | 72 | 185 | |
| 324 8" diameter | | | 38 | 90 | 275 | |
| 325 Discharge hose with coupling, 50' long, 2" diameter | | | 7 | 18 | 47 | |
| 326 3" diameter | | | 10 | 25 | 64 | |
| 327 4" diameter | | | 13 | 38 | 100 | |
| 328 6" diameter | | | 25 | 65 | 165 | |
| 329 8" diameter | | | 42 | 91 | 190 | |

Figure 22.5

The prices are based on quantities that would normally be purchased for electrical installations of $35,000 to $40,000 and up. Prices for small quantities must be adjusted accordingly. If more current costs for materials are available for the appropriate location, it is recommended that adjustments be made to the unit costs to reflect any cost difference.

*Installation*: The unit cost for installation of an item is calculated by multiplying the total hourly cost rate of the designated crew by the number of man-hours that the crew will require to install one unit of quantity. The resulting unit price for installation includes both labor and equipment costs (if any).

The labor rates used to determine the bare installation costs are shown for 35 standard trades in the Labor Rates chart (Figure 22.3), under the column "Base Rate Including Fringes". This rate includes a worker's actual hourly wage plus employer-paid benefits (health insurance, vacation, pension, etc.). The labor rates used are *national average* union rates based on trade union agreements (as of January 1 of the current year) from 30 major cities in the United States. As new wage agreements are negotiated within a calendar year, labor costs should be adjusted. The equipment portion of installation costs is discussed earlier in this chapter.

*Total Bare Costs*: This column simply represents the arithmetic sum of the bare material and installation costs. This total is the average cost to the installing contractor for the particular item of construction, supplied and installed, or "in place". No overhead and/or profit are included.

## Total Including Overhead and Profit

The price units in the "Total Including Overhead and Profit" column might also be called the "billing rate". These prices are, on the average, what the installing contractor would charge for the particular item of work.

The "Total Including Overhead and Profit" costs are determined by adding the following two calculations:

1. Bare materials – increased by 10% for handling. See Figure 22.2.
2. Bare installation costs – increased to include overhead and profit.

In order to increase the installation cost to include overhead and profit, labor and equipment costs are treated separately. Ten percent is added to the bare equipment cost for handling, management, etc. Labor costs are increased by percentages for overhead and profit, depending upon trade as shown in Figure 22.3. The resulting rates are listed in the right hand columns of the same figure. Note that the percentage increase for overhead and profit for the average electrician is 43.8% of the base rate. The following items are included in the increase for overhead and profit, also shown in Figure 22.3:

> **Workers' Compensation and Employer's Liability:** Workers' Compensation and Employer's Liability Insurance rates vary from state to state and are tied into the construction trade safety records in that particular state. Rates also vary by trade according to the hazard involved. (See Figures 22.6 and 22.7.) The proper authorities will most likely keep the contractor well informed of the rates and obligations.

**State and Federal Unemployment Insurance:** The employer's tax rate is adjusted by a merit-rating system according to the number of former employees applying for benefits. Contractors who find it possible to offer a maximum of steady employment can enjoy a reduction in the unemployment tax rate.

**Employer-Paid Social Security (FICA):** The tax rate is adjusted annually by the federal government. It is a percentage of an employee's salary up to a maximum annual contribution.

**Builder's Risk and Public Liability:** These insurance rates vary according to the trades involved and the state in which the work is done.

*Overhead*: The column listed as "Sub's Overhead" provides percentages to be added for office or operating overhead. This is the cost of doing business. The percentages are presented as national averages by trade as shown in Figure 22.3. Note that the operating overhead costs are applied to *labor only* in *Means Electrical Cost Data*.

### Workers' Compensation Rates (National Averages by Trade and Building Type)

| Trade | Insurance Rate (% of Labor Cost) | | % of Building Cost | | | % of Labor Cost | | |
|---|---|---|---|---|---|---|---|---|
| | Range | Average | Office Bldgs. | Schools & Apts. | Mfg. | Office Bldgs. | Schools & Apts. | Mfg. |
| Excavation, Grading, etc. | 2.1% to 27.8% | 7.2% | 4.8% | 4.9% | 4.5% | .35% | .35% | .32% |
| Piles & Foundations | 4.0 to 47.6 | 17.0 | 7.1 | 5.2 | 8.7 | 1.21 | .88 | 1.48 |
| Concrete | 2.3 to 27.1 | 9.2 | 5.0 | 14.8 | 3.7 | .46 | 1.36 | .34 |
| Masonry | 1.6 to 18.7 | 7.6 | 6.9 | 7.5 | 1.9 | .52 | .57 | .14 |
| Structural Steel | 3.1 to 42.9 | 19.3 | 10.7 | 3.9 | 17.6 | 2.07 | .75 | 3.40 |
| Misc. & Ornamental Metals | 1.3 to 15.7 | 6.8 | 2.8 | 4.0 | 3.6 | .19 | .27 | .24 |
| Carpentry & Millwork | 2.9 to 54.1 | 10.1 | 3.7 | 4.0 | 0.5 | .37 | .40 | .05 |
| Metal or Composition Siding | 2.7 to 23.3 | 7.8 | 2.3 | 0.3 | 4.3 | .18 | .02 | .34 |
| Roofing | 3.7 to 52.7 | 18.2 | 2.3 | 2.6 | 3.1 | .42 | .47 | .56 |
| Doors & Hardware | 2.0 to 15.9 | 5.9 | 0.9 | 1.4 | 0.4 | .05 | .08 | .02 |
| Sash & Glazing | 3.0 to 33.3 | 7.9 | 3.5 | 4.0 | 1.0 | .28 | .32 | .08 |
| Lath & Plaster | 2.3 to 22.9 | 7.7 | 3.3 | 6.9 | 0.8 | .25 | .53 | .06 |
| Tile, Marble & Floors | 1.1 to 18.7 | 5.4 | 2.6 | 3.0 | 0.5 | .14 | .16 | .03 |
| Acoustical Ceilings | 1.6 to 15.4 | 6.3 | 2.4 | 0.2 | 0.3 | .15 | .01 | .02 |
| Painting | 2.6 to 20.0 | 7.7 | 1.5 | 1.6 | 1.6 | .12 | .12 | .12 |
| Interior Partitions | 2.9 to 54.1 | 10.1 | 3.9 | 4.3 | 4.4 | .39 | .43 | .44 |
| Miscellaneous Items | 1.2 to 92.4 | 9.6 | 5.2 | 3.7 | 9.7 | .50 | .36 | .93 |
| Elevators | 1.4 to 15.6 | 5.5 | 2.1 | 1.1 | 2.2 | .12 | .06 | .12 |
| Sprinklers | 1.4 to 16.8 | 5.5 | 0.5 | — | 2.0 | .03 | — | .11 |
| Plumbing | 1.3 to 12.5 | 4.8 | 4.9 | 7.2 | 5.2 | .24 | .35 | .25 |
| Heat., Vent., Air Conditioning | 1.8 to 14.5 | 6.3 | 13.5 | 11.0 | 12.9 | .85 | .69 | .81 |
| Electrical | 1.0 to 13.1 | 4.0 | 10.1 | 8.4 | 11.1 | .40 | .34 | .44 |
| Total | 1.0% to 92.4% | — | 100.0% | 100.0% | 100.0% | 9.29% | 8.52% | 10.30% |
| | | | Overall Weighted Average | | | | 9.37% | |

Figure 22.6

*Profit*: This percentage is the fee added by the contractor to offer both a return on investment and an allowance to cover the risk involved in the type of construction being bid. The profit percentage may vary from 4% on large, straightforward projects to as much as 25% on smaller, high-risk jobs. Profit percentages are directly affected by economic conditions, the expected number of bidders, and the estimated risk involved in the project. For estimating purposes, *Means Electrical Cost Data* assumes 10% as a reasonable average profit factor.

## Square Foot and Cubic Foot Costs

Division 17 in *Means Electrical Cost Data* has been developed to facilitate the preparation of rapid preliminary budget estimates. The cost figures in this division are derived from more than 10,500 actual building projects contained in the R.S. Means data bank of construction costs and include the contractor's overhead and profit. The prices shown *do not* include architectural fees or land costs. The files are updated each year with costs for new projects. In no case are all subdivisions of a project listed.

These projects were located throughout the United States and reflect differences in square foot and cubic foot costs due to both the variations in labor and material costs, and the differences in the owners' requirements. For instance, a bank in a large city would have different features and costs than one in a rural area. This is true of all the different types of buildings analyzed. All individual cost items were computed and tabulated separately. Thus, the sum of the median figures for Plumbing, H.V.A.C. and Electrical will not normally add up to the total Mechanical and Electrical costs arrived at by separate analysis and tabulation of the projects.

### Workers' Compensation Rates. Average Rates by States.

| State | Weighted Average | State | Weighted Average | State | Weighted Average |
|---|---|---|---|---|---|
| Alabama | 6.4% | Kentucky | 5.7% | North Dakota | 6.9% |
| Alaska | 12.4 | Louisiana | 6.4 | Ohio | 7.0 |
| Arizona | 10.5 | Maine | 14.3 | Oklahoma | 8.1 |
| Arkansas | 6.4 | Maryland | 14.7 | Oregon | 21.5 |
| California | 11.5 | Massachusetts | 12.5 | Pennsylvania | 10.2 |
| Colorado | 10.9 | Michigan | 11.3 | Rhode Island | 12.2 |
| Connecticut | 15.0 | Minnesota | 13.6 | South Carolina | 7.4 |
| Delaware | 9.7 | Mississippi | 5.4 | South Dakota | 6.1 |
| District of Columbia | 16.5 | Missouri | 5.2 | Tennessee | 4.8 |
| Florida | 11.2 | Montana | 12.0 | Texas | 6.0 |
| Georgia | 5.4 | Nebraska | 6.1 | Utah | 5.3 |
| Hawaii | 25.1 | Nevada | 10.6 | Vermont | 6.1 |
| Idaho | 8.1 | New Hampshire | 13.3 | Virginia | 7.3 |
| Illinois | 11.4 | New Jersey | 6.6 | Washington | 6.8 |
| Indiana | 2.5 | New Mexico | 14.4 | West Virginia | 8.7 |
| Iowa | 6.3 | New York | 8.9 | Wisconsin | 6.5 |
| Kansas | 6.3 | North Carolina | 4.7 | Wyoming | 5.4 |

*Figure 22.7*

The data and prices presented on a Division 17 page (as shown in Figure 22.8) are listed both as square foot or cubic foot costs and as a percentage of total costs. Each category tabulates the data in a similar manner. The median, or middle figure, is listed. (This is *not* the average cost.) This means that 50% of all projects had a lower cost, and 50% had a higher cost per unit than the median figure. Figures in the "1/4" column indicate that 25% of the projects had lower costs and 75% had higher costs. Similarly, figures in the "3/4" column indicate that 75% had lower costs and 25% of the projects had higher costs.

The costs and figures represent all projects and do not take into account project size. As a rule, larger buildings (of the same type and relative location) will cost less to build per square foot than similar buildings of a smaller size. This cost difference is due to economies of scale as well as a lower exterior envelope-to-floor area ratio. From the electrical point of view, certain components are used only once within a wide range of building sizes. In such cases, the cost of these components will vary only slightly with the building size. A conversion is necessary to adjust project costs, when appropriate, based on size relative to the norm.

There are two stages of project development when square foot cost estimates are most useful. The first is during the conceptual stage when few, if any details are available. At this time, square foot costs are appropriate for ballpark budget purposes. As soon as details become available in the project design, the square foot approach should be discontinued and the project priced more accurately. After the detailed unit price estimate is completed, square foot costs can be used again — this time for verification and as a check against gross errors.

When using the figures in Division 17, it is recommended that the median cost column be consulted for preliminary figures if no additional information is available. When costs have been converted for location (see City Cost Indexes), the median numbers (as shown in Figure 22.8) should provide a fairly accurate base figure. This figure should then be adjusted according to the estimator's experience, local economic conditions, code requirements and the owner's particular project requirements. There is no need to factor the percentage figures, as these should remain relatively constant from city to city.

## Repair and Remodeling

Cost figures in *Means Electrical Cost Data* are based on new construction utilizing the most cost-effective combination of labor, equipment and material. The work is scheduled in the proper sequence to allow the various trades to accomplish their tasks in an efficient manner. Figure 22.9 (from Division 18 of *Means Electrical Cost Data*) shows factors that can be used to adjust figures in other sections of the book for repair and remodeling projects. For expanded coverage, see Means' *Repair and Remodeling Cost Data*.

## Section B — Systems Cost Tables

Means' systems data are divided into twelve "uniformat" divisions, which reorganize the components of construction into logical groupings. The Systems approach was devised to provide quick and easy methods for estimating even when only preliminary design data are available.

# 17.1 S.F., C.F. and % of Total Costs

| | | UNIT | UNIT COSTS | | | % OF TOTAL | | |
|---|---|---|---|---|---|---|---|---|
| | | | 1/4 | MEDIAN | 3/4 | 1/4 | MEDIAN | 3/4 |
| 310 800 | Total: Mechanical & Electrical | S.F. | 8.05 | 11.30 | 14.65 | 18.40% | 21.90% | 24.60% |
| 900 | Per rental unit, total cost | Unit | 38,800 | 45,900 | 50,500 | | | |
| 950 | Total: Mechanical & Electrical | " | 7,700 | 9,650 | 11,400 | | | |
| 50-001 | **HOUSING Public (low-rise)** | S.F. | 33.15 | 46.10 | 62.95 | | | |
| 002 | Total project costs | C.F. | 2.79 | 3.63 | 4.55 | | | |
| 272 | Plumbing | S.F. | 2.41 | 3.33 | 4.27 | 7.10% | 9% | 11.50% |
| 273 | Heating, ventilating, air conditioning | | 1.28 | 2.41 | 2.73 | 4.40% | 6% | 6.40% |
| 290 | Electrical | | 2.11 | 3.05 | 4.28 | 4.90% | 6.50% | 8.20% |
| 310 | Total: Mechanical & Electrical | | 6.45 | 9.30 | 12.15 | 15.70% | 19.20% | 22.10% |
| 900 | Per apartment, total cost | Apt. | 36,800 | 41,700 | 52,100 | | | |
| 950 | Total: Mechanical & Electrical | " | 6,175 | 8,475 | 10,600 | | | |
| 51-001 | **ICE SKATING RINKS** | S.F. | 31.40 | 43.90 | 72.05 | | | |
| 002 | Total project costs | C.F. | 1.78 | 2.23 | 2.63 | | | |
| 272 | Plumbing | S.F. | .94 | 1.39 | 2.13 | 3.10% | 3.20% | 4.60% |
| 290 | Electrical | | 2.48 | 3.25 | 4.51 | 5.70% | 7% | 10.10% |
| 310 | Total: Mechanical & Electrical | | 4.49 | 6.37 | 9.59 | 12.40% | 16.40% | 25.90% |
| 52-001 | **JAILS** | S.F. | 101 | 113 | 132 | | | |
| 002 | Total project costs | C.F. | 7.45 | 9.25 | 11.60 | | | |
| 272 | Plumbing | S.F. | 5.90 | 10.15 | 12.25 | 7% | 8.30% | 12% |
| 277 | Heating, ventilating, air conditioning | | 5.90 | 10.70 | 15.50 | 6.30% | 9.40% | 12.10% |
| 290 | Electrical | | 8.60 | 11.35 | 14.35 | 7.80% | 9.80% | 12.40% |
| 310 | Total: Mechanical & Electrical | | 22.15 | 31.80 | 41.25 | 23.20% | 29.60% | 35.30% |
| 53-001 | **LIBRARIES** | | 56.80 | 69.55 | 86.75 | | | |
| 002 | Total project costs | C.F. | 3.94 | 4.77 | 5.95 | | | |
| 272 | Plumbing | S.F. | 2.35 | 3.24 | 4.32 | 3.60% | 4.50% | 5.80% |
| 277 | Heating, ventilating, air conditioning | | 5.75 | 8.15 | 10.60 | 8.70% | 11% | 13.20% |
| 290 | Electrical | | 5.80 | 7.10 | 9.45 | 8.40% | 10.90% | 12.10% |
| 310 | Total: Mechanical & Electrical | | 12.25 | 17.15 | 24.25 | 19.40% | 25.50% | 29.60% |
| 55-001 | **MEDICAL CLINICS** | | 54.40 | 67.05 | 83.20 | | | |
| 002 | Total project costs | C.F. | 4.03 | 5.31 | 7.05 | | | |
| 272 | Plumbing | S.F. | 3.75 | 5.15 | 7.10 | 6.10% | 8.40% | 10.20% |
| 277 | Heating, ventilating, air conditioning | | 4.56 | 5.95 | 8.65 | 6.70% | 9% | 11.70% |
| 290 | Electrical | | 5.15 | 6.60 | 8.60 | 8.10% | 9.90% | 12% |
| 310 | Total: Mechanical & Electrical | | 11.85 | 15.20 | 21.30 | 19% | 24.20% | 30.10% |
| 57-001 | **MEDICAL OFFICES** | | 50.25 | 63.30 | 76.20 | | | |
| 002 | Total project costs | C.F. | 3.85 | 5.10 | 6.80 | | | |
| 272 | Plumbing | S.F. | 3.08 | 4.57 | 6.30 | 5.70% | 6.90% | 9.40% |
| 277 | Heating, ventilating, air conditioning | | 3.65 | 5.40 | 6.90 | 6.50% | 8% | 10.40% |
| 290 | Electrical | | 4.26 | 6.15 | 8 | 7.60% | 9.80% | 11.70% |
| 310 | Total: Mechanical & Electrical | | 9.80 | 13.75 | 18.25 | 17.20% | 22.40% | 27.40% |
| 59-001 | **MOTELS** | | 47.85 | 49.70 | 72.80 | | | |
| 002 | Total project costs | C.F. | 2.94 | 5.55 | 6.75 | | | |
| 272 | Plumbing | S.F. | 2.22 | 4.03 | 6.85 | 3.80% | 8.90% | 12.60% |
| 277 | Heating, ventilating, air conditioning | | 1.75 | 3.01 | 3.21 | 4.10% | 6.20% | 8.20% |
| 290 | Electrical | | 2.62 | 4.75 | 7.10 | 4.70% | 9.50% | 10.90% |
| 310 | Total: Mechanical & Electrical | | 8.35 | 14.05 | 16.40 | 17.30% | 27.70% | 33.30% |
| 900 | Per rental unit, total cost | Unit | 13,500 | 22,700 | 32,300 | | | |
| 950 | Total: Mechanical & Electrical | " | 3,750 | 5,025 | 5,525 | | | |
| 60-001 | **NURSING HOMES** | S.F. | 51.55 | 66.80 | 82.40 | | | |
| 002 | Total project costs | C.F. | 4.09 | 5.25 | 7 | | | |
| 272 | Plumbing | S.F. | 4.48 | 5.45 | 8.15 | 8.30% | 10.30% | 14.10% |
| 277 | Heating, ventilating, air conditioning | | 4.80 | 6.95 | 8.15 | 10.60% | 11.70% | 11.80% |
| 290 | Electrical | | 5.15 | 6.50 | 8 | 9.70% | 11% | 12.50% |
| 310 | Total: Mechanical & Electrical | | 11.75 | 16.05 | 23.15 | 22% | 28.10% | 33.20% |
| 900 | Per bed or person, total cost | Bed | 20,100 | 25,700 | 32,200 | | | |

Figure 22.8

| 17.1 S.F., C.F. and % of Total Costs | | UNIT | UNIT COSTS | | | % OF TOTAL | | |
|---|---|---|---|---|---|---|---|---|
| | | | 1/4 | MEDIAN | 3/4 | 1/4 | MEDIAN | 3/4 |
| 91-001 | THEATERS | S.F. | 41.95 | 53.25 | 81.55 | | | |
| 002 | Total project costs | C.F. | 2.19 | 3.06 | 4.53 | | | |
| 272 | Plumbing | S.F. | 1.39 | 1.62 | 4.87 | 2.90% | 4.60% | 6.10% |
| 277 | Heating, ventilating, air conditioning | | 3.88 | 5.20 | 6 | 7.30% | 11.60% | 13.30% |
| 290 | Electrical | | 3.93 | 5.30 | 8.20 | 8% | 9.90% | 11.80% |
| 310 | Total: Mechanical & Electrical | | 8.60 | 11 | 20.15 | 16% | 24.90% | 27.40% |
| 94-001 | TOWN HALLS City Halls & Municipal Buildings | ↓ | 50.70 | 63.70 | 83.40 | | | |
| 002 | Total project costs | C.F. | 3.42 | 5.05 | 6.45 | | | |
| 272 | Plumbing | S.F. | 1.78 | 3.48 | 6.10 | 4.20% | 5.90% | 7.90% |
| 277 | Heating, ventilating, air conditioning | | 3.78 | 7.50 | 8.60 | 7% | 9% | 13.20% |
| 290 | Electrical | | 4.02 | 6.10 | 8.50 | 7.90% | 9.40% | 11.60% |
| 310 | Total: Mechanical & Electrical | | 8 | 13.05 | 20.70 | 15.50% | 20.80% | 29.90% |
| 97-001 | WAREHOUSES And Storage Buildings | ↓ | 18.25 | 25.40 | 39 | | | |
| 002 | Total project costs | C.F. | .95 | 1.54 | 2.50 | | | |
| 272 | Plumbing | S.F. | .62 | 1.06 | 2.13 | 2.90% | 4.70% | 6.50% |
| 273 | Heating & ventilating | | .71 | 1.84 | 2.69 | 2.40% | 5% | 8.80% |
| 290 | Electrical | | 1.13 | 2.08 | 3.60 | 5.10% | 7.50% | 10.10% |
| 310 | Total: Mechanical & Electrical | | 2.09 | 3.55 | 8.10 | 9.60% | 15% | 21.40% |
| 99-001 | WAREHOUSE & OFFICES Combination | ↓ | 21.65 | 27.75 | 40 | | | |
| 002 | Total project costs | C.F. | 1.15 | 1.68 | 2.48 | | | |
| 272 | Plumbing | S.F. | .87 | 1.49 | 2.30 | 3.60% | 4.60% | 6.20% |
| 277 | Heating, ventilating, air conditioning | | 1.40 | 2.18 | 3.04 | 5% | 5.60% | 9.50% |
| 290 | Electrical | | 1.52 | 2.24 | 3.52 | 5.90% | 7.90% | 9.90% |
| 310 | Total: Mechanical & Electrical | ↓ | 2.88 | 4.90 | 7.55 | 11.50% | 16% | 21.30% |

For expanded coverage of these items see *Means' Square Foot Costs 1986*

| 18.1 Repair & Remodeling | | CREW | MAN-HOURS | UNIT | BARE COSTS | | | TOTAL INCL O&P |
|---|---|---|---|---|---|---|---|---|
| | | | | | EQUIP. | LABOR | TOTAL | |
| 20-001 | FACTORS To adjust figures in other sections of this | | | | | | | |
| 002 | book for repair and remodeling projects: | | | | | | | |
| 050 | Cut & patch to match existing construction, add, minimum | | | Costs | 2% | 3% | | |
| 055 | Maximum | | | | 5% | 9% | | |
| 080 | Dust protection, add, minimum | | | | 1% | 2% | | |
| 085 | Maximum | | | | 4% | 11% | | |
| 110 | Equipment usage curtailment, add, minimum | | | | 1% | 1% | | |
| 115 | Maximum | | | | 3% | 10% | | |
| 140 | Material handling & storage limitation, add, minimum | | | | 1% | 1% | | |
| 145 | Maximum | | | | 6% | 7% | | |
| 170 | Protection of existing work, add, minimum | | | | 2% | 2% | | |
| 175 | Maximum | | | | 5% | 7% | | |
| 200 | Shift work requirements, add, minimum | | | | | 5% | | |
| 205 | Maximum | | | | | 30% | | |
| 230 | Temporary shoring and bracing, add, minimum | | | | 2% | 5% | | |
| 235 | Maximum | | | ↓ | 5% | 12% | | |

*Figure* 22.9

The groupings, or systems, are presented in such a way that the estimator can easily vary components within the systems as well as substituting one system for another. This is extremely useful when adapting to budget, design, or other considerations. Figure 22.10 shows how the data are presented in the Systems section.

Each system is illustrated and accompanied by a detailed description. The book lists the components and sizes of each system, usually in the order of construction. Alternates for the most common variations are also listed. Each individual component is found in the Unit Price Section. If an alternate (not listed in Systems) is required, it can easily be derived.

*Quantity*: A unit of measure is established for each system. For example, lighting systems are measured by the square foot of floor area; motor systems are measured by "each"; generators are measured by Kilowatt. Within each system, the components are measured by industry standard, using the same units as in the Unit Price section.

*Material*: The cost of each component in the Material column is the "Bare Material Cost", plus 10% handling, for the unit and quantity as defined in the "Quantity" column.

*Installation*: Installation costs as listed in the Systems pages contain both labor and equipment costs. The labor rate includes the "Bare Labor Cost" plus the installing contractor's overhead and profit. These rates are shown in Figure 22.3. The equipment rate is the "Bare Equipment Cost", plus 10%.

## Section C — Estimating References

Throughout the Unit Price and Systems sections are reference numbers highlighted with bold squares. These numbers serve as footnotes, referring the reader to illustrations, charts, and estimating reference tables in Section C. Figure 22.11 shows an example reference number (for steel tubular scaffold) as it appears on a Unit Price page. Figure 22.12 shows the corresponding reference page from Section C. The development of unit costs for many items is explained in these reference tables. Design criteria for many types of electrical systems are also included to aid the designer/estimator in making appropriate choices.

## City Cost Indexes

The unit prices in *Means Electrical Cost Data* are national averages. When they are to be applied to a particular location, these prices must be adjusted to local conditions. R.S. Means has developed the City Cost Indexes for just that purpose. Section C of *Means Electrical Cost Data* contains tables of indexes for 162 U.S. and Canadian cities based on a 30 major city average of 100. The figures are broken down into material and installation for the Electrical and Mechanical UCI divisions, as shown in Figure 22.13. Please note that for each city there is a weighted average based on total project costs. This average is based on the relative contribution of each division to the construction process as a whole.

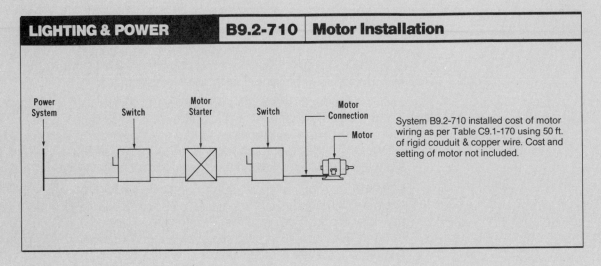

Power System | Switch | Motor Starter | Switch | Motor Connection — Motor

System B9.2-710 installed cost of motor wiring as per Table C9.1-170 using 50 ft. of rigid couduit & copper wire. Cost and setting of motor not included.

| System Components | QUANTITY | UNIT | COST EACH | | |
|---|---|---|---|---|---|
| | | | MAT. | INST. | TOTAL |
| SYSTEM 09.2-710-0200 | | | | | |
| MOTOR INSTALLATION, SINGLE PHASE, 115V, TO AND INCLUDING 1/3 HP MOTOR | | | | | |
| Wire 600V type THWN-THHN, copper solid #12 | 1.250 | C.L.F. | 4.54 | 29.21 | 33.75 |
| Steel intermediate conduit, (IMC) 1/2" diam | 50.000 | L.F. | 24 | 129 | 153 |
| Magnetic FVNR, 115V, 1/3 HP, size 00 starter | 1.000 | Ea. | 96.80 | 63.20 | 160 |
| Safety switch, fused, heavy duty, 240V 2P 30 amp | 1.000 | Ea. | 48.40 | 71.60 | 120 |
| Safety switch, non fused, heavy duty, 600V, 3 phase, 30 amp. | 1.000 | Ea. | 56.10 | 78.90 | 135 |
| Flexible metallic conduit, Greenfield 1/2" diam | 1.500 | L.F. | .42 | 1.92 | 2.34 |
| Connectors for flexible metallic conduit Greenfield 1/2" diam | 1.000 | Ea. | .37 | 3.23 | 3.60 |
| Coupling for Greenfield to conduit 1/2" diam | 1.000 | Ea. | 1.87 | 5.13 | 7 |
| Fuse cartridge nonrenewable, 250V 30 amp | 1.000 | Ea. | .55 | 5.15 | 5.70 |
| | | | | | |
| TOTAL | | | 233.05 | 387.34 | 620.39 |

| 9.2-710 | Motor Installation | COST EACH | | |
|---|---|---|---|---|
| | | MAT. | INST. | TOTAL |
| 0200 | Motor installation, single phase, 115V, to and including 1/3 HP motor | 235 | 385 | 620 |
| 0240 | To and incl. 1 HP motor | 245 | 390 | 635 |
| 0280 | To and incl. 2 HP motor | 265 | 410 | 675 |
| 0320 | To and incl. 3 HP motor | 305 | 420 | 725 |
| 0360 | 230V, to and including 1 HP motor | 240 | 395 | 635 |
| 0400 | To and incl. 2 HP motor | 255 | 395 | 650 |
| 0440 | To and incl. 3 HP motor | 285 | 425 | 710 |
| 0520 | Three phase, 200V, to and including 1-1/2 HP motor | 275 | 435 | 710 |
| 0560 | To and incl. 3 HP motor | 295 | 475 | 770 |
| 0600 | To and incl. 5 HP motor | 315 | 530 | 845 |
| 0640 | To and incl. 7-1/2 HP motor | 325 | 540 | 865 |
| 0680 | To and incl. 10 HP motor | 545 | 675 | 1220 |
| 0720 | To and incl. 15 HP motor | 730 | 750 | 1480 |
| 0760 | To and incl. 20 HP motor | 895 | 865 | 1760 |
| 0800 | To and incl. 25 HP motor | 910 | 870 | 1780 |
| 0840 | To and incl. 30 HP motor | 1475 | 1025 | 2500 |
| 0880 | To and incl. 40 HP motor | 1775 | 1225 | 3000 |
| 0920 | To and incl. 50 HP motor | 3150 | 1400 | 4550 |
| 0960 | To and incl. 60 HP motor | 3225 | 1475 | 4700 |
| 1000 | To and incl. 75 HP motor | 4075 | 1700 | 5775 |
| 1040 | To and incl. 100 HP motor | 8625 | 2000 | 10,625 |
| 1080 | To and incl. 125 HP motor | 8800 | 2200 | 11,000 |

*Figure 22.10*

| 1.1 Overhead | CREW | MAN-HOURS | UNIT | BARE COSTS MAT. | BARE COSTS INST. | BARE COSTS TOTAL | TOTAL INCL O&P |
|---|---|---|---|---|---|---|---|
| 005 C13.1 -100 Average | | | % | | 100% | | |
| 010 Maximum (Anchorage, AK) | | | " | | 141.30% | | |
| **24-001 MARK-UP** For General Contractors for change | | | | | | | |
| 010 of scope of job as bid | | | | | | | |
| 020 Extra work, by subcontractors, add | | | % | | | | 10% |
| 025 By General Contractor, add | | | | | | | 15% |
| 040 Omitted work, by subcontractors, deduct | | | | | | | 5% |
| 045 By General Contractor, deduct | | | | | | | 7.50% |
| 060 Overtime work, by subcontractor, add | | | | | | | 15% |
| 065 By General Contractor, add | | | | | | | 10% |
| 100 C10.3 -200 Subcontractors, on their own labor, minimum | | | | | 40% | | |
| 110 Maximum | | | ↓ | | 60.80% | | |
| **26-001 MATERIAL INDEX** (Div. 19) For 162 major U.S. and Canadian cities | | | | | | | |
| 002 Minimum (Greensboro, NC.) | | | % | 93.80% | | | |
| 004 Average | | | | 100% | | | |
| 006 Maximum (Anchorage, AK.) | | | | 125.30% | | | |
| **28-001 OVERHEAD** As percent of direct costs, minimum | | | | | | 5% | |
| 005 Average | | | | | | 12% | |
| 010 C10.3 -200 Maximum | | | ↓ | | | 22% | |
| **30-001 OVERHEAD & PROFIT** Allowance to add to items in this | | | | | | | |
| 002 book that do not include Subs O&P, average | | | % | | | 25% | |
| 010 C10.3 -200 Allowance to add to items in this book that | | | | | | | |
| 011 do include Subs O&P, minimum | | | % | | | | 5% |
| 015 Average | | | | | | | 10% |
| 020 Maximum | | | | | | | 15% |
| 030 Typical, by size of project, under $100,000 | | | | | | 30% | |
| 035 $500,000 project | | | | | | 25% | |
| 040 $2,000,000 project | | | | | | 20% | |
| 045 Over $10,000,000 project | | | ↓ | | | 15% | |
| **32-001 OVERTIME** C10.2 -400 For early completion of projects or where | | | | | | | |
| 002 labor shortages exist, add to usual labor, up to | | | Costs | | 118% | | |
| **34-001 PERFORMANCE BOND** C10.1 -302 For buildings, minimum | | | Job Cost | | | | .39% |
| 010 Maximum | | | | | | | 1.20% |
| **36-001 PERMITS** Rule of thumb, most cities, minimum | | | | | | | .50% |
| 010 Maximum | | | ↓ | | | | 2% |
| **38-001 PHOTOGRAPHS** 8" x 10", 4 shots, 2 prints ea., std. mounting | | | Set | 75 | | 75 | 83M |
| 010 Hinged linen mounts | | | | 87 | | 87 | 96M |
| 020 8" x 10", 4 shots, 2 prints each, in color | | | | 156 | | 156 | 170M |
| 030 For I.D. slugs, add to all above | | | | 2.40 | | 2.40 | 2.64M |
| 150 Time lapse equipment, camera and projector, buy | | | | 3,255 | | 3,255 | 3,575M |
| 155 Rent per month | | | ↓ | 420 | | 420 | 460M |
| 170 Cameraman and film, including processing, B.&W. | | | Day | 462 | | 462 | 510M |
| 172 Color | | | " | 525 | | 525 | 580M |
| **43-001 SCAFFOLD** Steel tubular, regular, buy | | | | | | | |
| 009 Building exterior, 1 to 5 stories | 3 Carp | 1.430 | C.S.F. | 12.50 | 29 | 41.50 | 55 |
| 020 C10.3 -300 6 to 12 stories | 4 Carp | 2.130 | | 11.20 | 43 | 54.20 | 74 |
| 031 13 to 20 stories | 5 Carp | 2.390 | | 9.60 | 48 | 57.60 | 80 |
| 046 Building interior walls, (area) up to 16' high | 3 Carp | 1.060 | | 11.60 | 21 | 32.60 | 43 |
| 056 16' to 40' high | | 1.280 | ↓ | 11.95 | 26 | 37.95 | 50 |
| 080 Building interior floor area, up to 30' high | ↓ | .282 | C.C.F. | 3.90 | 5.65 | 9.55 | 12.45 |
| 090 Over 30' high | 4 Carp | .332 | " | 4.40 | 6.65 | 11.05 | 14.45 |
| **44-001 SCAFFOLDING SPECIALTIES** | | | | | | | |
| 002 | | | | | | | |
| 150 Sidewalk bridge using tubular steel | | | | | | | |
| 151 scaffold frames, including planking | 3 Carp | .533 | L.F. | 3.50 | 10.65 | 14.15 | 19.30 |
| 160 For 2 uses per month, deduct from all above | | | | 50% | | | |
| 170 For 1 use every 2 months, add to all above | | | | 100% | | | |
| 190 Catwalks, 32" wide, no guardrails, 6' span, buy | | | Ea. | 104 | | 104 | 115M |
| 200 10' span, buy | | | " | 170 | | 170 | 185M |

Figure 22.11

## Steel Tubular Scaffolding (Div. 1.1-43)

On new construction, tubular scaffolding is efficient up to 60' high or five stories. Above this it is usually better to use a hung scaffolding if construction permits.

In repairing or cleaning the front of an existing building the cost of tubular scaffolding per S.F. of building front increases as the height increases above the first tier. The first tier cost is relatively high due to leveling and alignment. Swing scaffolding operations may interfere with tenants. In this case the tubular is more practical at all heights.

The minimum efficient crew for erection is three men. For heights over 50', a four-man crew is more efficient. Use two or more on top and two at the bottom for handing up or hoisting. Four men can erect and dismantle about nine frames per hour up to five stories. From five to eight stories they will average six frames per hour. With 7' horizontal spacing this will run about 300 S.F. and 200 S.F. of wall surface, respectively. Time for placing planks

must be added to the above. On heights above 50', five planks can be placed per man-hour.

The cost per 1,000 S.F. of building front in the table below was developed by pricing the materials required for a typical tubular scaffolding system eleven frames long and two frames high. Planks were figured five wide for standing plus two wide for materials.

Frames are 2', 4' and 5' wide and usually spaced 7' O.C. horizontally. Sidewalk frames are 6' wide. Rental rates will be lower for jobs over three months duration.

For jobs under twenty-five frames, figure rental at $6.00 per frame. For jobs over one hundred frames, rental can go as low as $2.25 per frame. These figures do not include accessories which are listed separately below. Large quantities for long periods can reduce rental rates by 20%.

| Item | Unit | Purchase, Each | | Monthly Rent, Each | | Per 1,000 S.F. of Building Front | |
|------|------|---------|------------|---------|------------|---------------|----------------|
| | | Regular | Heavy Duty | Regular | Heavy Duty | No. of Frames | Rental per Mo. |
| 5' Wide Frames, 3' High | Ea. | $45 | $52 | $3.10 | $3.40 | — | — |
| *5'-0" High | | 56 | 57 | 3.10 | 3.40 | — | — |
| *6'-6" High | | 71 | 76 | 3.10 | 3.40 | 24 | $ 74.40 |
| 2' & 4' Wide, 5' High | | — | 68 | — | 3.40 | — | — |
| 6'-0" High | | — | 79 | — | 3.40 | — | — |
| 6' Wide Frame, 7'-6" High | | 100 | 110 | 7.25 | 9.25 | — | — |
| Sidewalk Bracket, 20" | | 17 | 19 | .94 | 1.10 | 12 | 11.28 |
| Guardrail Post | | 13 | 18 | .53 | .63 | 12 | 6.36 |
| Guardrail, 7' section | | 5 | 6 | .47 | .52 | 11 | 5.17 |
| Cross Braces | | 15 | 16 | .47 | .58 | 44 | 20.68 |
| Screw Jacks & Plates | | 20 | 28 | 1.30 | 1.40 | 24 | 31.20 |
| 8" Casters | | 70 | — | 5.70 | 6.00 | — | — |
| 16' Plank, 2" x 10" | | 19 | — | 3.60 | 3.85 | 35 | 126.00 |
| 8' Plank, 2" x 10" | | 10 | — | 1.85 | 2.10 | 7 | 12.95 |
| 1' to 6' Extension Tube | | — | 65 | — | 2.10 | — | — |
| Shoring Stringers, steel, 10' to 12' long | L.F. | — | 6 | — | .35 | — | — |
| Aluminum, 12' to 16' long | | — | 12 | — | .58 | — | — |
| Aluminum joists with nailers, 10' to 22' long | | — | 7.60 | — | .35 | — | — |
| Flying Truss System, Aluminum | S.F.C.A. | — | 8.00 | — | .45 | — | — |
| | | | | | | Total | $288.04 |
| | | | | | | 2 Use/Mo. | $144.02 |

*Most commonly used

Scaffolding is often used as falsework over 15' high during construction of cast-in-place concrete beams and slabs. Two ft. wide scaffolding is generally used for heavy beam construction. The span between frames depends upon the load to be carried with a maximum span of 5'.

Heavy duty scaffolding with a capacity of 10,000#/leg can be spaced up to 10' O.C. depending upon form support design and loading.

Scaffolding used as horizontal shoring requires less than half the material required with conventional shoring.

On new construction, erection is done by carpenters.

Rolling towers supporting horizontal shores can reduce labor and speed the job. For maninetance work, catwalks with spans up to 70' can be supported by the rolling towers.

*Figure 22.12*

| DIVISION | ALABAMA | | | | | | | | | | | | ALASKA | | | ARIZONA | | |
|---|---|---|---|---|---|---|---|---|---|---|---|---|---|---|---|---|---|---|
| | BIRMINGHAM | | | HUNTSVILLE | | | MOBILE | | | MONTGOMERY | | | ANCHORAGE | | | PHOENIX | | |
| | MAT. | INST. | TOTAL | MAT. | INST. | TOTAL | MAT. | INST. | TOTAL | MAT. | INST. | TOTAL | MAT. | INST. | TOTAL | MAT. | INST. | TOTAL |
| MECHANICAL | 96.4 | 77.3 | 86.5 | 99.7 | 76.5 | 87.7 | 97.5 | 77.4 | 87.1 | 99.2 | 73.3 | 85.8 | 107.7 | 146.6 | 127.9 | 98.8 | 83.8 | 91.0 |
| ELECTRICAL | 94.0 | 77.3 | 82.1 | 91.9 | 77.1 | 81.4 | 89.7 | 78.2 | 81.5 | 90.7 | 72.2 | 77.5 | 110.8 | 146.8 | 136.5 | 105.1 | 89.4 | 93.9 |
| WEIGHTED PROJECT AVERAGE | 94.3 | 78.0 | 85.4 | 99.7 | 77.8 | 87.7 | 97.3 | 79.6 | 87.6 | 96.1 | 76.2 | 85.2 | 125.3 | 141.3 | 134.1 | 99.2 | 90.6 | 94.5 |

| DIVISION | ARIZONA | | | ARKANSAS | | | | | | CALIFORNIA | | | | | | | | |
|---|---|---|---|---|---|---|---|---|---|---|---|---|---|---|---|---|---|---|
| | TUCSON | | | FORT SMITH | | | LITTLE ROCK | | | ANAHEIM | | | BAKERSFIELD | | | FRESNO | | |
| | MAT. | INST. | TOTAL | MAT. | INST. | TOTAL | MAT. | INST. | TOTAL | MAT. | INST. | TOTAL | MAT. | INST. | TOTAL | MAT. | INST. | TOTAL |
| MECHANICAL | 98.8 | 83.8 | 91.0 | 97.5 | 72.5 | 84.5 | 97.0 | 73.1 | 84.6 | 96.7 | 125.7 | 111.8 | 94.9 | 116.6 | 106.1 | 92.4 | 125.2 | 109.4 |
| ELECTRICAL | 102.3 | 88.7 | 92.6 | 99.2 | 75.5 | 82.3 | 93.4 | 76.2 | 81.1 | 98.7 | 125.3 | 117.7 | 106.1 | 116.5 | 113.5 | 109.7 | 119.6 | 116.8 |
| WEIGHTED PROJECT AVERAGE | 98.9 | 90.7 | 94.4 | 96.7 | 76.4 | 85.5 | 98.7 | 77.8 | 87.2 | 100.6 | 123.9 | 113.4 | 98.5 | 118.8 | 109.6 | 98.6 | 120.3 | 110.5 |

| DIVISION | CALIFORNIA | | | | | | | | | | | | | | | | | |
|---|---|---|---|---|---|---|---|---|---|---|---|---|---|---|---|---|---|---|
| | LOS ANGELES | | | OXNARD | | | RIVERSIDE | | | SACRAMENTO | | | SAN DIEGO | | | SAN FRANCISCO | | |
| | MAT. | INST. | TOTAL | MAT. | INST. | TOTAL | MAT. | INST. | TOTAL | MAT. | INST. | TOTAL | MAT. | INST. | TOTAL | MAT. | INST. | TOTAL |
| MECHANICAL | 97.4 | 126.2 | 112.3 | 98.7 | 125.9 | 112.8 | 96.5 | 126.6 | 112.1 | 98.2 | 126.4 | 112.8 | 102.9 | 125.0 | 114.4 | 100.8 | 160.0 | 131.5 |
| ELECTRICAL | 101.8 | 133.2 | 124.2 | 98.7 | 125.1 | 117.5 | 98.2 | 124.5 | 117.0 | 109.7 | 125.9 | 121.3 | 106.1 | 113.8 | 111.6 | 110.7 | 152.7 | 140.6 |
| WEIGHTED PROJECT AVERAGE | 98.1 | 125.6 | 113.2 | 98.8 | 123.6 | 112.4 | 99.0 | 123.8 | 112.6 | 99.2 | 123.6 | 112.5 | 103.4 | 118.5 | 111.7 | 104.0 | 139.3 | 123.3 |

| DIVISION | CALIFORNIA | | | | | | | | | COLORADO | | | | | | CONNECTICUT | | |
|---|---|---|---|---|---|---|---|---|---|---|---|---|---|---|---|---|---|---|
| | SANTA BARBARA | | | STOCKTON | | | VALLEJO | | | COLORADO SPRINGS | | | DENVER | | | BRIDGEPORT | | |
| | MAT. | INST. | TOTAL | MAT. | INST. | TOTAL | MAT. | INST. | TOTAL | MAT. | INST. | TOTAL | MAT. | INST. | TOTAL | MAT. | INST. | TOTAL |
| MECHANICAL | 98.7 | 125.3 | 112.5 | 97.1 | 119.3 | 108.6 | 95.8 | 132.0 | 114.6 | 98.0 | 91.3 | 94.5 | 97.2 | 95.2 | 96.2 | 104.1 | 102.3 | 103.2 |
| ELECTRICAL | 97.9 | 123.2 | 116.0 | 100.3 | 121.9 | 115.7 | 110.0 | 132.1 | 125.8 | 97.9 | 89.9 | 92.2 | 95.2 | 91.4 | 92.5 | 102.8 | 101.9 | 102.2 |
| WEIGHTED PROJECT AVERAGE | 104.8 | 124.7 | 115.7 | 98.6 | 117.6 | 109.0 | 98.8 | 128.1 | 114.8 | 98.0 | 91.8 | 94.6 | 101.6 | 93.5 | 97.2 | 103.6 | 102.7 | 103.1 |

| DIVISION | CONNECTICUT | | | | | | | | | | | | DELAWARE | | | D.C. | | |
|---|---|---|---|---|---|---|---|---|---|---|---|---|---|---|---|---|---|---|
| | HARTFORD | | | NEW HAVEN | | | STAMFORD | | | WATERBURY | | | WILMINGTON | | | WASHINGTON | | |
| | MAT. | INST. | TOTAL | MAT. | INST. | TOTAL | MAT. | INST. | TOTAL | MAT. | INST. | TOTAL | MAT. | INST. | TOTAL | MAT. | INST. | TOTAL |
| MECHANICAL | 102.0 | 102.1 | 102.0 | 102.3 | 101.4 | 101.9 | 101.8 | 106.8 | 104.4 | 100.9 | 97.3 | 99.1 | 100.3 | 103.8 | 102.1 | 101.5 | 84.5 | 92.7 |
| ELECTRICAL | 98.6 | 104.8 | 103.0 | 92.9 | 101.9 | 99.4 | 92.9 | 102.9 | 100.1 | 91.6 | 100.7 | 98.1 | 105.8 | 102.8 | 103.7 | 99.0 | 90.3 | 92.8 |
| WEIGHTED PROJECT AVERAGE | 100.1 | 103.6 | 102.0 | 101.0 | 101.7 | 101.4 | 103.7 | 103.8 | 103.8 | 100.4 | 100.9 | 100.7 | 98.4 | 103.8 | 101.4 | 101.5 | 89.7 | 95.1 |

| DIVISION | FLORIDA | | | | | | | | | | | | | | | GEORGIA | | |
|---|---|---|---|---|---|---|---|---|---|---|---|---|---|---|---|---|---|---|
| | FT LAUDERDALE | | | JACKSONVILLE | | | MIAMI | | | ORLANDO | | | TAMPA | | | ATLANTA | | |
| | MAT. | INST. | TOTAL | MAT. | INST. | TOTAL | MAT. | INST. | TOTAL | MAT. | INST. | TOTAL | MAT. | INST. | TOTAL | MAT. | INST. | TOTAL |
| MECHANICAL | 101.2 | 88.9 | 94.8 | 100.4 | 81.1 | 90.4 | 97.8 | 94.3 | 96.0 | 97.1 | 79.7 | 88.1 | 97.4 | 81.5 | 89.1 | 102.8 | 78.3 | 90.1 |
| ELECTRICAL | 98.3 | 87.6 | 90.7 | 99.2 | 76.0 | 82.7 | 99.8 | 98.3 | 98.7 | 92.8 | 74.5 | 79.8 | 92.8 | 83.6 | 86.2 | 96.5 | 81.1 | 85.5 |
| WEIGHTED PROJECT AVERAGE | 97.1 | 91.4 | 93.9 | 98.1 | 79.0 | 87.6 | 95.6 | 94.0 | 94.8 | 94.6 | 76.7 | 84.8 | 98.8 | 85.5 | 91.5 | 98.9 | 80.3 | 88.7 |

| DIVISION | GEORGIA | | | | | | | | | HAWAII | | | IDAHO | | | ILLINOIS | | |
|---|---|---|---|---|---|---|---|---|---|---|---|---|---|---|---|---|---|---|
| | COLUMBUS | | | MACON | | | SAVANNAH | | | HONOLULU | | | BOISE | | | CHICAGO | | |
| | MAT. | INST. | TOTAL | MAT. | INST. | TOTAL | MAT. | INST. | TOTAL | MAT. | INST. | TOTAL | MAT. | INST. | TOTAL | MAT. | INST. | TOTAL |
| MECHANICAL | 98.6 | 68.5 | 83.0 | 98.9 | 72.0 | 85.0 | 98.6 | 71.8 | 84.7 | 111.9 | 112.9 | 112.4 | 96.6 | 88.1 | 92.2 | 97.5 | 95.9 | 96.7 |
| ELECTRICAL | 97.0 | 64.8 | 74.0 | 107.5 | 69.0 | 80.1 | 100.8 | 70.4 | 79.2 | 105.2 | 107.6 | 106.9 | 95.8 | 87.7 | 90.1 | 93.5 | 96.3 | 95.5 |
| WEIGHTED PROJECT AVERAGE | 98.5 | 68.6 | 82.1 | 97.1 | 75.5 | 85.3 | 99.4 | 75.9 | 86.5 | 115.1 | 111.7 | 113.3 | 98.5 | 90.6 | 94.2 | 97.5 | 100.2 | 99.0 |

| DIVISION | ILLINOIS | | | | | | | | | INDIANA | | | | | | | | |
|---|---|---|---|---|---|---|---|---|---|---|---|---|---|---|---|---|---|---|
| | PEORIA | | | ROCKFORD | | | SPRINGFIELD | | | EVANSVILLE | | | FORT WAYNE | | | GARY | | |
| | MAT. | INST. | TOTAL | MAT. | INST. | TOTAL | MAT. | INST. | TOTAL | MAT. | INST. | TOTAL | MAT. | INST. | TOTAL | MAT. | INST. | TOTAL |
| MECHANICAL | 96.1 | 95.8 | 95.9 | 102.0 | 99.9 | 100.9 | 97.1 | 92.1 | 94.5 | 98.2 | 95.6 | 96.8 | 98.7 | 91.7 | 95.1 | 97.7 | 100.5 | 99.2 |
| ELECTRICAL | 93.8 | 95.4 | 95.0 | 92.5 | 101.1 | 98.7 | 94.5 | 93.4 | 93.8 | 94.0 | 94.3 | 94.3 | 96.6 | 91.6 | 93.1 | 100.8 | 100.5 | 100.6 |
| WEIGHTED PROJECT AVERAGE | 97.9 | 94.0 | 95.8 | 102.2 | 100.9 | 101.5 | 100.3 | 94.0 | 96.9 | 99.1 | 96.9 | 97.9 | 97.7 | 93.3 | 95.3 | 98.0 | 101.4 | 99.9 |

| DIVISION | INDIANA | | | | | | | | | IOWA | | | | | | KANSAS | | |
|---|---|---|---|---|---|---|---|---|---|---|---|---|---|---|---|---|---|---|
| | INDIANAPOLIS | | | SOUTH BEND | | | TERRE HAUTE | | | DAVENPORT | | | DES MOINES | | | TOPEKA | | |
| | MAT. | INST. | TOTAL | MAT. | INST. | TOTAL | MAT. | INST. | TOTAL | MAT. | INST. | TOTAL | MAT. | INST. | TOTAL | MAT. | INST. | TOTAL |
| MECHANICAL | 100.6 | 96.2 | 98.3 | 98.1 | 92.6 | 95.2 | 98.1 | 94.2 | 96.1 | 100.0 | 90.7 | 95.2 | 96.6 | 88.5 | 92.4 | 99.5 | 84.5 | 91.7 |
| ELECTRICAL | 98.3 | 96.1 | 96.8 | 97.1 | 92.4 | 93.8 | 108.4 | 94.8 | 98.7 | 97.2 | 90.6 | 92.5 | 98.5 | 81.2 | 86.2 | 97.8 | 81.1 | 85.9 |
| WEIGHTED PROJECT AVERAGE | 99.4 | 96.5 | 97.8 | 98.1 | 94.3 | 96.0 | 97.2 | 94.5 | 95.7 | 98.0 | 91.1 | 94.2 | 98.8 | 85.2 | 91.3 | 98.1 | 85.9 | 91.4 |

*Figure 22.13*

In addition to adjusting the figures in *Means Electrical Cost Data* for particular locations, the City Cost Index can also be used to adjust costs from one city to another. For example, the price of the electrical work for a particular building type is known for City A. In order to budget the costs of the same building type in City B, the following calculation can be made:

$$\frac{\text{City B Index}}{\text{City A Index}} \times \text{City A Cost} = \text{City B Cost}$$

While City Cost Indexes provide a means to adjust prices for location, the Historical Cost Index (also included in *Means Electrical Cost Data* and shown in Figure 22.14), provides a means to adjust for time. Using the same principle as above, a time-adjustment factor can be calculated:

$$\frac{\text{Index for Year X}}{\text{Index for Year Y}} = \text{Time-adjustment Factor}$$

| Year | "Quarterly City Cost Index" Jan. 1, 1975 = 100 | | Current Index Based on Jan. 1, 1986 = 100 | | Year | "Quarterly City Cost Index" Jan. 1, 1975 = 100 | Current Index Based on Jan. 1, 1986 = 100 | | Year | "Quarterly City Cost Index" Jan. 1, 1975 = 100 | Current Index Based on Jan. 1, 1986 = 100 | |
|---|---|---|---|---|---|---|---|---|---|---|---|---|
| | Est. | Actual | Est. | Actual | | Actual | Est. | Actual | | Actual | Est. | Actual |
| Oct. 1986 | | | | | July 1973 | 86.3 | 44.9 | | July 1957 | 42.2 | 22.0 | |
| July 1986 | | | | | 1972 | 79.7 | 41.5 | | 1956 | 40.4 | 21.0 | |
| April 1986 | | | | | 1971 | 73.5 | 38.3 | | 1955 | 38.1 | 19.8 | |
| Jan. 1986 | 192.0 | | 100.0 | 100.0 | 1970 | 65.8 | 34.3 | | 1954 | 36.7 | 19.1 | |
| July 1985 | | 189.1 | 98.5 | | 1969 | 61.6 | 32.1 | | 1953 | 36.2 | 18.9 | |
| 1984 | | 187.6 | 97.7 | | 1968 | 56.9 | 29.6 | | 1952 | 35.3 | 18.4 | |
| 1983 | | 183.5 | 95.6 | | 1967 | 53.9 | 28.1 | | 1951 | 34.4 | 17.9 | |
| 1982 | | 174.3 | 90.8 | | 1966 | 51.9 | 27.0 | | 1950 | 31.4 | 16.4 | |
| 1981 | | 160.2 | 83.4 | | 1965 | 49.7 | 25.9 | | 1949 | 30.4 | 15.8 | |
| 1980 | | 144.0 | 75.0 | | 1964 | 48.6 | 25.3 | | 1948 | 30.4 | 15.8 | |
| 1979 | | 132.3 | 68.9 | | 1963 | 47.3 | 24.6 | | 1947 | 27.6 | 14.4 | |
| 1978 | | 122.4 | 63.8 | | 1962 | 46.2 | 24.1 | | 1946 | 23.2 | 12.1 | |
| 1977 | | 113.3 | 59.0 | | 1961 | 45.4 | 23.6 | | 1945 | 20.2 | 10.5 | |
| 1976 | | 107.3 | 55.9 | | 1960 | 45.0 | 23.4 | | 1944 | 19.3 | 10.1 | |
| 1975 | | 102.6 | 53.4 | | 1959 | 44.2 | 23.0 | | 1943 | 18.6 | 9.7 | |
| 1974 | | 94.7 | 49.3 | | 1958 | 43.0 | 22.4 | | 1942 | 18.0 | 9.4 | |

*Figure* 22.14

240

This time-adjustment factor can be used to determine the budget costs for a particular building type in Year X, based on costs for a similar building type known from Year Y. Used together, the two indexes allow for cost adjustments from one city during a given year to another city in another year (the present or otherwise). For example, an office building built in San Francisco in 1974, *originally cost* $1,000,000. How much will a similar building cost in Phoenix in 1986? Adjustment factors are developed as shown above using data from Figures 22.13 and 22.14:

$$\frac{\text{Phoenix index}}{\text{San Francisco index}} = \frac{94.5}{123.3} = 0.77$$

$$\frac{1986 \text{ index}}{1974 \text{ index}} = \frac{192.0}{94.7} = 2.03$$

Original cost x location adjustment x time adjustment = Proposed new cost

$$\$1,000,000 \text{ x } 0.77 \text{ x } 2.03 = \$1,563,000$$

Understanding how Means' prices are obtained and organized makes this annual cost book a more valuable tool for the estimating process. The Sample Estimate in the following chapter is based on prices in the 1986 edition of *Means Electrical Cost Data*.

# Chapter 23

# A SAMPLE ESTIMATE

This chapter contains a complete sample estimate for the electrical portion of an office building. In this step-by-step example, forms are filled out and calculations made according to the techniques described in Parts I and II. For this example, let us assume that an electrical contractor has received a request for quote. The request is from the construction management company of an office building to be built in the local area.

*Project Description*

The project selected for the sample estimate is a three-story office building with a basement parking garage. The building will have a composite steel frame with an aluminum panel and glass wall curtain system. Plans for the electrical work are shown in Figures 23.1 through 23.6. These plans, representing the major features of the electrical installation, are provided for illustrative purposes only. In actuality, such a building would require additional drawings with more plans, details, sections, and elevations. A full set of specifications would also be provided. The quantities, as given in the sample estimate, represent realistic conditions. Assumptions have been made for some items not shown that could normally be included.

This estimate may not necessarily be complete in every detail. Telephones, for example, are assumed to be outside the scope of the contract.

Nevertheless, the estimate demonstrates techniques and approaches which can be used to solve real estimating problems.

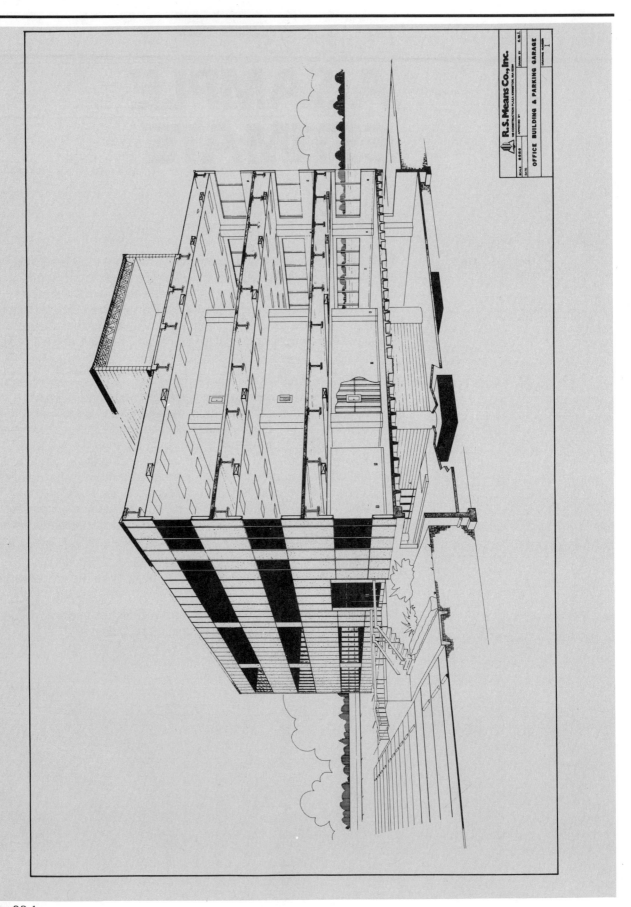

Figure 23.1

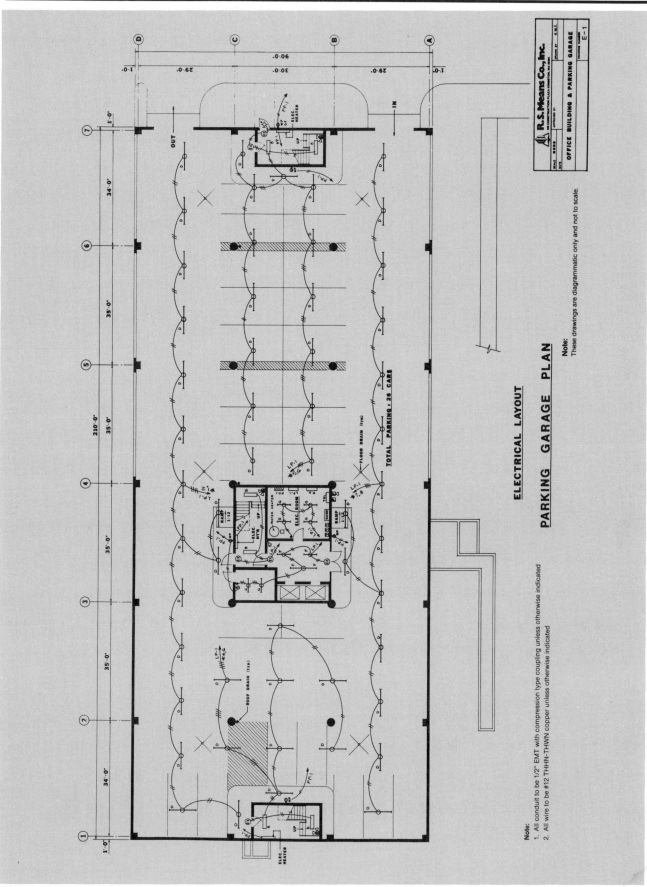

ELECTRICAL LAYOUT

PARKING GARAGE PLAN

Note:
These drawings are diagrammatic only and not to scale.

Note:
1. All conduit to be 1/2" EMT with compression type coupling unless otherwise indicated
2. All wire to be #12 THHN-THWN copper unless otherwise indicated

R.S. Means Co., Inc.
OFFICE BUILDING & PARKING GARAGE
E-1

TOTAL PARKING = 26 CARS

Figure 23.2

245

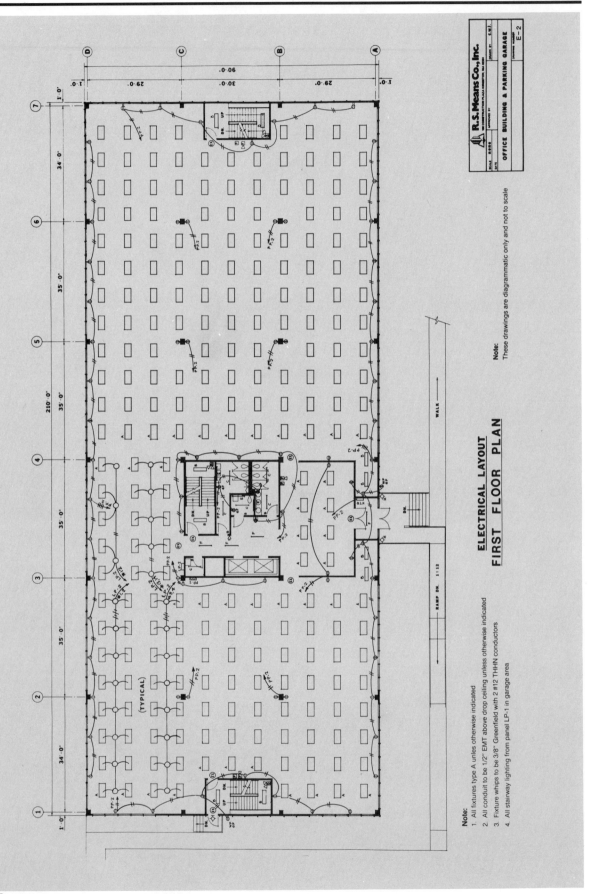

ELECTRICAL LAYOUT
FIRST FLOOR PLAN

Note:
These drawings are diagrammatic only and not to scale

Note:
1. All fixtures type A unless otherwise indicated
2. All conduit to be 1/2" EMT above drop ceiling unless otherwise indicated
3. Fixture whips to be 3/8" Greenfield with 2 #12 THHN conductors
4. All stairway lighting from panel LP-1 in garage area

OFFICE BUILDING & PARKING GARAGE

E-2

*Figure* 23.3

246

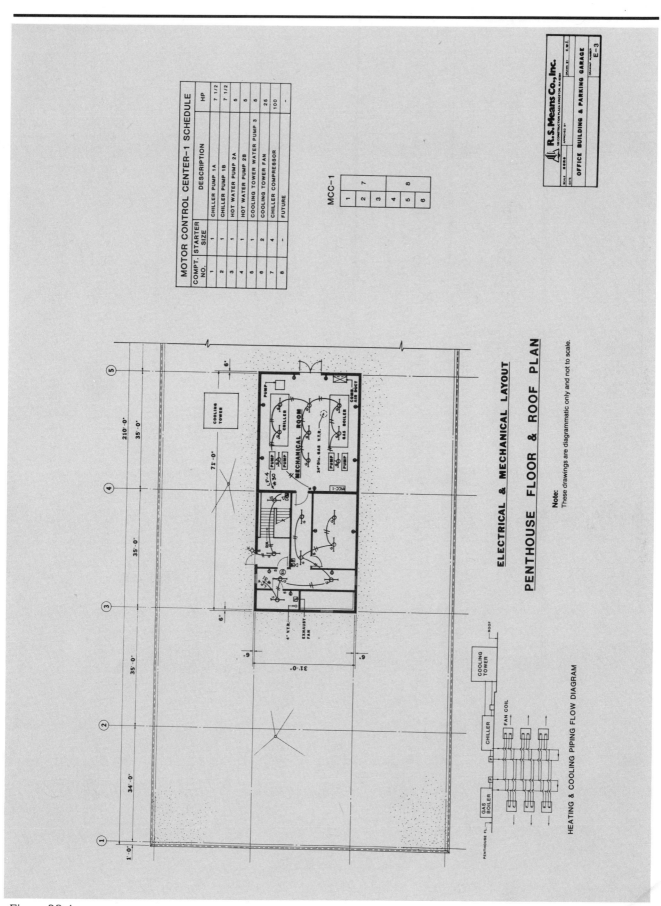

MOTOR CONTROL CENTER-1 SCHEDULE

| COMPT. NO. | STARTER SIZE | DESCRIPTION | HP |
|---|---|---|---|
| 1 | 1 | CHILLER PUMP 1A | 7 1/2 |
| 2 | 1 | CHILLER PUMP 1B | 7 1/2 |
| 3 | 1 | HOT WATER PUMP 2A | 5 |
| 4 | 1 | HOT WATER PUMP 2B | 5 |
| 5 | 1 | COOLING TOWER WATER PUMP 3 | 5 |
| 6 | 2 | COOLING TOWER FAN | 25 |
| 7 | 4 | CHILLER COMPRESSOR | 100 |
| 8 | - | FUTURE | - |

MCC-1

| 1 |  |
|---|---|
| 2 | 7 |
| 3 |  |
| 4 |  |
| 5 | 8 |
| 6 |  |

ELECTRICAL & MECHANICAL LAYOUT

PENTHOUSE FLOOR & ROOF PLAN

Note:
These drawings are diagrammatic only and not to scale.

HEATING & COOLING PIPING FLOW DIAGRAM

R.S. Means Co., Inc.

OFFICE BUILDING & PARKING GARAGE

E-3

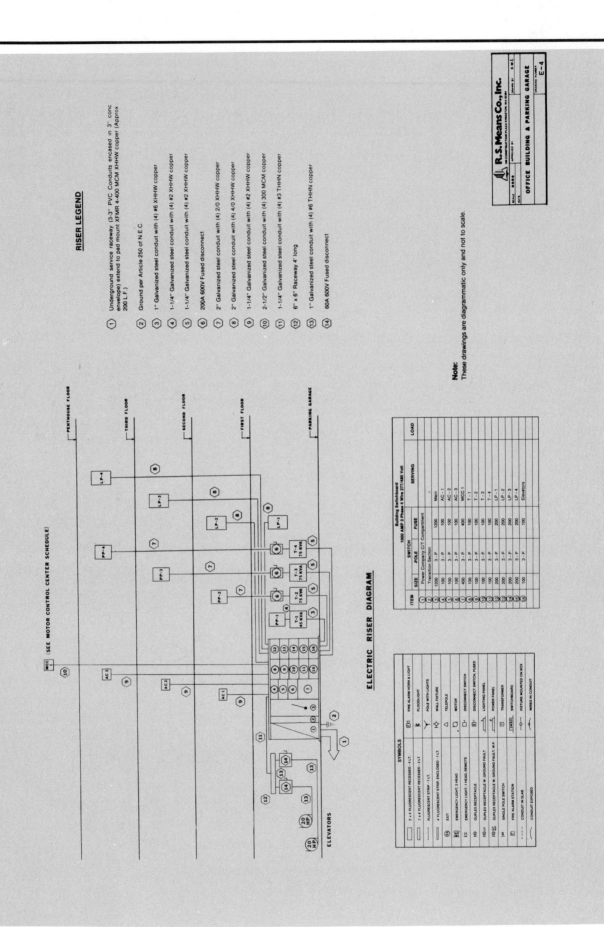

RISER LEGEND

① Underground service raceway (3-3" PVC Conduits encased in 3" conc envelope) extend to pad mount XFMR 4-400 MCM XHHW copper (Approx 200 L.F.)

② Ground per Article 250 of N.E.C.

③ 1" Galvanized steel conduit with (4) #6 XHHW copper

④ 1-1/4" Galvanized steel conduit with (4) #2 XHHW copper

⑤ 1-1/4" Galvanized steel conduit with (4) #2 XHHW copper

⑥ 200A 600V Fused disconnect

⑦ 2" Galvanized steel conduit with (4) 2/0 XHHW copper

⑧ 2" Galvanized steel conduit with (4) 4/0 XHHW copper

⑨ 1-1/4" Galvanized steel conduit with (4) #2 XHHW copper

⑩ 2-1/2" Galvanized steel conduit with (4) 300 MCM copper

⑪ 1-1/4" Galvanized steel conduit with (4) #3 THHN copper

⑫ 6" x 6" Raceway 4' long

⑬ 1" Galvanized steel conduit with (4) #6 THHN copper

⑭ 60A 600V Fused disconnect

**Note:**
These drawings are diagrammatic only and not to scale.

ELECTRIC RISER DIAGRAM

| ITEM | SIZE | POLE | FUSE | SERVING | LOAD |
|---|---|---|---|---|---|
| | | Power Company C/T Compartment | | | |
| | | Transition Section | | | |
| | 1200 | 3 - P | 1200 | Main | |
| | 100 | 3 - P | 100 | AC - 1 | |
| | 100 | 3 - P | 100 | AC - 2 | |
| | 100 | 3 - P | 100 | AC - 3 | |
| | 400 | 3 - P | 400 | MCC 1 | |
| | 100 | 3 - P | 100 | T - 1 | |
| | 100 | 3 - P | 100 | T - 2 | |
| | 100 | 3 - P | 100 | T - 3 | |
| | 100 | 3 - P | 100 | T - 4 | |
| | 200 | 3 - P | 200 | LP - 1 | |
| | 200 | 3 - P | 200 | LP - 2 | |
| | 200 | 3 - P | 200 | LP - 3 | |
| | 200 | 3 - P | 200 | LP - 4 | |
| | 100 | 3 - P | 100 | Elevators | |

Building Switchboard
1000 AMP 3 Phase 4 Wire 277/480 Volt

SYMBOLS

| | |
|---|---|
| 2 x 4 FLUORESCENT RECESSED - 4 LT | FIRE ALARM HORN & LIGHT |
| 1 x 4 FLUORESCENT RECESSED - 2 LT | FLOODLIGHT |
| FLUORESCENT STRIP - 1 LT | POLE WITH LIGHTS |
| 4' FLUORESCENT STRIP ENCLOSED - 1 LT | WALL FIXTURE |
| EXIT | TELEPOLE |
| EMERGENCY LIGHT. 2 HEAD | MOTOR |
| EMERGENCY LIGHT. 1 HEAD. REMOTE | DISCONNECT SWITCH |
| DUPLEX RECEPTACLE | DISCONNECT SWITCH. FUSED |
| DUPLEX RECEPTACLE W GROUND FAULT | LIGHTING PANEL |
| DUPLEX RECEPTACLE W GROUND FAULT. W.P. | POWER PANEL |
| SINGLE POLE SWITCH | TRANSFORMER |
| FIRE ALARM STATION | SWITCHBOARD |
| CONDUIT IN SLAB | FIXTURE MOUNTED ON BOX |
| CONDUIT EXPOSED | WIRES IN CONDUIT |

PENTHOUSE FLOOR
THIRD FLOOR
SECOND FLOOR
FIRST FLOOR
PARKING GARAGE

(SEE MOTOR CONTROL CENTER SCHEDULE)

ELEVATORS

Figure 23.5

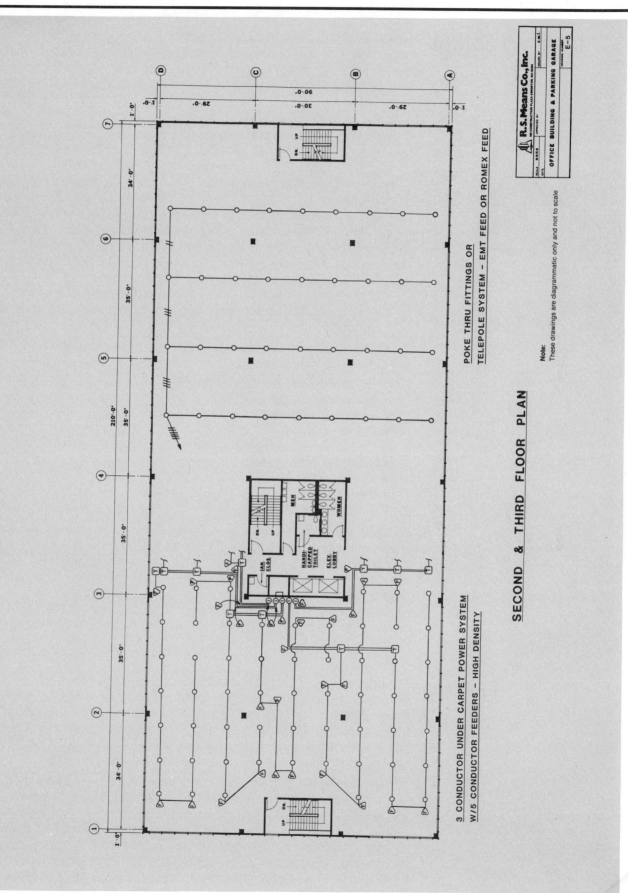

3 CONDUCTOR UNDER CARPET POWER SYSTEM
W/5 CONDUCTOR FEEDERS – HIGH DENSITY

POKE THRU FITTINGS OR
TELEPOLE SYSTEM – EMT FEED OR ROMEX FEED

SECOND & THIRD FLOOR PLAN

Note:
These drawings are diagrammatic only and not to scale

R.S. Means Co., Inc.

OFFICE BUILDING & PARKING GARAGE

E-5

*Figure 23.6*

249

## Getting Started

A good estimator must be able to visualize the proposed electrical installation from source (service) to end use (fixtures or devices). This process helps to identify each component needed to make the system work. Before starting the takeoff, the estimator should follow some basic steps:

- Read the electrical specifications thoroughly.
- Scan the electrical plans; check other sections of the plans and specifications for their potential effect on electrical work (special attention should be given to mechanical and site work).
- Check the architectural and structural plans for unique or atypical requirements.
- Check and become familiar with the fixture and power symbols.
- Clarify with the architect or engineer any unclear areas, making sure that the scope of work is understood. Addenda may be necessary to clarify certain items of work so that the responsibility for the performance of all work is defined.
- Immediately contact suppliers, manufacturers, and subsystem specialty contractors in order to get their quotations, sub-bids, and drawings.

Certain information should be taken from complete plans and specifications if the electrical portion of the work is to be properly estimated. In addition to floor plans, other sources of information are:

- Power riser diagram
- Panelboard schedule
- Fixture schedule
- Electrical symbol legend
- Reflected ceiling plans
- Branch circuit plans
- Fire alarm and telephone riser diagrams
- Special systems

Selecting a starting point can be difficult when a large scale project is involved. Lighting is often a good place to begin for the following reasons: First, lighting is an item common to nearly all projects. Secondly, it represents a large, if not the largest cost center for a project. Thirdly, when doing the lighting takeoff, the estimator is also taking a complete "tour" of the building. The resulting overview enables the estimator to select the order in which the remainder of the estimate should be carried out.

## Lighting

The lighting takeoff should begin with a counting of all the various fixtures. A fixture schedule may be provided in the drawings or listed in the specifications. Figure 23.7 shows a fixture schedule for this type of building. A copy of this schedule should be fixed to a wall or desk top for handy reference throughout the lighting takeoff. The legend of drawing symbols (see Figure 23.5) should be posted with the schedule.

To organize the lighting takeoff information, use a form, such as the "Lighting Schedule" form shown in Figure 23.8. Enter each fixture type with a brief description. Enter the symbol of each fixture as it will appear on the drawings. Note any special requirements or addenda; for example, an amendment to use Type C fixtures in lieu of Type F. Begin counting in an organized, systematic fashion. Arrange the schedule in such a way as to allow space for recording basement fixtures, followed by the fixtures for each of the other levels. Count carefully. Create columns for necessary components and count these components for each floor or area as you go. Included in this category are such items as raceways, fixture whips, boxes, and switches. Take advantage of any symmetry or repetition on different floors to save time. When all floors or areas are taken off, subtotal the columns. Make any needed adjustments and then calculate the total. It is a generally accepted practice to add 5% to conduit totals.

Wire is usually figured by multiplying the conduit footages by the overall average of the number of wires contained in all the conduit runs. In this example, allow three conductors per foot of conduit. Since wire length must exceed conduit lengths in order to make connections and terminations, add 10% to the wire totals.

Transfer the totals to a cost analysis form for pricing and extensions. "Pricing" refers to the process of assigning units of cost and/or hours to each activity code. "Extension" refers to the process of multiplying the value per unit times the total quantity of each item in order to get a total cost per item. See Figures 23.9 and 23.10.

Arrange the items on the cost analysis sheets into topical groupings, as much as is practical; this method can save time when researching the unit pricing. If changes must be made later, the affected items will be easier to locate.

| Type | Manufacturer & Catalog # | Fixture | Type | Lamps Qty | Volts | Watts | Mounting | Remarks |
|------|--------------------------|---------|------|-----------|-------|-------|----------|---------|
| A | Meansco  #7054 | 2'x4' Troffer | F-40 CW | 4 | 277 | 40 | Recessed | Acrylic Lens |
| B | Meansco  #7055 | 1'x4' Troffer | F-40 CW | 2 | 277 | 40 | Recessed | Acrylic Lens |
| C | Meansco  #7709 | 6"x4' | F-40 CW | 1 | 277 | 40 | Surface | Acrylic Wrap |
| D | Meansco  #7710 | 6"x8' Strip | F96T12 CW | 1 | 277 | 40 | Surface | |
| E | Meansco  #7900A | 6"x4' | F-40 | 1 | 277 | 40 | Surface | Mirror Light |
| F | Kingston  #100A | 6"x4' | F-40 | 1 | 277 | 40 | Surface | Acrylic Wrap |
| G | Kingston  #110C | ' Strip | F-40 CW | 1 | 277 | 40 | Surface | |
| H | Kingston  #3752 | Wallpack | HPS | 1 | 277 | 150 | Bracket | W/Photo Cell |
| I | Kingston  #201-202 | Floodlight | HPS | 1 | 277 | 400 | Surface | 2' Below Fascia |
| K | Kingston  #203 | | HPS | 1 | 277 | 100 | Wall Bracket | |
| L | Meansco  #8100 | Exit Light | 1-13W 20W | 1 | 120 | 13 | Surface | |
| | | | T6-1/2 | 2 | 6½ | 20 | | |
| M | Meansco  #9000 | Battery Unit | Sealed Beam | 2 | 12 | 18 | Wall Mount | 12 Volt Unit |

Figure 23.7

## Means Forms
### LIGHTING SCHEDULE

JOB NO. 86-002
JOB DESCRIPTION Office Bldg.
OWNER R. S. Means Company, Inc.
LOCATION Kingston, Ma

ESTIMATE NO. E-86-002
SHEET 1 OF 19
SCALE 1/4" = 1'
TAKE OFF BY PHD
CHECKED BY SM
DATE 1-3-86

FIXTURES Type F Deleted - Use Type C

| TYPE | A 2'x4' Troffer 4L | B 1'x4' Troffer 2L | C 6"x4" Acrylic | D 6"x8' Strip | E 6"x8' Mineral Light | G 4" Strip Surf. Mt. | H 150W Wall Pack | K 70W Wall Pack | L Incand. Exit Light | M Battery Emergency Lights | N Em. Lt. Remote Head | EMT 1/2" | Conn. 1/2" Set Screw | Conn. 1/2" Comp. Type | Whips 3/4"x6' 3w THHN | 4" Round Box | 4" Round Box Cover | 1 Gang Switch Box | S.P. Switch 20A | 4" Sq. Box | Sw. Cover Plate | 4" Plaster Rings | THHN #12 |
|---|---|---|---|---|---|---|---|---|---|---|---|---|---|---|---|---|---|---|---|---|---|---|---|
| Basmt | - | 6 | 4 | 53 | - | 6 | 1 | 1 | 6 | 3 | 1 | 1480 | 190 | 18 | 6 | 8 | 8 | 1 | 1 | 6 | 1 | 6 | 4,440 |
| 1st Fl. | 237 | 10 | 9 | - | 3 | 1 | 2 | 1 | 8 | 3 | 1 | 2070 | 264 | 26 | 247 | 128 | 128 | 5 | 5 | 8 | 5 | 8 | 6,210 |
| 2nd Fl. | 244 | 6 | 8 | - | 3 | 1 | 1 | - | 4 | 3 | 1 | 1880 | 242 | 24 | 250 | 129 | 129 | 5 | 5 | 4 | 5 | 4 | 5,640 |
| 3rd Fl. | 244 | 6 | 8 | - | 3 | 1 | - | - | 4 | 3 | 1 | 1880 | 242 | 24 | 250 | 129 | 129 | 5 | 5 | 4 | 5 | 4 | 5,640 |
| Pent Hs | - | - | - | - | - | 18 | - | 1 | 1 | 1 | - | 940 | 114 | 12 | 1 | 1 | 1 | 4 | 4 | 4 | 4 | 1 | 2,820 |
| Sub. Tot. | | | | | | | | | | | | 8250 +5% | | | | | | | | | | | 24,750 LF +10% |
| Adjust | | | | | | | | | | | | 413 | | | | | | | | | | | 2,475 |
| Total | 725 | 28 | 29 | 53 | 9 | 27 | 2 | 3 | 23 | 13 | 4 | 8663 | 1052 | 104 | 753 | 394 | 394 | 20 | 20 | 23 | 20 | 23 | 273 CLF |

Figure 23.8

252

## Means Forms
**COST ANALYSIS**

PROJECT **Office Bldg.** — Lighting

ESTIMATE NO. **E-86-002**

ARCHITECT **R.S. Means Company, Inc.**

DATE **1-3-86**

TAKE OFF BY: **PHD**  QUANTITIES BY: **PHD**  PRICES BY: **PHD**  EXTENSIONS BY: **SM**  CHECKED BY: **MG**

| DESCRIPTION | SOURCE/DIMENSIONS | | | QUANTITY | UNIT | man-Hours UNIT COST | man-Hours TOTAL | material UNIT COST | material TOTAL | Installation UNIT COST | Installation TOTAL |
|---|---|---|---|---|---|---|---|---|---|---|---|
| **Fixtures** | | | | | | | | | | | |
| Type A - 2'x4' Troffer, 4L | 16.6 | 10 | 060 | 725 | Ea. | 1.7 | 1233 | 49 | 35525 | 38 | 27550 |
| Type B - 1'x4' Troffer, 2L | | 10 | 020 | 28 | | 1.4 | 39 | 35 | 980 | 31 | 868 |
| Type C - 6"x4' Acrylic | | 10 | 202 | 29 | | 1.0 | 29 | 32 | 928 | 22 | 638 |
| Type D - 6"x8' strip | | 10 | 260 | 53 | | 1.19 | 63 | 32 | 1696 | 27 | 1431 |
| Type E - 6"x8' mirror Lt. | | 10 | 690 | 9 | | 1.0 | 9 | 39 | 351 | 22 | 198 |
| Type G - 4' strip Surf. mt. | | 10 | 220 | 27 | | .941 | 25 | 18 | 486 | 21 | 567 |
| Type H - 150w Wall Pack | | 50 | 117 | 2 | | 2.0 | 4 | 280 | 560 | 45 | 90 |
| Type K - 70w Wall Pack | | 50 | 116 | 3 | | 2.0 | 6 | 265 | 795 | 45 | 135 |
| Type L - Fluor. Exit Lt. | | 25 | 015 | 23 | | 1.0 | 23 | 89 | 2047 | 22 | 506 |
| Type M - Batt. Emergency Light | | 25 | 050 | 13 | | 2.0 | 26 | 195 | 2535 | 45 | 585 |
| Type N - Em. Lt. Remote Head | | 25 | 078 | 4 | | .3 | 1 | 16 | 64 | 6.70 | 27 |
| **Raceway** | | | | | | | | | | | |
| 1/2" EMT | 16.0 | 20 | 500 | 8,663 | LF. | .047 | 407 | .23 | 1992 | 1.05 | 9096 |
| 1/2" EMT - Conn. Set Screw | | 20 | 650 | 1,052 | Ea. | .067 | 70 | .35 | 368 | 1.49 | 1567 |
| 1/2" EMT - Conn. Comp. Type | | 20 | 880 | 104 | Ea. | .067 | 7 | .61 | 63 | 1.49 | 155 |
| **Fixture Whips** | | | | | | | | | | | |
| 3/8"x6' 3 wire, THHN #14 | 16.6 | 91 | 030 | 753 | Ea. | .25 | 188 | 5.20 | 3916 | 5.60 | 4217 |
| **Wire** | | | | | | | | | | | |
| #12 THHN | 16.1 | 10 | 120 | 273 | CLF. | .727 | 198 | 3.85 | 1051 | 16.30 | 4450 |
| Subtotals | | | | | | | 2328 | | 53357 | | 52080 |

*Figure 23.9*

253

## Means Forms

**COST ANALYSIS**

PROJECT **Office Bldg.**     **Lighting**

ARCHITECT **R.S. Means Company, Inc.**

TAKE OFF BY: **PHD**   QUANTITIES BY: **PHD**   PRICES BY: **PHD**

EXTENSIONS BY: **SM**   CHECKED BY: **MG**

| DESCRIPTION | SOURCE/DIMENSIONS | | | QUANTITY | UNIT | Man-Hours | | materials | | Installation | |
|---|---|---|---|---|---|---|---|---|---|---|---|
| | | | | | | UNIT COST | TOTAL | UNIT COST | TOTAL | UNIT COST | TOTAL |
| **Boxes and Devices** | | | | | | | | | | | |
| 4" Octagon Box | 16.2 | 20 | 002 | 394 | Ea. | .4 | 158 | .89 | 351 | 8.95 | 3526 |
| 4" Octagon Blank Cover | | 20 | 025 | 394 | | .125 | 49 | .41 | 462 | 2.80 | 1103 |
| 4" Square Box | | 20 | 015 | 23 | | .4 | 9 | 1.07 | 25 | 8.95 | 206 |
| Plaster Rings | | 20 | 050 | 23 | | .151 | 3 | 1.22 | 28 | 3.38 | 78 |
| Switch Box | | 20 | 065 | 20 | | .296 | 6 | .93 | 19 | 6.65 | 133 |
| S.P. Switch - 20A | | 30 | 050 | 20 | | .296 | 6 | 6.95 | 139 | 6.65 | 133 |
| Sw. Plate - 1 Gang | | 30 | 260 | 20 | | .1 | 2 | 2.10 | 42 | 2.24 | 45 |
| | | | | | | | | | | | |
| Subtotals | | | | | | | 233 | | 1166 | | 5224 |
| Subtotal Sheet 2 | | | | | | | 2328 | | 53357 | | 52080 |
| | | | | | | | | | | | |
| Lighting Totals | | | | | | | 2561 | | 54423 | | 57304 |

Figure 23.10

254

It is a good idea to note the source of the unit costs used for pricing. Since we will be using *Means Electrical Cost Data* (1986 edition) for this estimate, the line number from the book will be the source description.

From the perspective of an electrical contractor, the man-hours needed to install the work as well as the costs for material and for labor are listed. The column headings for each item include the man-hours, material cost (bare), and installation cost (bare). (Mark-ups for overhead and profit will be added at the "bottom line" of the estimate.) See Figure 23.9.

In the normal estimating process, the entire job takeoff is completed before the cost analysis sheets are priced. However, for the purpose of this example estimate, pricing and extending each segment of the job are discussed with that segment's takeoff.

Complete the pricing and extensions and subtotal each sheet. When each segment of the estimate is complete, combine the subtotals to get a category total. In this case, the "totals" (Figure 23.10) are for the lighting work. It is not necessary to combine material and labor costs as yet.

## Service and Distribution Equipment

After the lighting takeoff is complete, the next logical step is to take off the service entrance equipment and distribution feeders. This segment of the work is appropriately listed early in the takeoff because it represents the source of all power for the building. In addition, the equipment is located in the basement level and can be approached in the same bottom-to-top order as was used for lighting. Furthermore, service and distribution equipment usually represents the second largest cost segment (after lighting) for this type of project.

The equipment, conduit, cable, and destination equipment are represented in diagrams on the electric riser drawing (Figure 23.5). When using the riser diagram to organize the takeoff sheets, remember that it is a "not to scale" drawing. In order to estimate the conduit lengths, fittings, and cable lengths, it will be necessary to use the plan and elevation drawings. Relevant information that does not appear on the drawings might instead be found in the specifications. Some information may appear in both locations. For example, the specifications might inform the contractor that "the elevator equipment, including controls, will be furnished and installed by others. The electrical contractor shall furnish and install two feeders with safety disconnects to be mounted within sight of the elevator equipment." Items of this kind must be noted by the estimator when reading the specifications. In this way, the estimate can be done correctly as the takeoff is performed.

Begin the riser diagram takeoff by filling in a feeder schedule to tabulate each conduit and cable run (see Figures 23.11 and 23.12). Enter each circuit number, its origin, and destination. Using the riser legend (Figure 23.5), enter the given information for each circuit, such as conduit size, number of conductors, and size of conductors. Referring to the plan and elevation drawings, carefully determine the length of each conduit run. Count and list any extra items, such as wireway or conduit nipples. Add 5% for each conduit run as a tolerance and then total.

| JOB NO. 86-002 | Riser Diagram | | SHEET 4 OF 19 |

OWNER **R.S. means Company, Inc.** ADDRESS **Kingston, ma** — ESTIMATE NO. **E-86-002**

JOB DESCRIPTION **Office Bldg.** DRAWING NO. **ME-4** — DATE **1-3-86**

SCALE — TAKE OFF BY **PHD** — CHECKED BY **MG**

| PNL. NO. | CIRC. NO. | FROM | TO | AMPS | RACEWAY | | | | | | | WIRE | | | | | | |
|---|---|---|---|---|---|---|---|---|---|---|---|---|---|---|---|---|---|---|
| | | | | | SIZE | TYPE | L.F. | TERM | +5% | | | NO. | SIZE | TYPE | VOLTS | RUN | L.F. | +10% CLF |
| SWBD | 1 | Pad-mount XFMR | main Switch-board | 1200 | 3" | PVC | 3@ 200 600 | | 630 | | | 3 | 400 mcm | XHHW | 600 | 3 Quad @ 200' | 2400 | 26.4 |
| | 8 | Swbd. | T1 | 60 | 1" | RGS. | 15 | | 16 | | | 4 | #6 | | | 20 | 80 | .9 |
| | 8-A | T1 | PP-1 | 100 | 1¼" | | 5 | | 6 | | | 4 | #2 | | | 5 | 20 | .3 |
| | 9 | Swbd. | T2 | 100 | 1¼" | | 20 | | 21 | | | 4 | #2 | | | 25 | 100 | 1.10 |
| | 9-A | T2 | 200 A Disconn. PP-2 | 200 | 2" | | 50 | | 53 | | | 4 | 2/0 | | | 50 | 200 | 2.2 |
| | 10 | Swbd. | T3 | 100 | 1¼" | | 25 | | 27 | | | 4 | #2 | | | 30 | 120 | 1.4 |
| | 10-A | T3 | 200 A Disconn. PP3 | 200 | 2" | | 65 | | 69 | | | 4 | 2/0 | | | 65 | 260 | 2.9 |
| | 11 | Swbd. | T4 | 100 | 1¼" | | 30 | | 32 | | | 4 | #2 | | | 35 | 140 | 1.6 |
| | 11-A | T4 | 200 A Disconn. & PP4 | 200 | 2" | | 80 | | 84 | | | 4 | 2/0 | | | 80 | 320 | 3.6 |
| | 4 | Swbd. | AC1 | 100 | 1¼" | | 50 | | 53 | | | 4 | #2 | | | 55 | 220 | 2.5 |
| | 5 | Swbd. | AC2 | 100 | 1¼" | | 65 | | 69 | | | 4 | #2 | | | 70 | 280 | 3.1 |
| | 6 | Swbd. | AC3 | 100 | 1¼" | | 80 | | 84 | | | 4 | #2 | | | 85 | 340 | 3.8 |

NOTES

Figure 23.11

## Means Forms

**FEEDER SCHEDULE**

| JOB NO. 86-002 | Riser Diagram | SHEET 5 OF 19 |
|---|---|---|

| OWNER R.S. Means Company, Inc. | ADDRESS | ESTIMATE NO. E-86-002 |
|---|---|---|

| JOB DESCRIPTION Office Bldg. | DRAWING NO. ME-4 | DATE 1-3-86 |
|---|---|---|

| SCALE | TAKE OFF BY PHD | CHECKED BY MG |
|---|---|---|

| PNL. NO. | CIRC. NO. | FROM | TO | AMPS | RACEWAY SIZE | TYPE | L.F. | TERM | +5% L.F. | 6" nipple | 6"x6" Wire way | 6"x6" End cap | WIRE NO. | SIZE | TYPE | VOLTS | RUN | L.F. | +10% C.L.F. |
|---|---|---|---|---|---|---|---|---|---|---|---|---|---|---|---|---|---|---|---|
| Swbd | 7 | Swbd. | MCC | 400 | 1¼" | RGS | 85 | | 90 | | | | 4 | 300 mcm | XHHW | 600 | 90' | 320 | 3.6 |
| | 12 | | LP-1 | 200 | 2" | | 5 | | 6 | | | | 4 | 4/0 | | | 10 | 40 | .5 |
| | 13 | | LP-2 | 200 | 2" | | 50 | | 53 | | | | 4 | 4/0 | | | 55 | 220 | 2.5 |
| | 14 | | LP-3 | 200 | 2" | | 65 | | 69 | | | | 4 | 4/0 | | | 70 | 280 | 3.1 |
| | 15 | | LP-4 | 200 | 2" | | 80 | | 84 | | | | 4 | 4/0 | | | 85 | 340 | 3.8 |
| | 16 | | Elev. Disc. | 100 | 1¼" | | 25 | | 27 | 2 Ea | 6' | 2 Ea | 4 | #3 | | | 30 | 120 | 1.4 |
| | 16-A | Elev. Disc. A | A Elev. Eq. | 60 | 1" | | 10 | | 11 | | | | 4 | #6 | | | 15 | 60 | .7 |
| | 16-B | Elev. Disc. B | B Elev. Eq. | 60 | 1" | | 10 | | 11 | | | | 4 | #6 | | | 15 | 60 | .7 |

3" PVC - 630
1" RGS - 16, 11, 11 = (38)
1¼" RGS - 6, 21, 27, 32, 53, 69, 84, 90, 27 = (409)
2" RGS - 53, 69, 84, 6, 53, 69, 84 = (418)

NOTES
300 mcm - (3.6)
400 mcm - (26.4)
#6 - .88, .7, .7 = (2.3)
#2 - .3, 1.1, 1.4, 1.6, 2.5, 3.1, 3.8 = (13.8)
2/0 - 2.2, 2.9, 3.6 = (8.7)
4/0 - .5, 2.5, 3.1, 3.8 = (9.9)

#3 - (1.4)

*Figure 23.12*

257

The next step is calculating the "run" length of each circuit by using the measured conduit length – plus any length required at the equipment. For example, the wires entering the top of a switchboard may need to reach a circuit breaker located at the bottom; therefore add five feet to each cable run entering this board. Multiply the length of each run by the number of wires to arrive at the linear feet of each wire for each circuit. Finally, add 10% to the wire lengths as a tolerance.

Since the service and distribution equipment is clearly and simply represented on the riser drawing, it may be practical to take off this equipment directly, using a cost analysis sheet (see Figures 23.14, 23.15, and 23.16). Additional information, such as the panelboard schedule (Figure 23.13), may help to complete this takeoff.

Transfer the quantities from the feeder schedule to the cost analysis sheets. It is often convenient to summarize like quantities at the bottom of the feeder schedule before transferring those totals. Refer to the bottom of Figure 23.12. To complete this segment of the work, price and extend these cost analysis sheets.

Careful attention must be paid no matter what the price source, to assure that the unit of measure is compatible with those used in the takeoff. An example may be seen in Figure 23.14. The fourth item listed on that sheet reflects the fact that there are three distribution sections (or structures) in which the fusable switch components will be mounted. When this item is found in *Means Electrical Cost Data* on line 16.3-61-252 (see Figure 23.17), it is noted that the unit includes three sections. Therefore, the quantity multiplier is changed from three to one, and a note made on the takeoff sheet.

## Panelboard Schedule

| Panel | Main Breaker. | Main | Lugs | Amps | Circ. | Breakers # | Type | Location | Volts |
|-------|---------------|------|--------|------|-------|------------|---------|-----------|---------|
| PP-1  | Yes           | 3-P  | Wire   | 100  | 24    | (24)       | 1P 20 A | Basement  | 120/208 |
| PP-2  | No            | 3-P  | 4 Wire | 225  | 42    | (42)       | 1P 20 A | 1st Floor | 120/208 |
| PP-3  | No            | 3-P  | 4 Wire | 225  | 42    | (42)       | 1P 20 A | 2nd Floor | 120/208 |
| PP-4  | No            | 3-P  | 4 Wire | 225  | 42    | (42)       | 1P-20 A | 3rd Floor | 120/208 |
| LP-1  | No            | 3-P  | 4 Wire | 225  | 42    | (42)       | 1P-20 A | Basement  | 277/480 |
| LP-2  | No            | 3-P  | 4 Wire | 225  | 42    | (42)       | 1P-20 A | 1st Floor | 277/480 |
| LP-3  | No            | 3-P  | 4 Wire | 225  | 42    | (42)       | 1P-20 A | 2nd Floor | 277/480 |
| LP-4  | No            | 3-P  | 4 Wire | 225  | 42    | (42)       | 1P-20 A | 3rd Floor | 277/480 |

*Figure* 23.13

**COST ANALYSIS**

| | | | | | | | | | | |
|---|---|---|---|---|---|---|---|---|---|---|
| PROJECT Office Bldg. | Riser Diagram and Feeder Schedule | | | | | SHEET NO. 6 of 19 | | | | |
| ARCHITECT R.S. means Company, Inc. | | | | | | ESTIMATE NO. E-86-002 | | | | |
| | | | | | | DATE 1-3-86 | | | | |
| TAKE OFF BY: PHD | QUANTITIES BY: PHD | PRICES BY: PHD | | EXTENSIONS BY: SM | | CHECKED BY: MG | | | | |

| DESCRIPTION | SOURCE/DIMENSIONS | | | QUANTITY | UNIT | man-Hours | | material | | Installation | |
|---|---|---|---|---|---|---|---|---|---|---|---|
| | | | | | | UNIT COST | TOTAL | UNIT COST | TOTAL | UNIT COST | TOTAL |
| **Switchboard** | | | | | | | | | | | |
| main Switch, 1200 A | 16.3 | 61 | 104 | 1 | Ea. | 18.18 | 18 | 4350 | 4350 | 405 | 405 |
| C/T Compartment | | 61 | 194 | 1 | | 2.96 | 3 | 885 | 885 | 66 | 66 |
| Transition Section | | 80 | 180 | 1 | | 20 | 20 | 1675 | 1675 | 450 | 450 |
| Distribution Sections (For 3 Sections Add) | | 61 | 252 | 1 | | 40 | 40 | 2950 | 2950 | 895 | 895 |
| Fusable Sw, 3P, 100 A (240/480) | | 70 | 176 | 8 | | 3.2 | 26 | 250 | 2000 | 72 | 576 |
| Fusable Sw, 3P, 200 A (240/480) | | 70 | 180 | 4 | | 4.2 | 17 | 340 | 1360 | 94 | 376 |
| Fusable Sw, 3P, 400 A (240/480) | ↓ | 70 | 184 | 1 | ↓ | 6.15 | 6 | 775 | 775 | 140 | 140 |
| **Panelboards w/Bkrs.** | | | | | | | | | | | |
| 225 A, m. Lug, 42 Cir., 120/208 | 16.3 | 50 | 100 | 3 | Ea. | 23.53 | 71 | 705 | 2115 | 525 | 1575 |
| 225 A, m. Lug, 42 Cir, 277/480 | | 50 | 150 | 4 | | 26.67 | 107 | 1675 | 6700 | 595 | 2380 |
| 100 A, m.C.B., 24 Cir, 120/208 | ↓ | 50 | 205 | 1 | ↓ | 17.02 | 17 | 605 | 605 | 380 | 380 |
| **Safety Switches** | | | | | | | | | | | |
| H.D., Fused, 240 V, 3P, 200A | 16.3 | 55 | 350 | 3 | Ea. | 6.15 | 18 | 270 | 810 | 140 | 420 |
| H.D., Fused, 600 V, 3P, 60 A | 16.3 | 55 | 438 | 2 | Ea. | 3.48 | 7 | 120 | 240 | 78 | 156 |
| Subtotals | | | | | | | 350 | | 24465 | | 7819 |

Figure 23.14

259

# Means Forms

**COST ANALYSIS**

PROJECT **Office Bldg.  Riser Diagram and Feeder Schedule**  ESTIMATE NO. **E-86-002**

ARCHITECT **R.S. Means Company, Inc.**  DATE **1-3-86**

TAKE OFF BY: **PHD**  QUANTITIES BY: **PHD**  PRICES BY: **PHD**  EXTENSIONS BY: **SM**  CHECKED BY: **M6**

| DESCRIPTION | SOURCE/DIMENSIONS | | | QUANTITY | UNIT | man-Hours UNIT COST | TOTAL | material UNIT COST | TOTAL | Installation UNIT COST | TOTAL |
|---|---|---|---|---|---|---|---|---|---|---|---|
| **Transformers** | | | | | | | | | | | |
| 45 KVA, 3P, 120/208 | 16.4 | 10 | 350 | 1 | Ea. | 20.0 | 20 | 1225 | 1225 | 450 | 450 |
| 75 KVA, 3P, 120/208 | 16.4 | 10 | 370 | 3 | Ea. | 22.86 | 69 | 1850 | 5550 | 510 | 1530 |
| | | | | | | | | | | | |
| **Grounding & Wire** | | | | | | | | | | | |
| 4/0 Bare Copper | 16.1 | 80 | 100 | .20 | CLF | 2.81 | 1 | 95 | 19 | 63 | 13 |
| Cadweld - 4/0 to Bldg. Stl. | | 80 | 274 | 2 | Ea. | 1.14 | 2 | 3.55 | 7 | 26 | 52 |
| Cable Term - 4/0, 1h, Cu. | ↓ | 50 | 300 | 2 | Ea. | .737 | 1 | 6.50 | 13 | 16.30 | 33 |
| 400 mcm - XHHW Cu. | 16.1 | 10 | 330 | 26.4 | CLF | 4.71 | 124 | 230 | 6072 | 105 | 2772 |
| 300 mcm | | | | 326 | 3.6 | 4.21 | 15 | 155 | 558 | 94 | 338 |
| 4/0 | | | | 322 | 9.9 | 3.64 | 36 | 100 | 990 | 81 | 802 |
| 2/0 | | | | 318 | 8.7 | 2.76 | 24 | 66 | 574 | 62 | 539 |
| #2 | | | | 312 | 13.8 | 1.78 | 25 | 35 | 483 | 40 | 552 |
| #3 (THHN) | | | | 145 | 1.4 | 1.60 | 2 | 34 | 48 | 36 | 50 |
| #6 ↓ | ↓ | | | 308 | 2.3 | 1.23 | 3 | 15 | 34 | 28 | 64 |
| | | | | | | | | | | | |
| **Subtotals** | | | | | | | 322 | | 15573 | | 7195 |

*Figure 23.15*

**Means Forms**
COST ANALYSIS

PROJECT: **Office Bldg.   Riser Diagram and Feeder Schedule**

ARCHITECT: **R.S. means Company, Inc.**

DATE **1-3-86**

TAKE OFF BY: **PHD**   QUANTITIES BY: **PHD**   PRICES BY: **PHD**   EXTENSIONS BY: **Sm**   CHECKED BY: **mb**

| DESCRIPTION | SOURCE/DIMENSIONS | | | QUANTITY | UNIT | man-Hours | | material | | Installation | |
|---|---|---|---|---|---|---|---|---|---|---|---|
| | | | | | | UNIT COST | TOTAL | UNIT COST | TOTAL | UNIT COST | TOTAL |
| **Raceways** | | | | | | | | | | | |
| 3" PVC - Underground (3 @ 210') | 16.7 | 01 | 500/520 (Ave.) | 210 | LF. | .120 | 25 | 1.17 | 246 | 2.69 | 565 |
| 1" RGS | 16.0 | 20 | 180 | 38 | LF. | .123 | 5 | 1 | 38 | 2.76 | 105 |
| 1¼" RGS | | 20 | 183 | 409 | LF. | .133 | 54 | 1.35 | 552 | 2.99 | 1223 |
| 2" RGS | | 20 | 187 | 418 | LF. | .178 | 74 | 2.25 | 940 | 3.98 | 1664 |
| 6"x 6" Wireway | | 95 | 040 | 4 | LF. | .267 | 1 | 9.85 | 39 | 5.95 | 24 |
| 6"x 6" End Caps | | 95 | 400 | 2 | Ea. | .444 | 1 | 2.35 | 5 | 9.95 | 20 |
| 1¼"x 6" nipples RGS | | 57 | 094 | 2 | Ea. | .348 | 1 | 5 | 10 | 7.80 | 16 |
| 1"LB Fittings - Allow. | | 55 | 230 | 2 | Ea. | .727 | 1 | 5.30 | 11 | 16.30 | 33 |
| 1¼" LB Fittings - Allow. | | 55 | 233 | 10 | Ea. | 1.0 | 10 | 7.70 | 77 | 22 | 220 |
| 2" LB Fittings - Allow | | 55 | 237 | 10 | Ea. | 1.6 | 16 | 16.20 | 162 | 36 | 360 |
| | | | | | | | | | | | |
| Subtotals | | | | | | | 188 | | 2080 | | 4230 |
| Subtotal Sheet 6 | | | | | | | 350 | | 24465 | | 7819 |
| Subtotal Sheet 7 | | | | | | | 322 | | 15573 | | 7195 |
| Riser Diagram Totals | | | | | | | 860 | | 42118 | | 19244 |

Figure 23.16

# 16.3 Starters, Boards & Switches

| | | CREW | MAN-HOURS | UNIT | MAT. | INST. | TOTAL | TOTAL INCL O&P |
|---|---|---|---|---|---|---|---|---|
| 141 | 1200 amp | 1 Elec | 20.000 | Ea. | 5,125 | 450 | 5,575 | 6,275 |
| 142 | 1600 amp | | 21.050 | | 5,800 | 470 | 6,270 | 7,050 |
| 143 | 2000 amp | | 23.530 | | 7,225 | 525 | 7,750 | 8,700 |
| 150 | Main ground fault protector, 1200-2000 amp | | 2.960 | | 2,975 | 66 | 3,041 | 3,375 |
| 160 | Bus way connection, 200 amp | | 2.960 | | 170 | 66 | 236 | 280 |
| 161 | 400 amp | | 3.480 | | 200 | 78 | 278 | 330 |
| 162 | 600 amp | | 4.000 | | 260 | 90 | 350 | 415 |
| 163 | 800 amp | | 5.000 | | 285 | 110 | 395 | 475 |
| 164 | 1200 amp | | 6.150 | | 390 | 140 | 530 | 625 |
| 165 | 1600 amp | | 6.670 | | 460 | 150 | 610 | 720 |
| 166 | 2000 amp | | 8.000 | | 540 | 180 | 720 | 850 |
| 170 | Shunt trip for remote operation 200 amp | | 2.000 | | 570 | 45 | 615 | 690 |
| 171 | 400 amp | | 2.000 | | 920 | 45 | 965 | 1,075 |
| 172 | 600 amp | | 2.000 | | 1,425 | 45 | 1,470 | 1,625 |
| 173 | 800 amp | | 2.000 | | 1,825 | 45 | 1,870 | 2,075 |
| 174 | 1200-2000 amp | | 2.000 | | 4,050 | 45 | 4,095 | 4,525 |
| 180 | Motor operated main breaker 200 amp | | 2.000 | | 1,075 | 45 | 1,120 | 1,250 |
| 181 | 400 amp | | 2.000 | | 1,550 | 45 | 1,595 | 1,775 |
| 182 | 600 amp | | 2.000 | | 2,075 | 45 | 2,120 | 2,350 |
| 183 | 800 amp | | 2.000 | | 2,475 | 45 | 2,520 | 2,775 |
| 184 | 1200-2000 amp | | 2.000 | | 4,675 | 45 | 4,720 | 5,200 |
| 190 | Current/potential transformer metering compartment 200-800 amp | | 2.960 | | 700 | 66 | 766 | 865 |
| 194 | 1200 amp | | 2.960 | | 885 | 66 | 951 | 1,075 |
| 195 | 1600-2000 amp | | 2.960 | | 1,075 | 66 | 1,141 | 1,275 |
| 200 | With watt meter 200-800 amp | | 4.000 | | 2,150 | 90 | 2,240 | 2,500 |
| 204 | 1200 amp | | 4.000 | | 2,400 | 90 | 2,490 | 2,775 |
| 205 | 1600-2000 amp | | 4.000 | | 2,575 | 90 | 2,665 | 2,950 |
| 210 | Split bus 60-200 amp | | 1.510 | | 125 | 34 | 159 | 185 |
| 213 | 400 amp | | 3.480 | | 230 | 78 | 308 | 365 |
| 214 | 600 amp | | 4.440 | | 285 | 100 | 385 | 455 |
| 215 | 800 amp | | 6.150 | | 370 | 140 | 510 | 605 |
| 217 | 1200 amp | | 8.000 | | 425 | 180 | 605 | 725 |
| 225 | Contactor control 60 amp | | 4.000 | | 1,075 | 90 | 1,165 | 1,300 |
| 226 | 100 amp | | 5.330 | | 1,250 | 120 | 1,370 | 1,550 |
| 227 | 200 amp | | 8.000 | | 1,900 | 180 | 2,080 | 2,350 |
| 228 | 400 amp | | 16.000 | | 5,875 | 360 | 6,235 | 6,975 |
| 229 | 600 amp | | 19.050 | | 6,575 | 425 | 7,000 | 7,850 |
| 230 | 800 amp | | 22.220 | | 7,825 | 500 | 8,325 | 9,325 |
| 250 | Modifier for two distribution sections, add | | 20.000 | | 1,475 | 450 | 1,925 | 2,275 |
| 252 | Three distribution sections, add | | 40.000 | | 2,950 | 895 | 3,845 | 4,525 |
| 256 | Auxiliary pull section, 20", add | | 8.000 | | 920 | 180 | 1,100 | 1,275 |
| 258 | 24", add | | 8.890 | | 1,050 | 200 | 1,250 | 1,450 |
| 260 | 30", add | | 10.000 | | 1,175 | 225 | 1,400 | 1,625 |
| 262 | 36", add | | 11.430 | | 1,275 | 255 | 1,530 | 1,775 |
| 264 | Dog house, 12", add | | 6.670 | | 260 | 150 | 410 | 500 |
| 266 | 18", add | | 8.000 | | 305 | 180 | 485 | 595 |
| 65-001 | **DISTRIBUTION SECTION** | | | | | | | |
| 010 | Aluminum bus bars, not including breakers | | | | | | | |
| 016 | Subfeed lug-rated at 60 amp | 1 Elec | 12.310 | Ea. | 740 | 275 | 1,015 | 1,200 |
| 017 | 100 amp | | 12.700 | | 855 | 285 | 1,140 | 1,350 |
| 018 | 200 amp | | 13.330 | | 970 | 300 | 1,270 | 1,500 |
| 019 | 400 amp | | 14.550 | | 1,100 | 325 | 1,425 | 1,675 |
| 020 | 120/208 or 277/480 volt, 4 wire, 600 amp | | 16.000 | | 1,300 | 360 | 1,660 | 1,950 |
| 030 | 800 amp | | 18.180 | | 1,375 | 405 | 1,780 | 2,100 |
| 040 | 1000 amp | | 20.000 | | 1,500 | 450 | 1,950 | 2,300 |
| 050 | 1200 amp | | 22.220 | | 1,600 | 500 | 2,100 | 2,475 |
| 060 | 1600 amp | | 24.240 | | 1,825 | 545 | 2,370 | 2,800 |
| 070 | 2000 amp | | 25.810 | | 1,925 | 580 | 2,505 | 2,950 |
| 080 | 2500 amp | | 26.670 | | 2,325 | 595 | 2,920 | 3,425 |
| 090 | 3000 amp | | 28.570 | | 2,650 | 640 | 3,290 | 3,825 |

Figure 23.17

Occasionally, the exact work item may not appear in the reference source, and in such cases, common sense judgements must be made. For example, in Figure 23.16, notice the first item — 3″ PVC underground (3 conduits @ 210′). When searching for the corresponding line item, it is noted that an exact match does not exist, although cost units for 2 conduits and for four conduits are given. Refer to section 16.7-01, lines 500 and 520 (Figure 23.18). A suitable estimate may be made using the average value of these two lines.

## Motor Control Center and Motor Feeders

Since motor control centers and motor feeders are taken off in a manner similar to that used for service and distribution equipment, this segment of the work is a logical choice for the next category in the estimate. The motor control center (MCC) appears on the penthouse floor plan drawing (Figure 23.4). The same drawing also shows an elevation of the MCC and a schedule of the MCC's circuits. Again, use a feeder schedule form (Figure 23.19) to tabulate the conduit and wire requirements. Enter each circuit from the MCC and list its destination.

In this case, the designer has provided only the horsepower of the motor being fed. Conduit and wire sizes are not given; this information must be determined by the estimator.

To complete this information, proceed according to the following steps:

1. Knowing the motor's horsepower and operating voltage (460V, 3 phase), we can refer to a table to find the amperage required. The table (from *Means Electrical Cost Data*) is "Ampere Values Determined By Horsepower, Voltage, and Phase Values", and appears in the Appendix of this book. Enter the HP and AMP values on the feeder schedule.

2. With the amperages known, the size of the conductors can be found based on a table (from *Means Electrical Cost Data*), "Minimum Copper and Aluminum Wire Sizes Allowed for Various Types of Insulation", which also appears in the Appendix of this book. Enter the copper wire size under the proper column on the feeder schedule. (When sizing wire for motor feeders, it is often appropriate to use #12 AWG as a *minimum*).

3. The size of the conduit can now be listed from the table (*Means Electrical Cost Data*) "Maximum Number of Wires for Various Conduit Sizes" (see Appendix).

| 16.7 **Lighting Utilities** | CREW | MAN-HOURS | UNIT | BARE COSTS | | | TOTAL INCL O&P |
|---|---|---|---|---|---|---|---|
| | | | | MAT. | INST. | TOTAL | |
| 01-001 **ELECTRIC & TELEPHONE SITEWORK** Not including excavation, backfill | | | | | | | |
| 020     or cast in place concrete | | | | | | | |
| 040     Hand holes, precast concrete with concrete cover | | | | | | | |
| 060       2' x 2' x 3' deep | R-3 | 8.330 | Ea. | 230 | 225 | 455 | 560 |
| 080       3' x 3' x 3' deep | | 10.530 | | 315 | 280 | 595 | 735 |
| 100       4' x 4' x 4' deep | ▼ | 14.290 | ▼ | 680 | 385 | 1,065 | 1,275 |
| 120     Manholes, precast, with iron racks, pulling irons, C.I. frame | | | | | | | |
| 140       and cover, 4' x 6' x 7' deep | R-3 | 16.670 | Ea. | 1,150 | 445 | 1,595 | 1,875 |
| 160       6' x 8' x 7' deep | | 20.000 | | 1,475 | 535 | 2,010 | 2,375 |
| 180       6' x 10' x 7' deep | | 25.000 | | 1,675 | 670 | 2,345 | 2,775 |
| 200     Poles, wood, creosoted (see also division 16.6-50) 20' high | | 6.450 | | 80 | 175 | 255 | 325 |
| 240       25' high | | 6.900 | | 90 | 185 | 275 | 355 |
| 260       30' high | | 7.690 | | 100 | 205 | 305 | 395 |
| 280       35' high | | 8.330 | | 155 | 225 | 380 | 480 |
| 300       40' high | | 8.700 | | 175 | 235 | 410 | 515 |
| 320       45' high | ▼ | 11.760 | ▼ | 215 | 315 | 530 | 675 |
| 340     Cross arms with hardware & insulators | | | | | | | |
| 360       4' long | 1 Elec | 3.200 | Ea. | 22 | 72 | 94 | 125 |
| 380       5' long | | 3.330 | | 24 | 75 | 99 | 135 |
| 400       6' long | ▼ | 3.640 | ▼ | 30 | 81 | 111 | 150 |
| 420     Underground duct, banks ready for concrete fill, min. of 1-1/2" | | | | | | | |
| 440       between ducts. For wire & cable see division 16.1 | | | | | | | |
| 458       PVC, type EB, 1 @ 2" diameter | 1 Elec | .033 | L.F. | .29 | .75 | 1.04 | 1.39 |
| 460       2 @ 2" diameter | | .067 | | .58 | 1.49 | 2.07 | 2.79 |
| 480       4 @ 2" diameter | | .133 | | 1.16 | 2.99 | 4.15 | 5.55 |
| 490       1 @ 3" diameter | | .040 | | .39 | .90 | 1.29 | 1.72 |
| 500       2 @ 3" diameter | | .080 | | .78 | 1.79 | 2.57 | 3.43 |
| 520       4 @ 3" diameter | | .160 | | 1.56 | 3.58 | 5.14 | 6.85 |
| 530       1 @ 4" diameter | | .050 | | .62 | 1.12 | 1.74 | 2.29 |
| 540       2 @ 4" diameter | | .100 | | 1.24 | 2.24 | 3.48 | 4.59 |
| 560       4 @ 4" diameter | | .200 | | 2.48 | 4.48 | 6.96 | 9.15 |
| 580       6 @ 4" diameter | | .296 | | 3.72 | 6.65 | 10.37 | 13.65 |
| 581       1 @ 5" diameter | | .062 | | .93 | 1.38 | 2.31 | 3.01 |
| 582       2 @ 5" diameter | | .123 | | 1.86 | 2.76 | 4.62 | 6 |
| 584       4 @ 5" diameter | | .229 | | 3.72 | 5.10 | 8.82 | 11.45 |
| 586       6 @ 5" diameter | | .320 | | 5.58 | 7.15 | 12.73 | 16.45 |
| 587       1 @ 6" diameter | | .080 | | 1.32 | 1.79 | 3.11 | 4.03 |
| 588       2 @ 6" diameter | | .160 | | 2.64 | 3.58 | 6.22 | 8.05 |
| 590       4 @ 6" diameter | | .320 | | 5.28 | 7.15 | 12.43 | 16.10 |
| 592       6 @ 6" diameter | | .533 | | 7.92 | 11.95 | 19.87 | 26 |
| 620       Rigid galvanized steel, 2 @ 2" diameter | | .089 | | 3.80 | 1.99 | 5.79 | 7.05 |
| 640       4 @ 2" diameter | | .178 | | 7.60 | 3.98 | 11.58 | 14.10 |
| 680       2 @ 3" diameter | | .160 | | 8.50 | 3.58 | 12.08 | 14.50 |
| 700       4 @ 3" diameter | | .320 | | 17 | 7.15 | 24.15 | 29 |
| 720       2 @ 4" diameter | | .229 | | 13.20 | 5.10 | 18.30 | 22 |
| 740       4 @ 4" diameter | | .471 | | 26 | 10.55 | 36.55 | 44 |
| 760       6 @ 4" diameter | | .727 | | 40 | 16.30 | 56.30 | 67 |
| 762       2 @ 5" diameter | | .267 | | 28 | 5.95 | 33.95 | 39 |
| 764       4 @ 5" diameter | | .533 | | 56 | 11.95 | 67.95 | 79 |
| 766       6 @ 5" diameter | | .889 | | 84 | 19.90 | 103.90 | 120 |
| 768       2 @ 6" diameter | | .400 | | 40 | 8.95 | 48.95 | 57 |
| 770       4 @ 6" diameter | | .800 | | 80 | 17.90 | 97.90 | 115 |
| 772       6 @ 6" diameter | | 1.140 | ▼ | 120 | 26 | 146 | 170 |
| 800       Fittings, PVC type EB, elbow, 2" diameter | | .500 | Ea. | 2.60 | 11.20 | 13.80 | 18.95 |
| 820       3" diameter | | .571 | | 4.15 | 12.80 | 16.95 | 23 |
| 840       4" diameter | | .667 | | 8.65 | 14.95 | 23.60 | 31 |
| 842       5" diameter | | .800 | | 21 | 17.90 | 38.90 | 49 |
| 844       6" diameter | ▼ | .889 | | 27 | 19.90 | 46.90 | 58 |
| 850       Coupling, 2" diameter | | | | .78 | | .78 | .85M |
| 860       3" diameter | | | ▼ | .93 | | .93 | 1.02M |

*Figure* 23.18

264

**FEEDER
SCHEDULE**

| JOB NO. | | | MC motor Feeders | | | SHEET | 9 of 19 |
|---|---|---|---|---|---|---|---|

OWNER **R.S. means**    ADDRESS **Kingston, Ma**    ESTIMATE NO. **E-86-002**

JOB DESCRIPTION    DRAWING NO.    DATE **1-3-86**

SCALE    TAKE OFF BY **PHD**    CHECKED BY **JM**

| PNL. NO. | CIRC. NO. | FROM | TO | HP / AMPS | RACEWAY | | | | | | | | | WIRE | | | | | | | |
|---|---|---|---|---|---|---|---|---|---|---|---|---|---|---|---|---|---|---|---|---|---|
| | | | | | SIZE | TYPE | L.F. | TERM | FLEX | FLEX CONN. | LB | | | NO. | SIZE | TYPE | VOLTS | RUN | L.F. | LF +10% | TERM |
| | 1 | MCC 1 | Chiller Pp 1A | 7½ / 11 | ½" | RGS | 40 | | 3' | 2 | | | | 3 | #12 | XHHW | 600 | 40 | 120 | 140 | 6 |
| | 2 | | Chiller Pp 1B | 7½ / 11 | ½" | | 40 | | 3' | 2 | | | | 3 | #12 | | | | 40 | 120 | 140 | 6 |
| | 3 | | Hot wtr. Pp 2A | 5 / 7.6 | ½" | | 20 | | 3' | 2 | | | | 3 | #12 | | | | 20 | 60 | 70 | 6 |
| | 4 | | Hot wtr Pp 2B | 5 / 7.6 | ½" | | 20 | | 3' | 2 | | | | 3 | #12 | | | | 20 | 60 | 70 | 6 |
| | 5 | | Cooling Twr. Pp 3 | 5 / 7.6 | ½" | | 50 | | 3' | 2 | | | | 3 | #12 | | | | 50 | 150 | 170 | 6 |
| | 6 | | Cooling Twr Fan | 25 / 34 | ¾" | | 80 | | 3' | 2 | 2 | | | 3 | #8 | | | | 80 | 240 | 270 | 6 |
| | 7 | | Chiller Comp | 100 / 124 | 1¼" | | 30 | | 3' | 2 | 1 | | | 3 | #1 | | | | 30 | 90 | 100 | 6 |
| | 8 | | Future | | | | | | | | | | | | | | | | | | |

Conduit
½" - 40,40,20,20,50 = (170)
¾" - (80)
1¼" - (30)

NOTES
Wire   #12 - 140,140,70,70,170 = (5.90) CLF
#8 - (2.70) CLF
#1 - (1.00) CLF

*Figure 23.19*

Although sizing wires and conduits in this manner may not be precise, it is generally accurate enough for estimating purposes.

Conduit and wire lengths are measured and calculated in the same manner as the distribution feeders. The conduit connection to each motor should allow for three feet of flexible metallic conduit (or sealtite) with connectors. Also, the power leads to the motors will require termination lugs for each conductor.

Next, the MCC and its components can be taken off directly onto the cost analysis sheets (see Figures 23.20 and 23.21).

Price and extend these items after transferring the quantities from the feeder schedule to the cost analysis sheets.

## Branch Wiring

The "owner" in the sample estimate is considering two alternative methods for branch power receptacles: an undercarpet cable system (type FCC) and poke-through receptacles. A bid proposal is required for each system.

### Fixed Receptacles:
A portion of the branch wiring system is fixed to the structure walls and columns. The estimating process may be simplified if this portion is taken off separately from the undercarpet system or the poke-through system. The fixed receptacles are taken off as shown in Figure 23.22. Note that 5% is added to the conduit total and 10% to the wire total. The wire total is based upon three wires per foot of conduit. The wire total is then changed from linear feet (LF) to hundred linear feet (CLF) to match the cost units. The takeoff quantities are then transferred from the quantity sheet to the cost analysis sheet (Figure 23.23), priced, and extended.

### Undercarpet System:
A composite drawing showing typical representations of both systems is shown in Figure 23.6. The items for the undercarpet branch power system are taken off and extended according to the methods used for the fixed receptacles discussed above (see Figure 23.24). The quantities are transferred to the cost analysis sheet (Figure 23.25) and the items are priced and extended.

### Poke-through System — Alternate:
The poke-through system will be the option. Since the two systems are being considered as equal alternatives, it must be assumed that the same density of receptacles will be needed for each system. The poke-through devices come as an assembled unit; as a result, fewer components need to be listed and priced. The takeoff for this system is shown in Figure 23.26.

The cost analysis sheet for the poke-through system is shown in Figure 23.27. When pricing and extending this sheet note the third item — "Core 3 inch hole". Costs are taken from line 2.1-20-030 of *Means Electrical Cost Data* (see Figure 23.28). A substantial portion of the crew cost will be for equipment. See crew B-89 in Figure 23.29. This crew includes costs for the drill, bit, and associated equipment.

**Means Forms**

**COST ANALYSIS**

PROJECT **Office Bldg.** **MCC and MCC Feeders**

ESTIMATE NO. **E-86-002**

ARCHITECT **R.S. Means Company, Inc.**

DATE **1-3-86**

TAKE OFF BY: **PHD**   QUANTITIES BY: **PHD**   PRICES BY: **PHD**   EXTENSIONS BY: **SM**   CHECKED BY: **JM**

| DESCRIPTION | SOURCE/DIMENSIONS | | | QUANTITY | UNIT | Man-Hours | | Material | | Installation | |
|---|---|---|---|---|---|---|---|---|---|---|---|
| | | | | | | UNIT COST | TOTAL | UNIT COST | TOTAL | UNIT COST | TOTAL |
| **Motor Cont. Ctr.** | | | | | | | | | | | |
| Sections, 72" h | 16.3 | 25 | 090 | 2 | Ea. | 10 | 20 | 830 | 1660 | 225 | 450 |
| Copper Bus, Add | | | 110 | 2 | | — | — | 105 | 210 | — | — |
| Pilot Lights, Add | | | 170 | 7 | | .5 | 4 | 63 | 441 | 11.20 | 78 |
| Push Buttons | | | 180 | 7 | | .5 | 4 | 42 | 294 | 11.20 | 78 |
| Size 1 Starters | | | 010 | 5 | | 2.96 | 15 | 505 | 2525 | 66 | 330 |
| Size 2 Starters | | | 020 | 1 | | 4.0 | 4 | 610 | 610 | 90 | 90 |
| Size 4 Starters | ↓ | ↓ | 040 | 1 | ↓ | 11.43 | 11 | 1725 | 1725 | 255 | 255 |
| **Conduit** | | | | | | | | | | | |
| 1/2" RGS | 16.0 | 20 | 175 | 170 | LF. | .089 | 15 | .54 | 92 | 1.99 | 338 |
| 3/4" RGS | | | 177 | 80 | LF. | .10 | 8 | .66 | 53 | 2.24 | 179 |
| 1 1/4" RGS | | ↓ | 183 | 30 | LF. | .133 | 4 | 1.35 | 41 | 2.99 | 90 |
| 1/2" Flex. | | 60 | 010 | 15' | LF. | .04 | 1 | .25 | 4 | .90 | 14 |
| 1/2" Flex. Connectors | | | 043 | 10 | Ea. | .10 | 1 | .34 | 3 | 2.24 | 22 |
| 3/4" Flex. | | | 020 | 3' | LF. | .050 | — | .32 | 1 | 1.12 | 3 |
| 3/4" Flex. Connectors | | | 044 | 2 | Ea. | .114 | — | .85 | 2 | 2.56 | 5 |
| 1 1/4" Flex. | | | 030 | 3' | LF. | .114 | — | .83 | 2 | 2.56 | 8 |
| 1 1/4" Flex. Connectors | | ↓ | 050 | 2 | Ea. | .20 | — | 4.25 | 9 | 4.48 | 9 |
| 3/4" LB Fittings | | 55 | 229 | 2 | Ea. | .615 | 1 | 3.69 | 7 | 13.80 | 28 |
| 1 1/4" LB Fittings | ↓ | 55 | 233 | 1 | Ea. | 1.0 | 1 | 7.70 | 8 | 22 | 22 |
| **Subtotals** | | | | | | | 89 | | 7687 | | 1999 |

Figure 23.20

**Means Forms**

**COST ANALYSIS**

PROJECT Office Bldg.     MCC and MCC Feeders     ESTIMATE NO. E-86-002

ARCHITECT R.S. Means Company, Inc.     DATE 1-3-86

TAKE OFF BY: PHD   QUANTITIES BY: PHD   PRICES BY: PHD   EXTENSIONS BY: SM   CHECKED BY: JM

| DESCRIPTION | SOURCE/DIMENSIONS | | | QUANTITY | UNIT | Man-Hours | | Material | | Installation | |
|---|---|---|---|---|---|---|---|---|---|---|---|
| | | | | | | UNIT COST | TOTAL | UNIT COST | TOTAL | UNIT COST | TOTAL |
| **Wire** | | | | | | | | | | | |
| #12 XHHW, 600V, ST, Cu | 16.1 | 10 | 302 | 5.9 | CLF | .727 | 4 | 7.15 | 42 | 16.30 | 96 |
| #8 | | | 306 | 2.7 | CLF | 1.0 | 3 | 14.30 | 39 | 22 | 59 |
| #1 | ↓ | ↓ | 314 | 1.0 | CLF | 2.0 | 2 | 48 | 48 | 45 | 45 |
| #12 Termination Lugs | 16.1 | 50 | 163 | 30 | Ea. | .160 | 5 | .14 | 4 | 3.58 | 107 |
| #8 | | | 178 | 6 | Ea. | .222 | 1 | 1.60 | 10 | 4.98 | 30 |
| #1 | ↓ | ↓ | 240 | 6 | Ea. | .400 | 2 | 2.85 | 17 | 8.95 | 54 |
| Subtotals | | | | | | | 17 | | 160 | | 391 |
| Subtotal Sheet 10 | | | | | | | 89 | | 7687 | | 1999 |
| Motor Control Ctr. Totals | | | | | | | 106 | | 7847 | | 2390 |

Figure 23.21

268

# Means Forms
## QUANTITY SHEET

### Fixed Receptacles Branch Wiring

PROJECT: Office Bldg.
LOCATION: Kingston, Ma.
ARCHITECT: R.S. Means Company, Inc.
TAKE OFF BY: PHD
EXTENSIONS BY: PHD

ESTIMATE NO. E-86-002
DATE: 1-3-86
CHECKED BY: JM

**Branch Wiring for Recept.**

| DESCRIPTION | NO. | DIMENSIONS | Garage | | 1st Floor | | 2nd Floor | | 3rd Floor | | Penthouse | | Totals +5% Cond. +10% Wire | |
|---|---|---|---|---|---|---|---|---|---|---|---|---|---|---|
| | | | UNIT | | UNIT | | UNIT | | UNIT | | UNIT | | UNIT | |
| Cast Box-FD, 1/2" Hub, 1 Gang | WP | | 5 Ea. | 5 | 2 Ea. → | 2 | | | | | | | 7 Ea. | 7 |
| W.P. Recpt. Cover | | | | 5 | | 2 | | | | | | | | 7 |
| GFI Recpt. | GF | | | 1 | | 3 | | 1 | | 1 | | | | 6 |
| Duplex Recpt., 20 A | Ø | | | 5 | | 80 | 80 Ea. | 80 | 80 Ea. | 80 | 12 Ea. → | 12 | | 257 |
| Handy Box | | | | 1 | | 1 | | | | | | 12 | | 13 |
| Recpt. Cover | | | | 1 | | 1 | | 1 | | 1 | | 12 | | 13 |
| 4" Sq. Boxes | | | | 1 | | 86 | 81 Ea. | 81 | 81 Ea. | 81 | | | | 248 |
| 4" Sq. Plaster Ring | | | | | | 81 | | 75 | | 75 | | | | 231 |
| Recpt. Cover - S.S. | | | | → | | 81 | | 75 → | | 75 → | | | | 231 |
| 1/2" EMT | | | 310 LF. | | 1120 LF. | | 1060 LF. | | 1060 LF. | | 180 LF. | | 3920 LF. | |
| 1/2" EMT Connector-Compression | | | 5 Ea. | | 2 Ea. | | 1 | | 1 | | 1 | | 7 Ea. | |
| 1/2" EMT Connector-Set Screw | | | 5 Ea. | | 160 Ea. | | 148 Ea. | | 146 Ea. | | 24 Ea. | | 483 Ea. | |
| #12 Wire THHN-Solid | | | 930 LF. | | 3360 LF. | | 3180 LF. | | 3180 LF. | | 540 LF. | | 12310 LF. -OR- 124 CLF. | |

Figure 23.22

# Means Forms

COST
ANALYSIS

PROJECT **Office Bldg.**    **Branch Wiring Fixed Recpt.**    ESTIMATE NO. **E-86-002**

ARCHITECT **R.S. Means Company, Inc.**    DATE **1-3-86**

TAKE OFF BY: **PHD**    QUANTITIES BY: **PHD**    PRICES BY: **PHD**    EXTENSIONS BY: **MG**    CHECKED BY: **JM**

| DESCRIPTION | SOURCE/DIMENSIONS | | | QUANTITY | UNIT | Man-Hours | | material | | Installation | |
|---|---|---|---|---|---|---|---|---|---|---|---|
| | | | | | | UNIT COST | TOTAL | UNIT COST | TOTAL | UNIT COST | TOTAL |
| **Boxes & Wiring Devices** | | | | | | | | | | | |
| Cast Box, FD, ½" Hub, 1 Gang. | 16.2 | 20 | 142 | 7 | Ea. | .667 | 5 | 7.95 | 56 | 14.95 | 105 |
| W.P. Recpt. Cover | | 20 | 160 | 7 | | .125 | 1 | 3.30 | 23 | 2.80 | 20 |
| GFI Recpt., 15A | | 30 | 248 | 6 | | .296 | 2 | 36 | 216 | 6.65 | 40 |
| Duplex Recpt., 20A | | 30 | 247 | 257 | | .296 | 76 | 6.40 | 1645 | 6.65 | 1709 |
| Handy Box | | 20 | 055 | 13 | | .296 | 4 | .87 | 11 | 6.65 | 86 |
| Cover, Recpt. | | 30 | 056 | 13 | | .125 | 2 | .31 | 4 | 2.80 | 36 |
| 4" Sq. Boxes | | 20 | 015 | 248 | | .400 | 99 | 1.07 | 265 | 8.95 | 2220 |
| 4" Sq. Plaster Rings | | 20 | 030 | 231 | | .125 | 29 | .63 | 146 | 2.80 | 647 |
| Recpt. Cover - S.S. | ↓ | 30 | 260 | 231 | ↓ | .1 | 23 | 2.10 | 485 | 2.24 | 517 |
| | | | | | | | | | | | |
| **Conduit** | | | | | | | | | | | |
| ½" EMT | 16.0 | 20 | 500 | 3,920 | L.F. | .047 | 184 | .23 | 902 | 1.05 | 4116 |
| ½" EMT Conn. - Compr. | | | 880 | 7 | Ea. | .067 | — | .61 | 4 | 1.49 | 10 |
| ½" EMT Conn. - Set Scr. | ↓ | ↓ | 650 | 483 | Ea. | .067 | 32 | .35 | 169 | 1.49 | 720 |
| | | | | | | | | | | | |
| **Wire** | | | | | | | | | | | |
| #12 THHN - Solid | 16.1 | 10 | 094 | 124 | C.L.F. | .727 | 90 | 3.30 | 409 | 16.30 | 2021 |
| | | | | | | | | | | | |
| Fixed Recpt. Total | | | | | | | 547 | | 4335 | | 12247 |

Figure 23.23

## Means Forms
### QUANTITY SHEET

| | |
|---|---|
| PROJECT | Office Bldg. |
| LOCATION | Kingston, Ma |
| TAKE OFF BY | PHD |
| ARCHITECT | R. S. Means Company, Inc. |
| EXTENSIONS BY | PHD |
| CHECKED BY | JM |

Undercarpet System Branch Wiring

ESTIMATE NO. E-86-002
DATE 1-3-86

| DESCRIPTION | Symbol | DIMENSIONS | Garage | UNIT | 1st Floor | UNIT | 2nd Floor | UNIT | 3rd Floor | UNIT | Penthouse | UNIT | Totals UNIT+5% Wire | UNIT |
|---|---|---|---|---|---|---|---|---|---|---|---|---|---|---|
| Undercarpet System | | | | | | | | | | | | | | |
| Receptacle, Dir. Connect | -O- | | | | 143 | Ea. | 153 | Ea. | 153 | Ea. | | | 448 | Ea. |
| Cable Tap | T | | | | 13 | | 13 | | 13 | | | | 39 | |
| Cable Fold | F | | | | 39 | | 39 | | 39 | | | | 117 | |
| Transition Wall Box | ⊙ | | | | 5 | | 5 | | 5 | | | | 15 | |
| Transition Block | | | | | 5 | | 5 | | 5 | | | | 15 | |
| Flat Cable /3-w/#12 | — | | | | 1420 | LF | 1500 | LF | 1500 | LF | | | 4640 | LF |
| Flat Cable /5-w/#12 | = | | | | 300 | | 270 | | 270 | | | | 880 | |
| Top Shield (3-w) | | | | | 1420 | | 1500 | | 1500 | | | | 4640 | |
| Top Shield (5-w) | | | | | 300 | | 270 | | 270 | | | | 880 | |
| Drill Floor (2 Per Recpt.) | | | | | 284 | Ea. | 306 | Ea. | 306 | Ea. | | | 896 | Ea. |
| Mark Floor | | | | | 1720 | LF. | 1770 | LF. | 1770 | LF. | | | 5260 | LF. |
| Hold-Down Tape (Roll) | | | | | 4300 | LF. | 4425 | LF. | 4425 | LF. | | | 13800 | LF. |
| Tape Primer (Can) | | | | | 4 | Ea. | 4 | Ea. | 4 | Ea. | | | 12 | Ea. |
| Splicing Tool | | | | | 1 | Ea. | | | | | | | 1 | Ea. |

Figure 23.24

## Means Forms
**COST ANALYSIS**

PROJECT: *Office Bldg.*  Undercarpet System (Fcc) Branch Wiring  ESTIMATE NO. E-86-002

ARCHITECT: R.S. Means Company, Inc  DATE 1-3-86

TAKE OFF BY: PHD  QUANTITIES BY: PHD  PRICES BY: PHD  EXTENSIONS BY: Sm  CHECKED BY: Jm

| DESCRIPTION | SOURCE/DIMENSIONS | | | QUANTITY | UNIT | man-Hours | | material | | Installation | |
|---|---|---|---|---|---|---|---|---|---|---|---|
| | | | | | | UNIT COST | TOTAL | UNIT COST | TOTAL | UNIT COST | TOTAL |
| **Undercarpet Pwr. Syst.** | | | | | | | | | | | |
| Recept., Direct Conn. | 16.0 | 93 | 086 | 448 | Ea. | .32 | 143 | .46 | 20608 | 7.15 | 3203 |
| Cable Tap, 3/c | | | 035 | 39 | | .2 | 8 | 9 | 351 | 4.48 | 175 |
| Insul. Patch; Tap, End | | | 040 | 78 | | .167 | 13 | 21 | 1638 | 3.73 | 291 |
| Cable Fold | | | 045 | 117 | | .035 | 4 | — | — | .78 | 91 |
| Top Shield; Tap, Fold | | | 050 | 156 | | .083 | 13 | .54 | 84 | 1.87 | 292 |
| Transition Wall Box | | | 100 | 15 | | .4 | 6 | 28 | 420 | 8.95 | 134 |
| Transition Block | | | 070 | 15 | | .104 | 2 | 14.20 | 213 | 2.33 | 35 |
| Flat Conductor Cable, 3/w, #12 | | | 010 | 4,640 | LF. | .004 | 19 | 2.13 | 9883 | .10 | 464 |
| F.C.C., 5/w, #12 | | | 145 | 880 | | .008 | 7 | 3.54 | 3115 | .18 | 158 |
| Top Shield, 3/w | | | 020 | 4,640 | | .005 | 23 | 2.27 | 10533 | .10 | 464 |
| Top Shield, 5/w | | | 155 | 880 | | .005 | 4 | 3.50 | 3080 | .10 | 88 |
| Drill Floor | | | 810 | 896 | Ea. | .05 | 45 | .15 | 134 | 1.12 | 1004 |
| Mark Floor | | | 820 | 5,260 | LF. | .005 | 26 | — | — | .11 | 579 |
| Hold-Down Tape | | | 830 | 13,800 | LF. | .001 | 14 | .06 | 828 | .03 | 414 |
| Tape Primer- Can | | | 835 | 12 | Ea. | .083 | 1 | 10.75 | 129 | 1.87 | 22 |
| Splicing Tool | | | 840 | 1 | Ea. | — | — | 165 | 165 | | |
| **Undercarpet Syst. Totals** | | | | | | | 328 | | 51181 | | 7414 |

Figure 23.25

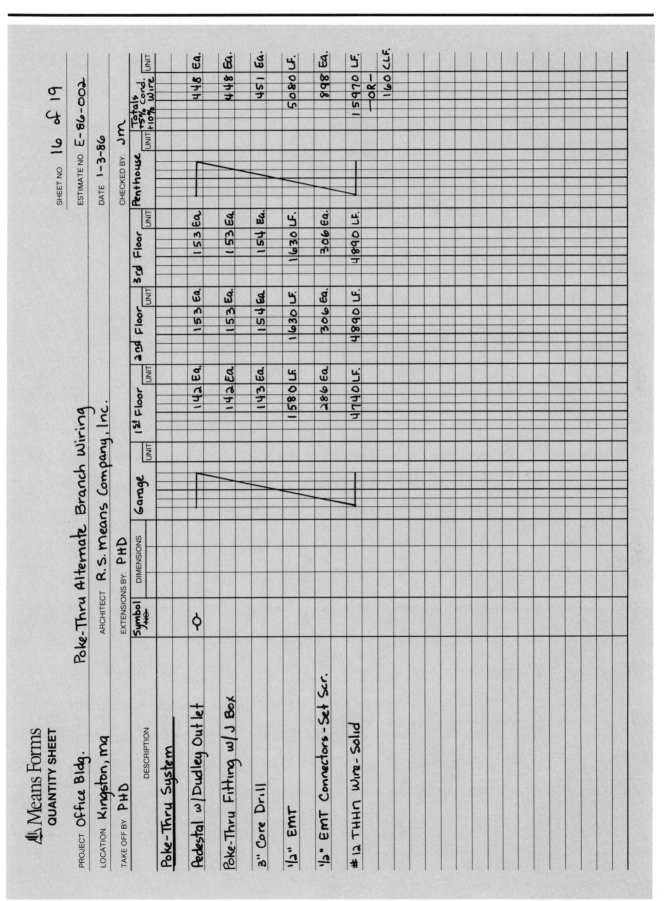

Figure 23.26

**COST ANALYSIS**

PROJECT **Office Bldg.** — **Poke-Thru-Alternate Branch Wiring**

ARCHITECT **R.S. Means Company, Inc.**

DATE **1-3-86**

TAKE OFF BY: **PHD**  QUANTITIES BY: **PHD**  PRICES BY: **PHD**  EXTENSIONS BY: **Sm**  CHECKED BY: **Jm**

| DESCRIPTION | SOURCE/DIMENSIONS | | | QUANTITY | UNIT | UNIT COST | TOTAL | UNIT COST | TOTAL | UNIT COST | TOTAL |
|---|---|---|---|---|---|---|---|---|---|---|---|
| Poke-Thru System | | | | | | | | | | | |
| Pedestal w/Duplex Recpt. | 16.2 | 20 | 214 | 448 | Ea. | .444 | 199 | 19.40 | 8691 | 9.95 | 4458 |
| Poke-Thru Fitting w/J.B. | 16.2 | 20 | 200 | 448 | Ea. | 1.18 | 529 | 4.90 | 2195 | 2.26 | 11648 |
| | | | | | | | | | | | |
| Cutting & Drilling | | | | | | | | | | | |
| Core 3" Hole | 2.1 | 20 | 030 | 451 | | .133 | 60 | | — | 16.10 | 7261 |
| | | | | | | | | | | | |
| Conduit | | | | | | | | | | | |
| 1/2" EMT | 16.0 | 20 | 500 | 5080 | LF. | .047 | 239 | .23 | 1168 | 1.05 | 5334 |
| 1/2" EMT Conn.- Set Scr. | 16.0 | 20 | 650 | 898 | Ea. | .067 | 60 | .35 | 314 | 1.49 | 1338 |
| | | | | | | | | | | | |
| Wire | | | | | | | | | | | |
| #12 THHN - Solid | 16.1 | 10 | 094 | 160 | C.LF. | .727 | 116 | 3.30 | 528 | 16.30 | 2608 |
| | | | | | | | | | | | |
| Poke-Thru Totals | | | | | | | 1203 | | 32653 | | 32647 |

Figure 23.27

| 2.1 Exploration & Clearing | CREW | MAN-HOURS | UNIT | BARE COSTS MAT. | BARE COSTS INST. | BARE COSTS TOTAL | TOTAL INCL O&P |
|---|---|---|---|---|---|---|---|
| 20-001 CORE DRILLING Reinforced concrete slab, up to 6" thick slab | | | | | | | |
| 002 | | | | | | | |
| 010   1" diameter core | B-89 | .107 | Ea. | | 12.85 | 12.85 | 14.85 |
| 015        Each added inch thick, add | | .013 | | | 1.61 | 1.61 | 1.86 |
| 030   3" diameter core | | .133 | | | 16.10 | 16.10 | 18.60 |
| 035        Each added inch thick, add | | .020 | | | 2.41 | 2.41 | 2.79 |
| 050   4" diameter core | | .178 | | | 21 | 21 | 25 |
| 055        Each added inch thick, add | | .027 | | | 3.22 | 3.22 | 3.72 |
| 070   6" diameter core | | .267 | | | 32 | 32 | 37 |
| 075        Each added inch thick, add | | .040 | | | 4.82 | 4.82 | 5.60 |
| 090   8" diameter core | | .333 | | | 40 | 40 | 46 |
| 095        Each added inch thick, add | | .053 | | | 6.45 | 6.45 | 7.45 |
| 110   10" diameter core | | .444 | | | 54 | 54 | 62 |
| 115        Each added inch thick, add | | .073 | | | 8.75 | 8.75 | 10.15 |
| 130   12" diameter core | | .533 | | | 64 | 64 | 74 |
| 135        Each added inch thick, add | | .089 | | | 10.70 | 10.70 | 12.40 |
| 150   14" diameter core | | .615 | | | 74 | 74 | 86 |
| 155        Each added inch thick, add | | .100 | | | 12.05 | 12.05 | 13.95 |
| 170   18" diameter core | | .762 | | | 92 | 92 | 105 |
| 175        Each added inch thick, add | ↓ | .133 | | | 16.10 | 16.10 | 18.60 |
| 176        For horizontal holes, add to above | | | ↓ | | 30% | | 30% |
| 177 Prestressed hollow core plank, 6" thick | | | | | | | |
| 178   1" diameter core | B-89 | .100 | Ea. | | 12.05 | 12.05 | 13.95 |
| 179        Each added inch thick, add | | .012 | | | 1.46 | 1.46 | 1.69 |
| 180   3" diameter core | | .123 | | | 14.85 | 14.85 | 17.15 |
| 181        Each added inch thick, add | | .018 | | | 2.19 | 2.19 | 2.54 |
| 182   4" diameter core | | .160 | | | 19.30 | 19.30 | 22 |
| 183        Each added inch thick, add | | .024 | | | 2.92 | 2.92 | 3.38 |
| 184   6" diameter core | | .229 | | | 28 | 28 | 32 |
| 185        Each added inch thick, add | | .036 | | | 4.38 | 4.38 | 5.05 |
| 186   8" diameter core | | .296 | | | 36 | 36 | 41 |
| 187        Each added inch thick, add | | .048 | | | 5.80 | 5.80 | 6.70 |
| 188   10" diameter core | | .400 | | | 48 | 48 | 56 |
| 189        Each added inch thick, add | | .067 | | | 8.05 | 8.05 | 9.30 |
| 190   12" diameter core | | .471 | | | 57 | 57 | 66 |
| 191        Each added inch thick, add | | .080 | | | 9.65 | 9.65 | 11.15 |
| 195 Minimum charge for above, 3" diameter core | | .889 | | | 105 | 105 | 125 |
| 200        4" diameter core | | .941 | | | 115 | 115 | 130 |
| 205        6" diameter core | | 1.030 | | | 125 | 125 | 145 |
| 210        8" diameter core | | 1.140 | | | 140 | 140 | 160 |
| 215        10" diameter core | | 1.330 | | | 160 | 160 | 185 |
| 220        12" diameter core | | 1.600 | | | 195 | 195 | 225 |
| 225        14" diameter core | ↓ | 1.850 | | | 225 | 225 | 260 |
| 230        18" diameter core | ↓ | 2.000 | ↓ | | 240 | 240 | 280 |

| | | | | BARE COSTS EQUIP. | BARE COSTS LABOR | BARE COSTS TOTAL | |
|---|---|---|---|---|---|---|---|
| 32-001 DUMP CHARGES Typical urban city, fees only | | | | | | | |
| 010        Building construction materials | | | C.Y. | | | | 3.85M |
| 020        Demolition lumber, trees, brush | | | | | | | 6.60M |
| 030        Rubbish only | | | ↓ | | | | 3.85M |
| 100   Reclamation station, usual charge | | | Ton | | | | 16.50M |
| 41-001 MOVING EQUIPMENT , remove and reset, 100' distance, | | | | | | | |
| 002        No obstructions, no assembly or leveling | | | | | | | |
| 010        Annealing furnace, 24' overall | B-67 | 4.000 | Ea. | 42 | 80 | 122 | 160 |
| 015 | | | | | | | |
| 020        Annealing oven, small | B-67 | 1.140 | Ea. | 12.10 | 23 | 35.10 | 46 |
| 024        Very large | | 16.000 | | 170 | 320 | 490 | 645 |
| 040        Band saw, small | ↓ | .640 | ↓ | 6.80 | 12.85 | 19.65 | 26 |
| 044        Large | | 1.140 | | 12.10 | 23 | 35.10 | 46 |

*Figure 23.28*

# CREWS

## Crew B-38

| Crew B-38 | Hr. | Daily | Hr. | Daily | Bare Costs | Incl. O&P |
|---|---|---|---|---|---|---|
| 2 Building Laborers | $15.90 | $254.40 | $23.05 | $368.80 | $17.08 | $24.76 |
| 1 Equip. Oper. (light) | 19.45 | 155.60 | 28.20 | 225.60 | | |
| 1 Backhoe Loader, 48 H.P. | | 156.00 | | 171.60 | | |
| 1 Demol. Hammer, Hyd. | | 212.45 | | 233.70 | 15.35 | 16.88 |
| 24 M.H., Daily Totals | | $778.45 | | $999.70 | $32.43 | $41.64 |

## Crew B-39

| Crew B-39 | Hr. | Daily | Hr. | Daily | Bare Costs | Incl. O&P |
|---|---|---|---|---|---|---|
| 1 Labor Foreman (outside) | $17.90 | $143.20 | $25.95 | $207.60 | $16.82 | $24.39 |
| 4 Building Laborers | 15.90 | 508.80 | 23.05 | 737.60 | | |
| 1 Equipment Oper. (light) | 19.45 | 155.60 | 28.20 | 225.60 | | |
| 1 Air Compr., 250 C.F.M. | | 111.40 | | 122.55 | | |
| Air Tools & Accessories | | 25.80 | | 28.40 | | |
| 2-50 Ft. Air Hoses, 1.5" Dia | | 13.20 | | 14.50 | 3.13 | 3.44 |
| 48 M.H., Daily Totals | | $958.00 | | $1336.25 | $19.95 | $27.83 |

## Crew B-53

| Crew B-53 | Hr. | Daily | Hr. | Daily | Bare Costs | Incl. O&P |
|---|---|---|---|---|---|---|
| 1 Equip. Oper. (light) | $19.45 | $155.60 | $28.20 | $225.60 | $19.45 | $28.20 |
| 1 Trencher, Chain, 12 H.P. | | 135.40 | | 148.95 | 16.92 | 18.61 |
| 8 M.H., Daily Totals | | $291.00 | | $374.55 | $36.37 | $46.81 |

## Crew B-54

| Crew B-54 | Hr. | Daily | Hr. | Daily | Bare Costs | Incl. O&P |
|---|---|---|---|---|---|---|
| 1 Equip. Oper. (light) | $19.45 | $155.60 | $28.20 | $225.60 | $19.45 | $28.20 |
| 1 Trencher, Chain, 40 H.P. | | 199.80 | | 219.80 | 24.97 | 27.47 |
| 8 M.H., Daily Totals | | $355.40 | | $445.40 | $44.42 | $55.67 |

## Crew B-67

| Crew B-67 | Hr. | Daily | Hr. | Daily | Bare Costs | Incl. O&P |
|---|---|---|---|---|---|---|
| 1 Millwright | $20.75 | $166.00 | $29.35 | $234.80 | $20.10 | $28.77 |
| 1 Equip. Oper. (light) | 19.45 | 155.60 | 28.20 | 225.60 | | |
| 1 Forklift | | 169.40 | | 186.35 | 10.58 | 11.64 |
| 16 M.H., Daily Totals | | $491.00 | | $646.75 | $30.68 | $40.41 |

## Crew B-68

| Crew B-68 | Hr. | Daily | Hr. | Daily | Bare Costs | Incl. O&P |
|---|---|---|---|---|---|---|
| 2 Millwrights | $20.75 | $332.00 | $29.35 | $469.60 | $20.31 | $28.96 |
| 1 Equip. Oper. (light) | 19.45 | 155.60 | 28.20 | 225.60 | | |
| 1 Forklift | | 169.40 | | 186.35 | 7.05 | 7.76 |
| 24 M.H., Daily Totals | | $657.00 | | $881.55 | $27.36 | $36.72 |

## Crew B-89

| Crew B-89 | Hr. | Daily | Hr. | Daily | Bare Costs | Incl. O&P |
|---|---|---|---|---|---|---|
| 1 Equip. Oper. (light) | $19.45 | $155.60 | $28.20 | $225.60 | $19.45 | $28.20 |
| 1 Stake Body, 3 Ton | | 648.00 | | 712.80 | | |
| 1 Air Compressor | | 51.40 | | 56.55 | | |
| 1 Water Tank, 65 Gal. | | 18.45 | | 20.30 | | |
| 1 Generator, 10 K.W. | | 54.60 | | 60.05 | | |
| 1 Concrete Saw | | 36.60 | | 40.25 | 101.13 | 111.24 |
| 8 M.H., Daily Totals | | $964.65 | | $1115.55 | $120.58 | $139.44 |

## Crew D-1

| Crew D-1 | Hr. | Daily | Hr. | Daily | Bare Costs | Incl. O&P |
|---|---|---|---|---|---|---|
| 1 Bricklayer | $20.50 | $164.00 | $29.20 | $233.60 | $18.25 | $26.00 |
| 1 Bricklayer Helper | 16.00 | 128.00 | 22.80 | 182.40 | | |
| 16 M.H., Daily Totals | | $292.00 | | $416.00 | $18.25 | $26.00 |

## Crew E-4

| Crew E-4 | Hr. | Daily | Hr. | Daily | Bare Costs | Incl. O&P |
|---|---|---|---|---|---|---|
| 1 Struc. Steel Foreman | $23.70 | $189.60 | $37.25 | $298.00 | $22.20 | $34.88 |
| 3 Struc. Steel Workers | 21.70 | 520.80 | 34.10 | 818.40 | | |
| 1 Gas Welding Machine | | 56.80 | | 62.50 | 1.77 | 1.95 |
| 32 M.H., Daily Totals | | $767.20 | | $1178.90 | $23.97 | $36.83 |

## Crew F-2

| Crew F-2 | Hr. | Daily | Hr. | Daily | Bare Costs | Incl. O&P |
|---|---|---|---|---|---|---|
| 2 Carpenters | $20.00 | $320.00 | $29.00 | $464.00 | $20.00 | $29.00 |
| Power Tools | | 17.20 | | 18.90 | 1.07 | 1.18 |
| 16 M.H., Daily Totals | | $337.20 | | $482.90 | $21.07 | $30.18 |

## Crew L-1

| Crew L-1 | Hr. | Daily | Hr. | Daily | Bare Costs | Incl. O&P |
|---|---|---|---|---|---|---|
| 1 Electrician | $22.40 | $179.20 | $32.20 | $257.60 | $22.47 | $32.40 |
| 1 Plumber | 22.55 | 180.40 | 32.60 | 260.80 | | |
| 16 M.H., Daily Totals | | $359.60 | | $518.40 | $22.47 | $32.40 |

## Crew L-3

| Crew L-3 | Hr. | Daily | Hr. | Daily | Bare Costs | Incl. O&P |
|---|---|---|---|---|---|---|
| 1 Carpenter | $20.00 | $160.00 | $29.00 | $232.00 | $21.27 | $30.83 |
| .5 Electrician | 22.40 | 89.60 | 32.20 | 128.80 | | |
| .5 Sheet Metal Worker | 22.70 | 90.80 | 33.15 | 132.60 | | |
| 16 M.H., Daily Totals | | $340.40 | | $493.40 | $21.27 | $30.83 |

## Crew L-4

| Crew L-4 | Hr. | Daily | Hr. | Daily | Bare Costs | Incl. O&P |
|---|---|---|---|---|---|---|
| 2 Skilled Workers | $20.50 | $328.00 | $29.90 | $478.40 | $18.85 | $27.53 |
| 1 Helper | 15.55 | 124.40 | 22.80 | 182.40 | | |
| 24 M.H., Daily Totals | | $452.40 | | $660.80 | $18.85 | $27.53 |

## Crew L-6

| Crew L-6 | Hr. | Daily | Hr. | Daily | Bare Costs | Incl. O&P |
|---|---|---|---|---|---|---|
| 1 Plumber | $22.55 | $180.40 | $32.60 | $260.80 | $22.50 | $32.46 |
| .5 Electrician | 22.40 | 89.60 | 32.20 | 128.80 | | |
| 12 M.H., Daily Totals | | $270.00 | | $389.60 | $22.50 | $32.46 |

## Crew L-7

| Crew L-7 | Hr. | Daily | Hr. | Daily | Bare Costs | Incl. O&P |
|---|---|---|---|---|---|---|
| 2 Carpenters | $20.00 | $320.00 | $29.00 | $464.00 | $19.17 | $27.75 |
| 1 Building Laborer | 15.90 | 127.20 | 23.05 | 184.40 | | |
| .5 Electrician | 22.40 | 89.60 | 32.20 | 128.80 | | |
| 28 M.H., Daily Totals | | $536.80 | | $777.20 | $19.17 | $27.75 |

## Crew L-9

| Crew L-9 | Hr. | Daily | Hr. | Daily | Bare Costs | Incl. O&P |
|---|---|---|---|---|---|---|
| 1 Labor Foreman (inside) | $16.40 | $131.20 | $23.75 | $190.00 | $18.02 | $26.67 |
| 2 Building Laborers | 15.90 | 254.40 | 23.05 | 368.80 | | |
| 1 Struc. Steel Worker | 21.70 | 173.60 | 34.10 | 272.80 | | |
| .5 Electrician | 22.40 | 89.60 | 32.20 | 128.80 | | |
| 36 M.H., Daily Totals | | $648.80 | | $960.40 | $18.02 | $26.67 |

## Crew Q-1

| Crew Q-1 | Hr. | Daily | Hr. | Daily | Bare Costs | Incl. O&P |
|---|---|---|---|---|---|---|
| 1 Plumber | $22.55 | $180.40 | $32.60 | $260.80 | $20.29 | $29.35 |
| 1 Plumber Apprentice | 18.04 | 144.32 | 26.10 | 208.80 | | |
| 16 M.H., Daily Totals | | $324.72 | | $469.60 | $20.29 | $29.35 |

## Crew Q-2

| Crew Q-2 | Hr. | Daily | Hr. | Daily | Bare Costs | Incl. O&P |
|---|---|---|---|---|---|---|
| 2 Plumbers | $22.55 | $360.80 | $32.60 | $521.60 | $21.04 | $30.43 |
| 1 Plumber Apprentice | 18.04 | 144.32 | 26.10 | 208.80 | | |
| 24 M.H., Daily Totals | | $505.12 | | $730.40 | $21.04 | $30.43 |

## Crew Q-3

| Crew Q-3 | Hr. | Daily | Hr. | Daily | Bare Costs | Incl. O&P |
|---|---|---|---|---|---|---|
| 1 Plumber Foreman (ins) | $23.05 | $184.40 | $33.35 | $266.80 | $21.54 | $31.16 |
| 2 Plumbers | 22.55 | 360.80 | 32.60 | 521.60 | | |
| 1 Plumber Apprentice | 18.04 | 144.32 | 26.10 | 208.80 | | |
| 32 M.H., Daily Totals | | $689.52 | | $997.20 | $21.54 | $31.16 |

## Crew Q-5

| Crew Q-5 | Hr. | Daily | Hr. | Daily | Bare Costs | Incl. O&P |
|---|---|---|---|---|---|---|
| 1 Steamfitter | $22.75 | $182.00 | $32.90 | $263.20 | $20.47 | $29.60 |
| 1 Steamfitter Apprentice | 18.20 | 145.60 | 26.30 | 210.40 | | |
| 16 M.H., Daily Totals | | $327.60 | | $473.60 | $20.47 | $29.60 |

*Figure 23.29*

## Fire Detection System

There are several special systems which could be included in a building of this type. Telephone, burglar alarm, heat tracing, and sound systems are among them. For the purpose of this sample estimate, all of these systems are either furnished and installed by others or not used at all. Only the fire detection system is included in the scope of work.

Although some items of a fire detection system are shown in Figure 23.3, a full set of drawings would have additional details and specifications. Figure 23.30 shows the takeoff for this system. To maintain consistency, this takeoff sheet is arranged in a fashion similar to that of the branch wiring takeoff, i.e., by floors from the basement to the penthouse.

The quantities are transferred to the cost analysis sheet in Figure 23.31 where they are priced and extended. These special systems conclude the takeoff phase of the sample estimate.

## The Estimate Summary

**The Recap Sheet:** It is generally a good practice to list the cost totals from each segment of the work onto one sheet of paper often called a "Recap". This procedure will allow the estimator to review all the segments and to add them into a total contractor's cost. Any alternates are usually listed and calculated on this sheet as shown in Figure 23.32. The scheduling estimate can be drawn up from this recap sheet.

**The Project Schedule:** When the work for all sections is priced, the estimator should complete the Project Schedule so that time-related costs in the Project Overhead Summary can be determined. When preparing the schedule, the estimator must visualize the entire construction process in order to determine the correct sequence of work. Certain tasks must be completed before others are begun. Different trades will work simultaneously. Material deliveries will also affect scheduling. All such variables must be incorporated into the Project Schedule. An example is shown in Figure 23.33. The man-hour figures, which have been calculated for each section are used to assist with scheduling. The estimator must be careful to not only use the man-hours for each section independently, but to coordinate each section with related work.

The schedule shows that the project will last approximately one year. The duration of the electrical work will be about six months. Time-dependent items, such as equipment rental and superintendent costs, can be included in the Project Overhead Summary. Some items, such as permits and insurance, are dependent on total job costs. The total direct costs for the project can be determined as shown in the Estimate Summary (Figures 23.34 and 23.35). All costs can now be included.

**The Bottom Line:** The estimator is now able to complete the Estimate Summary as shown. Appropriate contingency, sales tax, and overhead and profit costs must be added to the direct costs of the project. Ten percent is added to material. The overhead and profit percentage of 43.8% for labor is obtained from Figure 6.1 as the average mark-up for electricians. Contractors should determine appropriate mark-ups for their own companies as discussed in Part 1, Chapter 7 of this book. Finally, the totals represent national average prices, and should be adjusted with the City Cost Index for your locality.

# Means Forms
## QUANTITY SHEET

PROJECT: Office Bldg.    Fire Detection Alarm System

LOCATION: Kingston, Ma

ARCHITECT: R.S. Means Company, Inc

TAKE OFF BY: PHD    EXTENSIONS BY: PHD

ESTIMATE NO. E-86-002

DATE 1-3-86

CHECKED BY: JM

| DESCRIPTION | Symbol No. | DIMENSIONS | Basement | UNIT | 1st Floor | UNIT | 2nd Floor | UNIT | 3rd Floor | UNIT | Penthouse | UNIT | Totals +5% Cond. +10% Wire | UNIT |
|---|---|---|---|---|---|---|---|---|---|---|---|---|---|---|
| **Fire Alarm Components** | | | | | | | | | | | | | | |
| Control Panel | ⬭ | | 1 | EA. | | | | | | | | | 1 | EA. |
| Break-Glass Station | F | | 1 | EA. | 3 | EA. | 3 | EA. | 3 | EA. | 1 | EA. | 11 | EA. |
| Light & Horn | FX | | 1 | EA. | 3 | EA. | 3 | EA. | 3 | EA. | 1 | EA. | 11 | EA. |
| Master Box | E | | 1 | | | | | | | | | | 1 | EA. |
| Annunciator | | | 1 | | 1 | EA. | 1 | | 1 | | 1 | | 1 | EA. |
| Smoke Detector | | | 6 | EA. | 24 | EA. | 24 | EA. | 24 | EA. | 5 | EA. | 83 | EA. |
| **Raceways and Boxes** | | | | | | | | | | | | | | |
| 1/2" EMT | | | 250 | LF. | 920 | LF. | 920 | LF. | 920 | LF. | 110 | LF. | 3280 | LF. |
| 1/2" EMT Connectors - Compr. | | | 18 | EA. | 64 | EA. | 60 | EA. | 60 | EA. | 14 | EA. | 216 | EA. |
| 4" Sq. Box | | | 9 | EA. | 32 | EA. | 30 | EA. | 30 | EA. | 7 | EA. | 108 | EA. |
| 4" Sq. Plaster Rings | | | 9 | EA. | 32 | EA. | 30 | EA. | 30 | EA. | 7 | EA. | 108 | EA. |
| **Conductors** | | | | | | | | | | | | | | |
| Fire Alarm, #18, 1 Pr. | | | 30 | LF. | 1000 | LF. | 1000 | LF. | 1000 | LF. | 350 | LF. | 3720 LF (or 37 CLF) | LF. |
| Fire Alarm, #18, 6 Pr. | | | 280 | LF. | 500 | LF. | 250 | LF. | 280 | LF. | 60 | LF. | 1510 LF (or 15 CLF) | LF. |

Figure 23.30

# Means Forms
**COST ANALYSIS**

| | | | |
|---|---|---|---|
| PROJECT: Office Building — Fire Detection System | | ESTIMATE NO. E-86-002 | |
| ARCHITECT: R.S. means Company, Inc | | DATE 1-3-86 | |
| TAKE OFF BY: PHD | QUANTITIES BY: PHD | PRICES BY: PHD | EXTENSIONS BY: Sm | CHECKED BY: Jm |

| DESCRIPTION | SOURCE/DIMENSIONS | | | QUANTITY | UNIT | man-Hours | | material | | Installation | |
|---|---|---|---|---|---|---|---|---|---|---|---|
| | | | | | | UNIT COST | TOTAL | UNIT COST | TOTAL | UNIT COST | TOTAL |
| **Fire Alarm Comp.** | | | | | | | | | | | |
| Control Panel | 16.8 | 15 | 400 | 1 | Ea. | 12.12 | 12 | 1325 | 1325 | 270 | 270 |
|   Battery & Rack | | | 420 | 1 | | 2.0 | 2 | 530 | 530 | 45 | 45 |
| Break-Glass Station | | | 700 | 11 | | 1.0 | 11 | 27 | 297 | 22 | 242 |
|   Light & Horn | | | 560 | 11 | | 1.51 | 17 | 75 | 825 | 34 | 374 |
| Master Box | | | 680 | 1 | | 2.96 | 3 | 1150 | 1150 | 66 | 66 |
| Annunciator, 12 zone | | | 800 | 1 | | 6.15 | 6 | 210 | 210 | 140 | 140 |
| Smoke Detector | ↓ | ↓ | 520 | 83 | ↓ | 1.29 | 107 | 64 | 5312 | 29 | 2407 |
| **Raceways & Boxes** | | | | | | | | | | | |
| ½" EMT | 16.0 | 20 | 500 | 3280 | LF. | .047 | 154 | .23 | 754 | 1.05 | 3444 |
| ½" EMT Conn-Compr. | 16.0 | 20 | 890 | 216 | Ea. | .067 | 14 | .78 | 168 | 1.49 | 322 |
| 4" Sq. Box | 16.2 | 20 | 015 | 108 | | .4 | 43 | 1.07 | 116 | 8.95 | 967 |
| 4" Sq. Plaster Rings | 16.2 | 20 | 030 | 108 | ↓ | .125 | 14 | .63 | 68 | 2.80 | 302 |
| **Conductors** | | | | | | | | | | | |
| Fire Alarm, #18, 1 Pr. | 16.1 | 72 | 185 | 37 | C.L.F. | 1.0 | 37 | 57 | 2109 | 22 | 814 |
| Fire Alarm, #18, 6 Pr. | 16.1 | 72 | 200 | 15 | C.L.F. | 2.0 | 30 | 220 | 3300 | 45 | 675 |
| Fire Protection Total | | | | | | | 450 | | 16164 | | 10068 |

Figure 23.31

## Means Forms
**COST ANALYSIS**

PROJECT **Office Bldg.**

ESTIMATE NO. **E-86-002**

ARCHITECT **R.S. Means Company, Inc.**

DATE **1-3-86**

TAKE OFF BY: **PHD**    QUANTITIES BY: **PHD**    PRICES BY: **PHD**    EXTENSIONS BY: **SM**    CHECKED BY: **JM**

| Segment DESCRIPTION | SOURCE/DIMENSIONS | | | Sheet no. QUANTITY | UNIT | man-Hours | | material | | Installation | |
|---|---|---|---|---|---|---|---|---|---|---|---|
| | | | | | | UNIT COST | TOTAL | UNIT COST | TOTAL | UNIT COST | TOTAL |
| Lighting | | | | 3 | | | 2561 | | 54123 | | 57304 |
| Service & Distribution (Riser) | | | | 8 | | | 860 | | 42118 | | 19244 |
| MCC and Feeders | | | | 11 | | | 106 | | 7847 | | 2390 |
| Branch Power Wiring Fixed Recpt. | | | | 13 | | | 547 | | 4335 | | 12247 |
| Undercarpet Recpt. | | | | 15 | | | 328 | | 51181 | | 7414 |
| Fire Detection | | | | 19 | | | 450 | | 16164 | | 10068 |
| Totals w/Undercarpet | | | | | | | 4852 | | 175768 | | 108667 |
| Alternate w/Poke-Thru | | | | | | | | | | | |
| Deduct- Undercarpet | | | | 15 | | | ⟨328⟩ | | ⟨51181⟩ | | ⟨7414⟩ |
| Add- Poke-Thru | | | | 17 | | | 1203 | | 32653 | | 32647 |
| Totals w/Poke-Thru | | | | | | | 5727 | | 157240 | | 133900 |

Figure 23.32

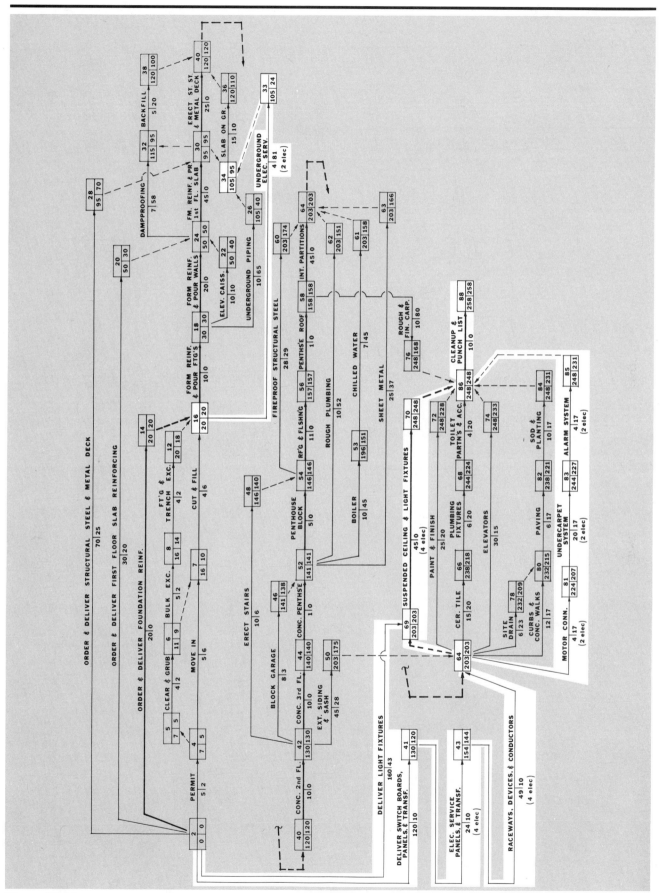

Figure 23.33

281

## Means Forms

**COST ANALYSIS**

PROJECT **Office Bldg.**

ESTIMATE NO. **E-86-002**

ARCHITECT **R.S. Means Company, Inc.**

DATE **1-3-86**

TAKE OFF BY: **PHD**    QUANTITIES BY: **PHD**    PRICES BY: **PHD**    EXTENSIONS BY: **Sm**    CHECKED BY: **Jm**

| DESCRIPTION | SOURCE/DIMENSIONS | | | QUANTITY | UNIT | material | | Installation | | Total | |
|---|---|---|---|---|---|---|---|---|---|---|---|
| | | | | | | UNIT COST | TOTAL | UNIT COST | TOTAL | UNIT COST | TOTAL |
| With Undercarpet | | | | | | | | | | | |
| | | | | | | | | | | | |
| Direct Cost Totals From Recap Sheet | | | | | | | 175768 | | 108667 | | |
| | | | | | | | | | | | |
| Storage Tlr. | 1.1 | 58 | 552 | 6 | mo. | 75 | 450 | | | | |
| Small Tools For Labor | 1.1 | 48 | 010 | Labor | % | 1 | 1087 | | | | |
| Toilets | 1.5 | 15 | 641 | 2@6 | mo. | 73 | 876 | | | | |
| Pick-Up Truck | 1.5 | 15 | 710 | 6 | mo. | 380 | 2280 | | | | |
| | | | | | | | | | | | |
| material Subtotal | | | | | | | 180461 | | | | |
| | | | | | | | | | | | |
| Sales Tax | | | | mtl. | % | 5 | 9023 | | | | |
| | | | | | | | 189484 | | | | |
| | | | | | | | | | | | |
| material Handling | | | | mtl. | % | 10 | 18948 | | | | |
| | | | | | | | | | | | |
| material Total | | | | | | | 208432 | | | 95.5% | 199053 |
| | | | | | | | | | | | |
| Contractor's OH & P | | | | Labor | % | | | 43.8 | 47596 | | |
| | | | | | | | | | | | |
| Labor Total | | | | | | | | | 156263 | 116.8% | 182515 |
| | | | | | | | | | | | |
| Bid Price with Undercarpet | | | | | | | | | | | 381568 |

*(margin note, vertical: Adjust To Boston ↓)*

Figure 23.34

## Means Forms

**COST ANALYSIS**

PROJECT **Office Building**

ARCHITECT **R.S. Means Company, Inc.**

TAKE OFF BY: **PHD**  QUANTITIES BY: **PHD**  PRICES BY: **PHD**  EXTENSIONS BY: **SM**  CHECKED BY: **JM**

SHEET NO. **Alternate Estimate Summary**

ESTIMATE NO. **E-86-002**

DATE **1-3-86**

| DESCRIPTION | SOURCE/DIMENSIONS | | | QUANTITY | UNIT | Material UNIT COST | Material TOTAL | Installation UNIT COST | Installation TOTAL | Total UNIT COST | Total TOTAL |
|---|---|---|---|---|---|---|---|---|---|---|---|
| **With Poke-Thru** | | | | | | | | | | | |
| Direct Cost Totals From Recap Sheet | | | | | | | 157240 | | 133900 | | |
| Storage | 1.1 | 58 | 552 | 6 | mo. | 75 | 450 | | | | |
| Small Tools for Labor | 1.1 | 48 | 010 | Labor | % | 1 | 1339 | | | | |
| Toilets | 1.5 | 15 | 641 | 2@6 | mo. | 73 | 876 | | | | |
| Pick-Up Truck | 1.5 | 15 | 710 | 6 | mo. | 380 | 2280 | | | | |
| Material Subtotal | | | | | | | 162185 | | | | |
| Sales Tax | | | | mtl. | % | 5 | 8109 | | | | |
| | | | | | | | 170294 | | | | |
| Material Handling | | | | mtl. | % | 10 | 17029 | | | | |
| Material Total | | | | | | | 187323 | | | 9557 | 178893 |
| Contractor's OH&P | | | | Labor | % | | | 43.8 | 58648 | | |
| Labor Total | | | | | | | | | 192548 | 116.87 | 224896 |
| Bid Price with Poke-Thru Alternate | | | | | | | | | | | 403789 |

Figure 23.35

283

# APPENDIX

# APPENDIX

Frequently, drawings and specifications do not contain complete, detailed information for the estimator. The design must therefore be interpreted as to sizes, capacities, and/or equipment requirements in order to identify and select unit prices for the estimate.

This section contains the following reference information as an aid to the electrical estimator:

- An extensive list of legend symbols and abbreviations. These designations are commonly used and accepted on electrical plan drawings, schematic diagrams, wiring diagrams, and takeoff forms.
- Charts and tables to determine weights and volumes, ampacities, and conduit and wire sizes.
- A listing of NEMA enclosure requirement designations for various environments.
- The electrical pages from R.S. Means' "Spec-Aid", reproduced as an example form for surveying the scope of work.
- A comprehensive list of abbreviations used throughout the construction industry.

## Lighting Outlets

○    Ceiling Surface Incandescent Fixture

⊢○    Wall Surface Incandescent Fixture

Ⓡ    Ceiling Recess Incandescent Fixture

⊢Ⓡ    Wall Recess Incandescent Fixture

ᴬ○₃ᵦ    Standard Designation for All Lighting Fixtures – A = Fixture Type, 3 = Circuit Number, b = Switch Control

Ⓑ    Ceiling Blanked Outlet

⊢Ⓑ    Wall Blanked Outlet

Ⓔ    Ceiling Electrical Outlet

⊢Ⓔ    Wall Electrical Outlet

Ⓙ    Ceiling Junction Box

⊢Ⓙ    Wall Junction Box

Ⓛ₍ₚₛ₎    Ceiling Lamp Holder with Pull Switch

⊢Ⓛ₍ₚₛ₎    Wall Lamp Holder with Pull Switch

Ⓛ    Ceiling Outlet Controlled by Low Voltage Switching when Relay is Installed in Outlet Box

⊢Ⓛ    Wall Outlet – Same as Above

◇    Outlet Box with Extension Ring

[EX]→    Exit Sign with Arrow as Indicated

[○]    Surface Fluorescent Fixture

[○]ₚ    Pendant Fluorescent Fixture

[OR]    Recessed Fluorescent Fixture

[○]    Wall Surface Fluorescent Fixture

⊢[---]⊣    Channel Mounted Fluorescent Fixture

[○ | ]    Surface or Pendant Continous Row Fluorescent Fixtures

[OR | ]    Recessed Continuous Row Fluorescent Fixtures

●    Incandescent Fixture on Emergency Circuit

[ ● ]    Fluorescent Fixture on Emergency Circuit

## Receptacle Outlets

⊢◯    Single Receptacle Outlet

⊢◉    Duplex Receptacle Outlet

⊢◉ˣ    Duplex Receptacle Outlet "X" Indicates above Counter Max. Height = 42" or above Counter

⊢◉_WP    Weatherproof Receptacle Outlet

⊢⊕    Triplex Receptacle Outlet

⊢⊕    Quadruplex Receptacle Outlet

⊢⊖    Duplex Receptacle Outlet – Split Wired

⊢⊕    Triplex Receptacle Outlet – Split Wired

⊢◬    Single Special Purpose Receptacle Outlet

⊢◬    Duplex Special Purpose Receptacle Outlet

⊢◉_R    Range Outlet

⊢▲_DW    Special Purpose Connection – Dishwasher

⊢◉_XP    Explosion-proof Receptacle Outlet Max. Height = 36" to ℄

⊢◉    Multi-outlet Assembly

Ⓒ    Clock Hanger Receptacle

Ⓕ    Fan Hanger Receptacle

⊖    Floor Single Receptacle Outlet

## Receptacle Outlets (Cont).

⊖ Floor Duplex Receptacle Outlet

△ Floor Special Purpose Outlet

◄ Floor Telephone Outlet – Public

◁ Floor Telephone Outlet – Private

Underfloor Duct and Junction Box for Triple, Double, or Single Duct System as Indicated by Number of Parallel Lines

Cellular Floor Header Duct

## Switch Outlets

$S$ — Single Pole Switch
Max. Height = 42" to ℄

$S_2$ — Double Pole Switch

$S_3$ — Three-way Switch

$S_4$ — Four-way Switch

$S_D$ — Automatic Door Switch

$S_K$ — Key Operated Switch

$S_P$ — Switch & Pilot Lamp

$S_{CB}$ — Circuit Breaker

$S_{WCB}$ — Weatherproof Circuit Breaker

$S_{MC}$ — Momentary Contact Switch

$S_{RC}$ — Remote Control Switch (Receiver)

$S_{WP}$ — Weatherproof Switch

$S_F$ — Fused Switch

$S_L$ — Switch for Low Voltage Switching System

$S_{LM}$ — Master Switch for Low Voltage Switching System

$S_T$ — Time Switch

$S_{TH}$ — Thermal Rated Motor Switch

$S_{DM}$ — Incandescent Dimmer Swtich

$S_{FDM}$ — Fluorescent Dimmer Switch

⊖$_S$ — Switch & Single Receptacle

⊖$_S$ — Switch & Double Receptacle

⊖$_A$
$S_A$ — Special Outlet Circuits

## Institutional, Commercial, & Industrial System Outlets

⬡ Nurses Call System Devices – any Type

◇ Paging System Devices – any Type

□ Fire Alarm System Devices – any Type

F Fire Alarm Manual Station – Max. Height = 48" to ℄

F Fire Alarm Horn with Integral Warning Light

⊗ Fire Alarm Thermodetector, Fixed Temperature

Ⓢ Smoke Detector

⊗ Fire Alarm Thermodetector, Rate of Rise

F Fire Alarm Master Box – Max. Height per Fire Department

H Magnetic Door Holder

ANN Fire Alarm Annunciator

◇ Staff Register System – any Type

🕐 Electrical Clock System Devices – any Type

◄ Public Telephone System Devices

289

## Institutional, Commercial, & Industrial System Outlets (Cont).

—◁ Private Telephone System Devices

—⬠ Watchman System Devices

—◁ᴸ Sound System, L = Speaker, V = Volume Control

—◉ Other Signal System Devices – CTV = Television Antenna, DP = Data Processing

⟦SC⟧ Signal Central Station

▱ Telephone Interconnection Box

(PE) Pneumatic/Electric Switch

(EP) Electric/Pneumatic Switch

⟦GP⟧ Operating Room Grounding Plate

⟦P⟧₆ Patient Ground Point – 6 = Number of Jacks

## Panelboards

▭ Flush Mounted Panelboard & Cabinet

▭ Surface Mounted Panelboard & Cabinet

▬ Lighting Panel

▨ Power Panel

▭ Heating Panel

⊠ Controller (Starter)

⊐ Externally Operated Disconnect Switch

## Busducts & Wireways

⟦T⟧⟦T⟧⟦T⟧ Trolley Duct

⟦B⟧⟦B⟧⟦B⟧ Busway (Service, Feeder, or Plug-in)

⟦C⟧⟦C⟧⟦C⟧ Cable through Ladder or Channel

⟦W⟧⟦W⟧⟦W⟧ Wireway

⟦J⟧ Bus Duct Junction Box

## Electrical Distribution or Lighting System, Aerial, Lightning Protection

○ Pole

⊗— Street Light & Bracket

△ Transformer

——— Primary Circuit

– – – Secondary Circuit

— – — Auxiliary System Circuits

——→ Down Guy

—•— Head Guy

—○→ Sidewalk Guy

⊂— Service Weather Head

⊘ Lightning Rod

—L— Lightning Protection System Conductor

## Residential Signaling System Outlets

▣ Push Botton

▢↘ Buzzer

◖▢ Bell

◖▢↘ Bell and Buzzer Combination

◇ Annunciator

◀ Outside Telephone

◁ Interconnecting Telephone

▐◀ Telephone Switchboard

⟦BT⟧ Bell Ringing Transformer

## Residential Signaling System Outlets (Cont).

|  |  |
|---|---|
| D | Electric Door Opener |
| M | Maid's Signal Plug |
| R | Radio Antenna Outlet |
| CH | Chime |
| TV | Television Antenna Outlet |
| T | Thermostat |

## Underground Electrical Distribution or Lighting System

|  |  |
|---|---|
| M | Manhole |
| H | Handhole |
| TM | Transformer – Manhole or Vault |
| TP | Transformer Pad |
| – – – | Underground Direct Burial Cable |
| | Underground Duct Line |
| | Street Light Standard Fed from Underground Circuit |

## Panel Circuits & Miscellaneous

|  |  |
|---|---|
| – – – – | Conduit Concealed in Floor or Walls |
| – – – – – | Wiring Exposed |
| | Home Run to Panelboard – Number of Arrows Indicate Number of Circuits |
| | Home Run to Panelboard – Two-Wire Circuit |
| | Home Run to Panelboard – Number of Slashes Indicate Number of Wires (When more than Two) |
| LS-LI,3,5 | Home Run to Panelboard – 'LS' Indicates Panel Designation; LI, 3, 5, Indicates Circuit Breaker No. |

|  |  |
|---|---|
| —C— | Clock Circuit, Conduit and Wire |
| —E— | Emergency Conduit and Wiring |
| —T— | Telephone Conduit and Wiring |
| —— | Feeders |
| ——o | Conduit Turned Up |
| ——● | Conduit Turned Down |
| G | Generator |
| M | Motor |
| 5 | Motor – Numeral Indicates Horsepower |
| I | Instrument (Specify) |
| T | Transformer |
| 8 | Remote Start-Stop Push Button Station |
| 8/8 | Remote Start-Stop Push Button Station w/Pilot Light |
| HTR | Electric Heater Wall Unit |

## Common Abbreviations on Drawings

|  |  |
|---|---|
| EWC | Electric Water Cooler |
| EDH | Electric Duct Heater |
| AFF | Above Finished Floor |
| UH | Unit Heater |
| GFI | Ground Fault Interrupter |
| GFP | Ground Fault Protector |
| GFCB | Ground Fault Circuit Breaker |
| EC | Empty Conduit |
| WP | Weatherproof |
| VP | Vaporproof |

Common Abbreviations on Drawings (Cont).

| | | | | |
|---|---|---|---|---|
| **EXP** | Explosion-proof | | **ATS** | Automatic Transfer Switch |
| **AD** | Auto Damper | | **IS** | Isolating Switch |
| **LP** | Lighting Panel | | **FATC** | Fire Alarm Terminal Cabinet |
| **PP** | Power Panel | | **FA** | Fire Alarm |
| **IP** | Isolation Panel | | **CAM** | Closed Circuit TV Camera |
| **MC** | Motor Controller | | **MON** | Closed Circuit TV Monitor |
| **MCC** | Motor Control Center | | **MG** | Motor-Generator Set |

| Amperes | Size Wire | Maximum Circuit Length | | | | |
|---|---|---|---|---|---|---|
| | | 2 Wire, 1 Phase | | 3 Wires, 3 Phase | | |
| | | 115V | 230V | 230V | 460V | 575V |
| 15 | 14 | 45 ft. | 90 ft. | 100 ft. | 200 ft. | 250 ft. |
| 20 | 12 | 50 | 100 | 112 | 224 | 280 |
| 25 | 10 | 65 | 130 | 145 | 290 | 363 |
| 35 | 8 | 75 | 150 | 168 | 336 | 420 |
| 50 | 6 | 90 | 180 | 201 | 402 | 503 |
| 70 | 4 | 90 | 180 | 201 | 402 | 503 |
| 90 | 2 | 115 | 230 | 257 | 514 | 643 |
| 100 | 1 | 130 | 260 | 291 | 582 | 728 |
| 125 | 1/0 | 130 | 260 | 291 | 582 | 728 |
| 150 | 2/0 | 135 | 270 | 302 | 604 | 755 |
| 175 | 3/0 | 145 | 290 | 325 | 650 | 812 |
| 225 | 4/0 | 145 | 290 | 325 | 650 | 812 |
| 250 | 250 MCM | 150 | 300 | 336 | 672 | 840 |
| 275 | 300 | 165 | 330 | 370 | 740 | 925 |
| 300 | 350 | 180 | 360 | 403 | 806 | 1007 |
| 400 | 500 | 190 | 380 | 425 | 850 | 1060 |

Maximum Circuit Length in Feet for Various Power Requirements

# Minimum Copper and Aluminum Wire Size Allowed for Various Types of Insulation

## Minimum Wire Sizes

| Amperes | Copper THW THWN or XHHW | Copper THHN XHHW * | Aluminum THW XHHW | Aluminum THHN XHHW * |
|---|---|---|---|---|
| 15A | #14 | #14 | #12 | #12 |
| 20 | #12 | #12 | #12 | |
| 25 | | #10 | #10 | #10 |
| 30 | #10 | #10 | #10 | |
| 40 | | | #8 | |
| 45 | | | #8 | #8 |
| 50 | #8 | #8 | #6 | |
| 55 | | | | |
| 60 | | | | #6 |
| 65 | #6 | #6 | #4 | #4 |
| 75 | #4 | | #3 | #3 |
| 85 | | | | |
| 90 | #3 | #4 | #2 | #2 |
| 95 | | | | |
| 100 | | | #1 | |
| 110 | #2 | #3 | #1 | #1 |
| 115 | | | | |
| 120 | | | 1/0 | |
| 130 | #1 | #2 | 2/0 | 1/0 |
| 135 | 1/0 | #1 | | 2/0 |
| 150 | | | | |
| 155 | | | 3/0 | |
| 170 | | 1/0 | 3/0 | 3/0 |
| 175 | 2/0 | | | |
| 180 | | | 4/0 | |
| 195 | 3/0 | 2/0 | | |
| 200 | | | | |
| 205 | | | | |
| 225 | 4/0 | 3/0 | 250MCM | 4/0 |
| 230 | | | 300MCM | |
| 250 | | | 350MCM | |
| 255 | 250MCM | 4/0 | | 250MCM |
| 260 | | | | |
| 270 | | | | |
| 280 | 300MCM | 250MCM | 400MCM | 300MCM |
| 285 | | | | |
| 290 | | | | |
| 305 | 350MCM | 300MCM | 500MCM | 350MCM |
| 310 | | | | |
| 320 | | | | |
| 335 | 400MCM | 350MCM | 600MCM | 400MCM |
| 340 | | | | |
| 350 | | | | |
| 375 | 500MCM | 400MCM | 700MCM | 500MCM |
| 380 | | | 750MCM | |
| 385 | | | | |
| 420 | 600MCM | 500MCM | | 600MCM |
| 430 | | | | 700MCM |
| 435 | | | | 750MCM |
| 475 | | 600MCM | | |

*Dry Locations Only

Notes:
1. Size #14 to 4/0 is in AWG units (American Wire Gauge)
2. Size 250 to 750 is in MCM units (Thousand Circular Mils)
3. Use next higher ampere value if exact value is not listed in table.
4. For loads that operate continuously increase ampere value by 25% to obtain proper wire size.

# Maximum Number of Wires (Insulations Noted) for Various Conduit Sizes

| Copper Wire Size | 1/2" TW | 1/2" THW | 1/2" THWN | 3/4" TW | 3/4" THW | 3/4" THWN | 1" TW | 1" THW | 1" THWN | 1-1/4" TW | 1-1/4" THW | 1-1/4" THWN | 1-1/2" TW | 1-1/2" THW | 1-1/2" THWN | 2" TW | 2" THW | 2" THWN | 2-1/2" THW | 2-1/2" THWN | 3" THW | 3" THWN | 3-1/2" THW | 3-1/2" THWN | 4" THW | 4" THWN |
|---|---|---|---|---|---|---|---|---|---|---|---|---|---|---|---|---|---|---|---|---|---|---|---|---|---|---|
| #14 | 9 | 6 | 13 | 15 | 10 | 24 | 25 | 16 | 39 | 44 | 29 | 69 | 60 | 40 | 94 | 99 | 65 | 154 | 93 |  | 143 |  | 192 |  |  |  |
| #12 | 7 | 4 | 10 | 12 | 8 | 18 | 19 | 13 | 29 | 35 | 24 | 51 | 47 | 32 | 70 | 78 | 53 | 114 | 76 | 164 | 117 |  | 157 |  |  |  |
| #10 | 5 | 4 | 6 | 9 | 6 | 11 | 15 | 11 | 18 | 26 | 19 | 32 | 36 | 26 | 44 | 60 | 43 | 73 | 61 | 104 | 95 | 160 | 127 |  | 163 |  |
| #8 | 2 | 1 | 3 | 4 | 3 | 5 | 7 | 5 | 9 | 12 | 10 | 16 | 17 | 13 | 22 | 28 | 22 | 36 | 32 | 51 | 49 | 79 | 66 | 106 | 85 | 136 |
| #6 |  | 1 | 1 |  | 2 | 4 |  | 4 | 6 |  | 7 | 11 |  | 10 | 15 |  | 16 | 26 | 23 | 37 | 36 | 57 | 48 | 76 | 62 | 98 |
| #4 |  | 1 | 1 |  | 1 | 2 |  | 3 | 4 |  | 5 | 7 |  | 7 | 9 |  | 12 | 16 | 17 | 22 | 27 | 35 | 36 | 47 | 47 | 60 |
| #3 |  | 1 | — |  | 1 | 1 |  | 2 | 3 |  | 4 | 6 |  | 6 | 8 |  | 10 | 13 | 15 | 19 | 23 | 29 | 31 | 39 | 40 | 51 |
| #2 |  | 1 | 1 |  | 1 | 1 |  | 2 | 3 |  | 4 | 5 |  | 5 | 7 |  | 9 | 11 | 13 | 16 | 20 | 25 | 27 | 33 | 34 | 43 |
| #1 |  | 1 | 1 |  | 1 | 1 |  | 1 | 1 |  | 3 | 3 |  | 4 | 5 |  | 6 | 8 | 9 | 12 | 14 | 18 | 19 | 25 | 25 | 32 |
| 1/0 |  | — | — |  | — | — |  | — | — |  | 2 | 3 |  | 3 | 4 |  | 5 | 7 | 8 | 10 | 12 | 15 | 16 | 21 | 21 | 27 |
| 2/0 |  | — | — |  | — | — |  | — | 1 |  | 1 | 2 |  | 3 | 3 |  | 5 | 6 | 7 | 8 | 10 | 13 | 14 | 17 | 18 | 22 |
| 3/0 |  | 1 | — |  | 1 | 1 |  | 1 | 1 |  | 1 | 1 |  | 2 | 3 |  | 4 | 5 | 6 | 7 | 9 | 11 | 12 | 14 | 15 | 18 |
| 4/0 |  |  |  |  |  | — |  | — | 1 |  | 1 | 1 |  | 1 | 2 |  | 3 | 4 | 5 | 6 | 7 | 9 | 10 | 12 | 13 | 15 |
| 250MCM |  |  |  |  |  | 1 |  | — | — |  |  | 1 |  |  | 1 |  | 2 | 3 | 4 | 4 | 6 | 7 | 8 | 10 | 10 | 12 |
| 300 |  |  |  |  |  |  |  | — | — |  |  | — |  |  | 1 |  | 2 | 3 | 3 | 4 | 5 | 6 | 7 | 8 | 9 | 11 |
| 350 |  |  |  |  |  |  |  | 1 | 1 |  |  | 1 |  |  | 1 |  | 1 | 2 | 3 | 3 | 4 | 5 | 6 | 7 | 8 | 9 |
| 400 |  |  |  |  |  |  |  |  |  |  |  |  |  |  | 1 |  | 1 | 1 | 2 | 3 | 4 | 5 | 5 | 6 | 7 | 8 |
| 500 |  |  |  |  |  |  |  |  |  |  |  |  |  |  |  |  | 1 | 1 | 1 | 2 | 3 | 4 | 4 | 5 | 6 | 7 |
| 600 |  |  |  |  |  |  |  |  |  |  |  |  |  |  |  |  | 1 | 1 | 1 | 1 | 3 | 3 | 4 | 4 | 5 | 5 |
| 700 |  |  |  |  |  |  |  |  |  |  |  |  |  |  |  |  | — | 1 | 1 | 1 | 2 | 3 | 3 | 4 | 4 | 5 |
| 750 |  |  |  |  |  |  |  |  |  |  |  |  |  |  |  |  | — | 1 | 1 | 1 | 2 | 2 | 3 | 3 | 4 | 4 |

## Metric Equivalent, Conduit and Wire (THW)

| Conduit Size | | Wire | Wire Size | | Wire | Wire Size | | Wire | Wire Size | |
|---|---|---|---|---|---|---|---|---|---|---|
| Inches | Millimeters | Amp. | AWG | Millimeters | Amp. | AWG | Millimeters | Amp. | AWG MCM | Millimeters |
| 3/4 | 20 | 20 | #12 | 2.5 | 80 | #3 | 25 | 235 | 250 | 120 |
| 1 | 25 | 25 | 10 | 4 | 100 | 2 | 35 | 280 | 300 | 150 |
| 1-1/4 | 30 | 30 | 8 | 6 | 130 | 1/0 | 50 | 325 | 400 | 185 |
| 1-1/2 | 40 | 45 | 6 | 10 | 165 | 3/0 | 70 | 360 | 500 | 240 |
| 2 | 50 | 60 | 4 | 16 | 200 | 4/0 | 95 | 415 | 600 | 300 |

# Concrete for Conduit Encasement

Table below lists C.Y. of concrete for 100 L.F. of trench with 3" of concrete cover on all sides.

## Trench Layout and Number of Conduits in Trench

| Conduit Separation | Conduit Diameter | 2 | 3 | 4 | 6 | 8 | 9 | 12 | 16 | 20 |
|---|---|---|---|---|---|---|---|---|---|---|
| 1" | 2" | 2.4 C.Y. | 3.1 C.Y. | 3.3 C.Y. | 4.1 C.Y. | 4.9 C.Y. | 5.1 C.Y. | 6.1 C.Y. | 7.3 C.Y. | 8.5 C.Y. |
| | 3" | 3.0 | 3.9 | 4.2 | 5.3 | 6.4 | 6.7 | 8.1 | 9.8 | 11.3 |
| | 3½" | 3.3 | 4.2 | 4.6 | 5.9 | 7.2 | 7.4 | 9.1 | 11.0 | 12.9 |
| | 4" | 3.6 | 4.7 | 5.0 | 6.5 | 7.9 | 8.3 | 10.1 | 12.2 | 14.5 |
| | 4½" | 3.9 | 5.1 | 5.5 | 7.1 | 8.7 | 9.2 | 11.3 | 13.6 | 16.0 |
| | 5" | 4.2 | 5.5 | 6.0 | 7.8 | 9.6 | 10.0 | 12.2 | 15.0 | 17.8 |
| 1½" | 2" | 2.6 | 3.3 | 3.6 | 4.6 | 5.7 | 5.9 | 7.3 | 8.9 | 10.5 |
| | 3" | 3.2 | 4.1 | 4.5 | 5.9 | 7.3 | 7.6 | 9.4 | 11.8 | 13.7 |
| | 3½" | 3.4 | 4.5 | 4.9 | 6.5 | 8.0 | 8.5 | 10.6 | 13.0 | 15.4 |
| | 4" | 3.7 | 5.0 | 5.4 | 7.2 | 8.9 | 9.5 | 11.7 | 14.4 | 17.2 |
| | 4½" | 4.0 | 5.3 | 6.0 | 7.9 | 9.8 | 10.3 | 12.9 | 16.0 | 19.0 |
| | 5" | 4.4 | 5.8 | 6.5 | 8.6 | 10.7 | 11.3 | 14.1 | 17.5 | 21.0 |
| 2" | 2" | 2.7 | 3.5 | 3.9 | 5.2 | 6.4 | 6.8 | 8.5 | 10.5 | 12.5 |
| | 3" | 3.3 | 4.4 | 4.9 | 6.5 | 8.1 | 8.7 | 10.8 | 13.5 | 16.2 |
| | 3½" | 3.5 | 4.8 | 5.4 | 7.2 | 9.0 | 9.6 | 12.0 | 15.1 | 18.1 |
| | 4" | 3.8 | 5.2 | 5.9 | 7.9 | 9.9 | 10.6 | 13.3 | 16.7 | 20.1 |
| | 4½" | 4.2 | 5.6 | 6.4 | 8.7 | 10.9 | 12.6 | 14.6 | 18.3 | 22.0 |
| | 5" | 4.5 | 6.1 | 7.0 | 9.4 | 11.8 | 12.7 | 16.0 | 20.1 | 24.2 |
| 3" | 2" | 2.9 | 4.4 | 4.6 | 6.4 | 8.1 | 8.9 | 11.1 | 14.1 | 17.2 |
| | 3" | 3.5 | 4.9 | 5.7 | 7.8 | 10.0 | 10.9 | 13.9 | 17.8 | 21.6 |
| | 3½" | 3.8 | 5.3 | 6.2 | 8.6 | 11.1 | 12.0 | 15.3 | 19.6 | 24.0 |
| | 4" | 4.2 | 5.8 | 6.8 | 9.5 | 12.1 | 13.1 | 16.8 | 21.6 | 26.3 |
| | 4½" | 4.5 | 6.2 | 7.4 | 10.3 | 13.2 | 14.2 | 18.3 | 23.5 | 28.7 |
| | 5" | 4.8 | 6.7 | 7.9 | 11.1 | 14.2 | 15.5 | 19.9 | 25.5 | 32.5 |
| Trench Layout | | | | | | | | | | |

| Weight (Lbs./L.F.) of 4 Pole Aluminum and Copper Bus Duct by Ampere Load | | | | |
|---|---|---|---|---|
| Amperes | Aluminum Feeder | Copper Feeder | Aluminum Plug-In | Copper Plug-In |
| 225 | | | 7 | 7 |
| 400 | | | 8 | 13 |
| 600 | 10 | 10 | 11 | 14 |
| 800 | 10 | 19 | 13 | 18 |
| 1000 | 11 | 19 | 16 | 22 |
| 1350 | 14 | 24 | 20 | 30 |
| 1600 | 17 | 26 | 25 | 39 |
| 2000 | 19 | 30 | 29 | 46 |
| 2500 | 27 | 43 | 36 | 56 |
| 3000 | 30 | 48 | 42 | 73 |
| 4000 | 39 | 67 | | |
| 5000 | | 78 | | |

| Size Required and Weight (Lbs./1000 L.F.) of Aluminum and Copper THW Wire by Ampere Load | | | | |
|---|---|---|---|---|
| Amperes | Copper Size | Aluminum Size | Copper Weight | Aluminum Weight |
| 15 | 14 | 12 | 24 | 11 |
| 20 | 12 | 10 | 33 | 17 |
| 30 | 10 | 8 | 48 | 39 |
| 45 | 8 | 6 | 77 | 52 |
| 65 | 6 | 4 | 112 | 72 |
| 85 | 4 | 2 | 167 | 101 |
| 100 | 3 | 1 | 205 | 136 |
| 115 | 2 | 1/0 | 252 | 162 |
| 130 | 1 | 2/0 | 324 | 194 |
| 150 | 1/0 | 3/0 | 397 | 233 |
| 175 | 2/0 | 4/0 | 491 | 282 |
| 200 | 3/0 | 250 | 608 | 347 |
| 230 | 4/0 | 300 | 753 | 403 |
| 255 | 250 | 400 | 899 | 512 |
| 285 | 300 | 500 | 1068 | 620 |
| 310 | 350 | 500 | 1233 | 620 |
| 335 | 400 | 600 | 1396 | 772 |
| 380 | 500 | 750 | 1732 | 951 |

| Conduit Weight Comparisons (Lbs. per 100 ft.) | | | | | | | | | | | | |
|---|---|---|---|---|---|---|---|---|---|---|---|---|
| Type | 1/2" | 3/4" | 1" | 1-1/4" | 1-1/2" | 2" | 2-1/2" | 3" | 3-1/2" | 4" | 5" | 6" |
| Rigid Aluminum | 28 | 37 | 55 | 72 | 89 | 119 | 188 | 246 | 296 | 350 | 479 | 630 |
| Rigid Steel | 79 | 105 | 153 | 201 | 249 | 332 | 527 | 683 | 831 | 972 | 1314 | 1745 |
| Intermediate Steel (IMC) | 60 | 82 | 116 | 150 | 182 | 242 | 401 | 493 | 573 | 638 | | |
| Electrical Metallic Tubing (EMT) | 29 | 45 | 65 | 96 | 111 | 141 | 215 | 260 | 365 | 390 | | |
| Polyvinyl Chloride, Schedule 40 | 16 | 22 | 32 | 43 | 52 | 69 | 109 | 142 | 170 | 202 | 271 | 350 |
| Polyvinyl Chloride Encased Burial | | | | | | 38 | | 67 | 88 | 105 | 149 | 202 |

| H.P. | Amperes | | | | | |
|---|---|---|---|---|---|---|
| | Single Phase | | Three Phase | | | |
| | 115V | 230V | 200V | 230V | 460V | 575V |
| 1/6 | 4.4A | 2.2A | | | | |
| 1/4 | 5.8 | 2.9 | | | | |
| 1/3 | 7.2 | 3.6 | | | | |
| 1/2 | 9.8 | 4.9 | 2.3A | 2.0A | 1.0A | 0.8A |
| 3/4 | 13.8 | 6.9 | 3.2 | 2.8 | 1.4 | 1.1 |
| 1 | 16 | 8 | 4.1 | 3.6 | 1.8 | 1.4 |
| 1-1/2 | 20 | 10 | 6.0 | 5.2 | 2.6 | 2.1 |
| 2 | 24 | 12 | 7.8 | 6.8 | 3.4 | 2.7 |
| 3 | 34 | 17 | 11.0 | 9.6 | 4.8 | 3.9 |
| 5 | | | 17.5 | 15.2 | 7.6 | 6.1 |
| 7-1/2 | | | 25.3 | 22 | 11 | 9 |
| 10 | | | 32.2 | 28 | 14 | 11 |
| 15 | | | 48.3 | 42 | 21 | 17 |
| 20 | | | 62.1 | 54 | 27 | 22 |
| 25 | | | 78.2 | 68 | 34 | 27 |
| 30 | | | 92.0 | 80 | 40 | 32 |
| 40 | | | 119.6 | 104 | 52 | 41 |
| 50 | | | 149.5 | 130 | 65 | 52 |
| 60 | | | 177 | 154 | 77 | 62 |
| 75 | | | 221 | 192 | 96 | 77 |
| 100 | | | 285 | 248 | 124 | 99 |
| 125 | | | 359 | 312 | 156 | 125 |
| 150 | | | 414 | 360 | 180 | 144 |
| 200 | | | 552 | 480 | 240 | 192 |

Ampere Values Determined by Horsepower, Voltage and Phase Values

# NEMA Enclosures

Electrical enclosures serve two basic purposes; they protect people from accidental contact with enclosed electrical devices and connections, and they protect the enclosed devices and connections from specified external conditions. The National Electrical Manufacturers Association (NEMA) has established the following standards. Because these descriptions are not intended to be complete representations of NEMA listings, consultation of NEMA literature is advised for detailed information.

The following definitions and descriptions pertain to NONHAZARDOUS locations:

NEMA Type 1: General purpose enclosures intended for use indoors, primarily to prevent accidental contact of personnel with the enclosed equipment in areas that do not involve unusual conditions.

NEMA Type 2: Dripproof indoor enclosures intended to protect the enclosed equipment against dripping noncorrosive liquids and falling dirt.

NEMA Type 3: Dustproof, raintight and sleet-resistant (ice-resistant) enclosures intended for use outdoors to protect the enclosed equipment against wind-blown dust, rain, sleet, and external ice formation.

NEMA Type 3R: Rainproof and sleet-resistant (ice-resistant) enclosures which are intended for use outdoors to protect the enclosed equipment against rain. These enclosures are constructed so that the accumulation and melting of sleet (ice) will not damage the enclosure and its internal mechanisms.

NEMA Type 3S: Enclosures intended for outdoor use to provide limited protection against wind-blown dust, rain, and sleet (ice) and to allow operation of external mechanisms when ice-laden.

NEMA Type 4: Watertight and dust-tight enclosures intended for use indoors and out – to protect the enclosed equipment against splashing water, seepage of water, falling or hose-directed water, and severe external condensation.

NEMA Type 4X: Watertight, dust-tight, and corrosion-resistant indoor and outdoor enclosures featuring the same provisions as Type 4 enclosures, plus corrosion resistance.

NEMA Type 5: Indoor enclosures intended primarily to provide limited protection against dust and falling dirt.

NEMA Type 6: Enclosures intended for indoor and outdoor use – primarily to provide limited protection against the entry of water during occasional temporary submersion at a limited depth.

NEMA Type 6R: Enclosures intended for indoor and outdoor use – primarily to provide limited protection against the entry of water during prolonged submersion at a limited depth.

NEMA Type 11: Enclosures intended for indoor use – primarily to provide, by means of oil immersion, limited protection to enclosed equipment against the corrosive effects of liquids and gases.

NEMA Type 12: Dust-tight and driptight indoor enclosures intended for use indoors in industrial locations to protect the enclosed equipment against fibers, flyings, lint, dust, and dirt, as well as light splashing, seepage, dripping, and external condensation of noncorrosive liquids.

NEMA Type 13: Oiltight and dust-tight indoor enclosures intended primarily to house pilot devices, such as limit switches, foot switches, push buttons, selector switches, and pilot lights, and to protect these devices against lint and dust, seepage, external condensation, and sprayed water, oil, and noncorrosive coolant.

The following definitions and descriptions pertain to HAZARDOUS, or CLASSIFIED, locations:

NEMA Type 7: Enclosures intended for use in indoor locations classified as Class 1, Groups A, B, C, or D, as defined in the National Electrical Code.

NEMA Type 9: Enclosures intended for use in indoor locations classified as Class 2, Groups E, F, or G, as defined in the National Electrical Code.

## Man-Hours

| Description | m/hr | Unit |
|---|---|---|
| NEMA 1 | | |
| 12"L x 12"W x 4"D | 1.330 | Ea. |
| NEMA 3R | | |
| 12"L x 12"W x 6"D | 1.600 | Ea. |
| NEMA 4 | | |
| 12"L x 12"W x 6"D | 4.000 | Ea. |
| NEMA 7 | | |
| 12"L x 12"W x 6"D | 8.000 | Ea. |
| NEMA 9 | | |
| 12"L x 12"W x 6"D | 5.000 | Ea. |
| NEMA 12 | | |
| 12"L x 14"W x 6"D | 1.510 | Ea. |

Screw Cover - NEMA 1

Hinged Cover - NEMA 1

Rainproof and Weatherproof, Screw cover - NEMA 3R

Sheet Metal Pull Boxes

 **Means Forms**
**SPEC-AID**

## DIVISION 16: ELECTRICAL

PROJECT                                LOCATION

**Incoming Service:** ☐ Overhead    ☐ Underground

                                        Primary                              Secondary

    Voltage

    Unit Sub-station & Size

    Number of Manholes

    Feeder Size

    Length

    Conduit

    Duct

    Concrete: ☐ No    ☐ Yes

    Other

**Building Service:** Size _____ Amps    Switchboard

    Panels: ☐ Distribution _____ Lighting _____ Power

    Describe

**Motor Control Center:** Furnished by

    Describe

**Bus Duct:** ☐ No    ☐ Yes    Size _____ Amps    Application

    Describe

**Cable Tray:** ☐ No    ☐ Yes    Describe

**Emergency System:** ☐ No    ☐ Yes    Allowance _____ ☐ Separate Contract

    Generator: ☐ No    ☐ Diesel    ☐ Gas    ☐ Gasoline _____ Size _____ KW

    Transfer Switch: ☐ No    ☐ Yes    Number _____ Size _____ Amps

    Area Protection Relay Panels: ☐ No    ☐ Yes

    Other

**Conduit:** ☐ No    ☐ Yes    ☐ Aluminum

    ☐ Electric Metallic Tubing

    ☐ Galvanized Steel

    ☐ Plastic

**Wire:** ☐ No    ☐ Yes    ☐ Type Installation

    ☐ Armored Cable

    ☐ Building Wire

    ☐ Metallic Sheath Cable

    ☐

**Underfloor Duct:** ☐ No    ☐ Yes    Describe

**Header Duct:** ☐ No    ☐ Yes    Describe

**Trench Duct:** ☐ No    ☐ Yes    Describe

**Underground Duct:** ☐ No    ☐ Yes    Describe

**Explosion Proof Areas:** ☐ No    ☐ Yes    Describe

**Motors:** ☐ No    ☐ Yes    Total H.P. _____ No. of Fractional H.P. _____ Voltage

    ☐ 1/2 to 5 H.P. _____ ☐ 7-1/2 to 25 H.P. _____ ☐ Over 25 H.P.

    Describe

    Starters: Type

    Supplied by:

**DIVISION 16: ELECTRICAL**

**Telephone System:** ☐ No ☐ Yes  Service Size _____ Length _____
  Manhole: ☐ No ☐ Yes  Number _____ Termination _____
  Concrete Encased: ☐ No ☐ Yes ☐ Rigid Galv. ☐ Duct ☐ _____

**Fire Alarm System:** ☐ No ☐ Yes  Service Size _____ Length _____ Wire Type _____
  Concrete Encased: ☐ No ☐ Yes ☐ Rigid Galv. ☐ Duct ☐ _____
  ☐ Stations _____ ☐ Horns _____ ☐ Lights _____ ☐ Combination _____
  Detectors: ☐ Rate of Rise _____ ☐ Fixed _____ ☐ Smoke _____
    Describe _____ Insulation _____ Wire Size _____
  ☐ Zones _____ ☐ Conduit _____ ☐ E.M.T. _____ ☐ Empty _____
    Describe _____

**Watchmans Tour:** ☐ No ☐ Yes ☐ Stations _____ ☐ Door Switches _____
  ☐ Alarm Bells _____ ☐ Key Re-sets _____ ☐ _____
  ☐ Conduit _____ ☐ E.M.T. _____ ☐ Wire _____ ☐ Empty _____
    Describe _____

**Clock System:** ☐ No ☐ Yes ☐ Electronic ☐ Wired ☐ _____
  ☐ Single Dial _____ ☐ Double Dial _____ ☐ Program Bell _____
  ☐ Conduit _____ ☐ E.M.T. _____ ☐ Empty _____
    Describe _____

**Sound System:** ☐ No ☐ Yes  Type _____ Speakers _____
  ☐ Conduit _____ ☐ Cable _____ ☐ E.M.T. _____ ☐ Empty _____
    Describe _____

**Television System:** ☐ No ☐ Yes  Describe _____
  ☐ Antenna _____ ☐ Closed Circuit _____ ☐ Teaching _____ ☐ Security _____
  ☐ Learning Laboratory _____ ☐ _____
  ☐ Conduit _____ ☐ E.M.T. _____ ☐ Wire _____ ☐ Empty _____

**Lightning Protection:** ☐ No ☐ Yes  Describe _____
**Low Voltage Switching:** ☐ No ☐ Yes  Describe _____
**Scoreboards:** ☐ No ☐ Yes  Describe _____ Number _____
**Comfort Systems:** ☐ No ☐ Electric Heat ☐ Snow Melting ☐ _____
    Describe _____
**Other Systems:** _____
_____
_____

**Lighting Fixtures:** ☐ No ☐ Yes ☐ Allowance _____ ☐ Separate Contract
  ☐ Economy ☐ Commercial ☐ Deluxe ☐ Explosion Proof ☐ _____
  ☐ Incandescent _____
  _____ Foot Candles _____
  ☐ Fluorescent _____
  _____ Foot Candles _____
  ☐ Mercury Vapor _____
  _____ Foot Candles _____
  ☐ _____ Foot Candles _____
  ☐ Step Lighting _____ ☐ Planter Lighting _____ ☐ Fountain Lighting _____
  ☐ Site Lighting _____ ☐ Poles _____ ☐ Area Lighting _____ ☐ Flood Lighting _____
  Dimming System: ☐ No ☐ Yes ☐ Incandescent ☐ Fluorescent _____
  Ceilings: ☐ T Bar ☐ Concealed Spline ☐ _____
  Emergency Battery Units: ☐ No ☐ Lead Acid ☐ Nickel Cadmium ☐ 6 Volt _____ 12 Volt _____
    Describe _____
**Special Considerations:** _____
_____
_____

# Abbreviations

| | | | | | | | |
|---|---|---|---|---|---|---|---|
| A | Area Square Feet; Ampere | Calc | Calculated | D.H. | Double Hung |
| ABS | Acrylonitrile Butadiene Styrene; Asbestos Bonded Steel | Cap. | Capacity | DHW | Domestic Hot Water |
| | | Carp. | Carpenter | Diag. | Diagonal |
| A.C. | Alternating Current ; Air Conditioning; Asbestos Cement | C.B. | Circuit Breaker | Diam. | Diameter |
| | | C.C.F. | Hundred Cubic Feet | Distrib. | Distribution |
| | | cd | Candela | Dk. | Deck |
| A.C.I. | American Concrete Institute | cd/sf | Candela per Square Foot | D.L. | Dead Load; Diesel |
| Addit. | Additional | CD | Grade of Plywood Face & Back | Do. | Ditto |
| Adj. | Adjustable | CDX | Plywood, grade C&D, exterior glue | Dp. | Depth |
| af | Audio-frenquency | Cefi. | Cement Finisher | D.P.S.T. | Double Pole, Single Throw |
| A.G.A. | American Gas Association | Cem. | Cement | Dr. | Driver |
| Agg. | Aggregate | CF | Hundred Feet | Drink. | Drinking |
| A.H. | Ampere Hours | C.F. | Cubic Feet | D.S. | Double Strength |
| A hr | Ampere-hour | CFM | Cubic Feet per Minute | D.S.A. | Double Strength A Grade |
| A.I.A. | American Institute of Architects | c.g. | Center of Gravity | D.S.B. | Double Strength B Grade |
| AIC | Ampere Interrupting Capacity | CHW | Commercial Hot Water | Dty. | Duty |
| Allow. | Allowance | C.I. | Cast Iron | DWV | Drain Waste Vent |
| alt. | Altitude | C.I.P. | Cast in Place | DX | Deluxe White, Direct Expansion |
| Alum. | Aluminum | Circ. | Circuit | dyn | Dyne |
| a.m. | ante meridiem | C.L. | Carload Lot | e | Eccentricity |
| Amp. | Ampere | Clab. | Common Laborer | E | Equipment Only; East |
| Approx. | Approximate | C.L.F. | Hundred Linear Feet | Ea. | Each |
| Apt. | Apartment | CLF | Current Limiting Fuse | Econ. | Economy |
| Asb. | Asbestos | CLP | Cross Linked Polyethylene | EDP | Electronic Data Processing |
| A.S.B.C. | American Standard Building Code | cm | Centimeter | E.D.R. | Equiv. Direct Radiation |
| Asbe. | Asbestos Worker | CMP | Corr. Metal Pipe | Eq. | Equation |
| A.S.H.R.A.E. | American Society of Heating, Refrig. & AC Engineers | C.M.U. | Concrete Masonry Unit | Elec. | Electrician; Electrical |
| | | Col. | Column | Elev. | Elevator; Elevating |
| A.S.M.E. | American Society of Mechanical Engineers | $CO_2$ | Carbon Dioxide | EMT | Electrical Metallic Conduit; Thin Wall Conduit |
| | | Comb. | Combination | | |
| A.S.T.M. | American Society for Testing and Materials | Compr. | Compressor | Eng. | Engine |
| | | Conc. | Concrete | EPDM | Ethylene Propylene Diene Monomer |
| Attchmt. | Attachment | Cont. | Continuous; | | |
| Avg. | Average | | Continued | Eqhv. | Equip. Oper., heavy |
| Bbl. | Barrel | Corr. | Corrugated | Eqlt. | Equip. Oper., light |
| B.&B. | Grade B and Better; Balled & Burlapped | Cos | Cosine | Eqmd. | Equip. Oper., medium |
| | | Cot | Cotangent | Eqmm. | Equip. Oper., Master Mechanic |
| B.&S. | Bell and Spigot | Cov. | Cover | Eqol. | Equip. Oper., oilers |
| B.&W. | Black and White | CPA | Control Point Adjustment | Equip. | Equipment |
| b.c.c. | Body-centered Cubic | Cplg. | Coupling | ERW | Electric Resistance Welded |
| B.F. | Board Feet | C.P.M. | Critical Path Method | Est. | Estimated |
| Bg. Cem. | Bag of Cement | CPVC | Chlorinated Polyvinyl Chloride | esu | Electrostatic Units |
| BHP | Brake Horse Power | C. Pr. | Hundred Pair | E.W. | Each Way |
| B.I. | Black Iron | CRC | Cold Rolled Channel | EWT | Entering Water Temperature |
| Bit.; | | Creos. | Creosote | Excav. | Excavation |
| Bitum. | Bituminous | Crpt. | Carpet & Linoleum Layer | Exp. | Expansion |
| Bk. | Backed | CRT | Cathode-ray Tube | Ext. | Exterior |
| Bkrs. | Breakers | CS | Carbon Steel | Extru. | Extrusion |
| Bldg. | Building | Csc | Cosecant | f. | Fiber stress |
| Blk. | Block | C.S.F. | Hundred Square Feet | F | Fahrenheit; Female; Fill |
| Bm. | Beam | C.S.I. | Construction Specification Institute | Fab. | Fabricated |
| Boil. | Boilermaker | | | FBGS | Fiberglass |
| B.P.M. | Blows per Minute | C.T. | Current Transformer | F.C. | Footcandles |
| BR | Bedroom | CTS | Copper Tube Size | f.c.c. | Face-centered Cubic |
| Brg. | Bearing | Cu | Cubic | f'c. | Compressive Stress in Concrete; Extreme Compressive Stress |
| Brhe. | Bricklayer Helper | Cu. Ft. | Cubic Foot | | |
| Bric. | Bricklayer | cw | Continuous Wave | F.E. | Front End |
| Brk. | Brick | C.W. | Cool White | FEP | Fluorinated Ethylene Propylene (Teflon) |
| Brng. | Bearing | Cwt. | 100 Pounds | | |
| Brs. | Brass | C.W.X. | Cool White Deluxe | F.G. | Flat Grain |
| Brz. | Bronze | C.Y. | Cubic Yard (27 cubic feet) | F.H.A. | Federal Housing Administration |
| Bsn. | Basin | C.Y./Hr. | Cubic Yard per Hour | Fig. | Figure |
| Btr. | Better | Cyl. | Cylinder | Fin. | Finished |
| BTU | British Thermal Unit | d | Penny (nail size) | Fixt. | Fixture |
| BTUH | BTU per Hour | D | Deep; Depth; Discharge | Fl. Oz. | Fluid Ounces |
| BX | Interlocked Armored Cable | Dis.; | | Flr. | Floor |
| c | Conductivity | Disch. | Discharge | F.M. | Frequency Modulation; Factory Mutual |
| C | Hundred; Centigrade | Db. | Decibel | | |
| | | Dbl. | Double | Fmg. | Framing |
| C/C | Center to Center | DC | Direct Current | Fndtn. | Foundation |
| Cab. | Cabinet | Demob. | Demobilization | Fori. | Foreman, inside |
| Cair. | Air Tool Laborer | d.f.u. | Drainage Fixture Units | Foro. | Foreman, outside |

# Abbreviations

| Abbrev. | Definition |
|---|---|
| Fount. | Fountain |
| FPM | Feet per Minute |
| FPT | Female Pipe Thread |
| Fr. | Frame |
| F.R. | Fire Rating |
| FRK | Foil Reinforced Kraft |
| FRP | Fiberglass Reinforced Plastic |
| FS | Forged Steel |
| FSC | Cast Body; Cast Switch Box |
| Ft. | Foot; Feet |
| Ftng. | Fitting |
| Ftg. | Footing |
| Ft. Lb. | Foot Pound |
| Furn. | Furniture |
| FVNR | Full Voltage Non Reversing |
| FXM | Female by Male |
| Fy. | Minimum Yield Stress of Steel |
| g | Gram |
| G | Gauss |
| Ga. | Gauge |
| Gal. | Gallon |
| Gal./Min. | Gallon Per Minute |
| Galv. | Galvanized |
| Gen. | General |
| Glaz. | Glazier |
| GPD | Gallons per Day |
| GPH | Gallons per Hour |
| GPM | Gallons per Minute |
| GR | Grade |
| Gran. | Granular |
| Grnd. | Ground |
| H | High; High Strength Bar Joist; Henry |
| H.C. | High Capacity |
| H.D. | Heavy Duty; High Density |
| H.D.O. | High Density Overlaid |
| Hdr. | Header |
| Hdwe. | Hardware |
| Help. | Helper average |
| HEPA | High Efficiency Particulate Air Filter |
| Hg | Mercury |
| H.O. | High Output |
| Horiz. | Horizontal |
| H.P. | Horsepower; High Pressure |
| H.P.F. | High Power Factor |
| Hr. | Hour |
| Hrs./Day | Hours Per Day |
| HSC | High Short Circuit |
| Ht. | Height |
| Htg. | Heating |
| Htrs. | Heaters |
| HVAC | Heating, Ventilating & Air Conditioning |
| Hvy. | Heavy |
| HW | Hot Water |
| Hyd.; Hydr. | Hydraulic |
| Hz. | Hertz (cycles) |
| I. | Moment of Inertia |
| I.C. | Interrupting Capacity |
| ID | Inside Diameter |
| I.D. | Inside Dimension; Identification |
| I.F. | Inside Frosted |
| I.M.C. | Intermediate Metal Conduit |
| In. | Inch |
| Incan. | Incandescent |
| Incl. | Included; Including |
| Int. | Interior |
| Inst. | Installation |
| Insul. | Insulation |
| I.P. | Iron Pipe |
| I.P.S. | Iron Pipe Size |
| I.P.T. | Iron Pipe Threaded |
| J | Joule |
| J.I.C. | Joint Industrial Council |
| K. | Thousand; Thousand Pounds |
| K.D.A.T. | Kiln Dried After Treatment |
| kg | Kilogram |
| kG | Kilogauss |
| kgf | Kilogram force |
| kHz | Kilohertz |
| Kip. | 1000 Pounds |
| KJ | Kiljoule |
| K.L. | Effective Length Factor |
| Km | Kilometer |
| K.L.F. | Kips per Linear Foot |
| K.S.F. | Kips per Square Foot |
| K.S.I. | Kips per Square Inch |
| K.V. | Kilo Volt |
| K.V.A. | Kilo Volt Ampere |
| K.V.A.R. | Kilovar (Reactance) |
| KW | Kilo Watt |
| KWh | Kilowatt-hour |
| L | Labor Only; Length; Long |
| Lab. | Labor |
| lat | Latitude |
| Lath. | Lather |
| Lav. | Lavatory |
| lb.; # | Pound |
| L.B. | Load Bearing; L Conduit Body |
| L. & E. | Labor & Equipment |
| lb./hr. | Pounds per Hour |
| lb./L.F. | Pounds per Linear Foot |
| lbf/sq in. | Pound-force per Square Inch |
| L.C.L. | Less than Carload Lot |
| Ld. | Load |
| L.F. | Linear Foot |
| Lg. | Long; Length; Large |
| L. & H. | Light and Heat |
| L.H. | Long Span High Strength Bar Joist |
| L.J. | Long Span Standard Strength Bar Joist |
| L.L. | Live Load |
| L.L.D. | Lamp Lumen Depreciation |
| lm | Lumen |
| lm/sf | Lumen per Square Foot |
| lm/W | Lumen Per Watt |
| L.O.A. | Length Over All |
| log | Logarithm |
| L.P. | Liquefied Petroleum; Low Pressure |
| L.P.F. | Low Power Factor |
| Lt. | Light |
| Lt. Ga. | Light Gauge |
| L.T.L. | Less than Truckload Lot |
| Lt. Wt. | Lightweight |
| L.V. | Low Voltage |
| M | Thousand; Material; Male; Medium Wall Copper |
| m/hr | Manhour |
| mA | Milliampere |
| Mach. | Machine |
| Mag. Str. | Magnetic Starter |
| Maint. | Maintenance |
| Marb. | Marble Setter |
| Mat. | Material |
| Mat'l. | Material |
| Max. | Maximum |
| MBF | Thousand Board Feet |
| MBH | Thousand BTU's per hr. |
| M.C.F. | Thousand Cubic Feet |
| M.C.F.M. | Thousand Cubic Feet per Minute |
| M.C.M. | Thousand Circular Mils |
| M.C.P. | Motor Circuit Protector |
| MD | Medium Duty |
| M.D.O. | Medium Density Overlaid |
| Med. | Medium |
| MF | Thousand Feet |
| M.F.B.M. | Thousand Feet Board Measure |
| Mfg. | Manufacturing |
| Mfrs. | Manufacturers |
| mg | Milligram |
| MGD | Million Gallons per Day |
| MGPH | Thousand Gallons per Hour |
| MH | Manhole; Metal Halide; Man Hour |
| MHz | Megahertz |
| Mi. | Mile |
| MI | Malleable Iron; Mineral Insulated |
| mm | Millimeter |
| Mill. | Millwright |
| Min. | Minimum |
| Misc. | Miscellaneous |
| ml | Milliliter |
| M.L.F. | Thousand Linear Feet |
| Mo. | Month |
| Mobil. | Mobilization |
| Mog. | Mogul Base |
| MPH | Miles per Hour |
| MPT | Male Pipe Thread |
| MRT | Mile Round Trip |
| ms | millisecond |
| M.S.F. | Thousand Square Feet |
| Mstz. | Mosaic & Terrazzo Worker |
| M.S.Y. | Thousand Square Yards |
| Mtd. | Mounted |
| Mthe. | Mosaic & Terrazzo Helper |
| Mtng. | Mounting |
| Mult. | Multi; Multiply |
| MV | Megavolt |
| MW | Megawatt |
| MXM | Male by Male |
| MYD | Thousand yards |
| N | Natural; North |
| nA | nanoampere |
| NA | Not Available; Not Applicable |
| N.B.C. | National Building Code |
| NC | Normally Closed |
| N.E.M.A. | National Electrical Manufacturers Association |
| NEHB | Bolted Circuit Breaker to 600V. |
| N.L.B. | Non-Load-Bearing |
| nm | nanometer |
| No. | Number |
| NO | Normally Open |
| N.O.C. | Not Otherwise Classified |
| Nose. | Nosing |
| N.P.T. | National Pipe Thread |
| NQOB | Bolted Circuit Breaker to 240V. |
| N.R.C. | Noise Reduction Coefficient |
| N.R.S. | Non Rising Stem |
| ns | nanosecond |
| nW | nanowatt |
| OB | Opposing Blade |
| OC | On Center |
| OD | Outside Diameter |
| O.D. | Outside Dimension |
| ODS | Overhead Distribution System |
| O & P | Overhead and Profit |
| Oper. | Operator |
| Opng. | Opening |
| Orna. | Ornamental |
| O.S.&Y. | Outside Stem and Yoke |
| Ovhd | Overhead |
| Oz. | Ounce |

# Abbreviations

| | | | | | |
|---|---|---|---|---|---|
| P. | Pole; Applied Load; Projection | S. | Suction; Single Entrance; South | T.S. | Trigger Start |
| p. | Page | | | Tr. | Trade |
| Pape. | Paperhanger | Scaf. | Scaffold | Transf. | Transformer |
| PAR | Weatherproof Reflector | Sch.; | | Trhv. | Truck Driver, Heavy |
| Pc. | Piece | Sched. | Schedule | Trlr. | Trailer |
| P.C. | Portland Cement; Power Connector | S.C.R. | Modular Brick | Trlt. | Truck Driver, Light |
| | | S.D.R. | Standard Dimension Ratio | TV | Television |
| P.C.F. | Pounds per Cubic Foot | S.E. | Surfaced Edge | T.W. | Thermoplastic Water Resistant Wire |
| P.E. | Professional Engineer; Porcelain Enamel; Polyethylene; Plain End | S.E.R.; | | | |
| | | S.E.U. | Service Entrance Cable | UCI | Uniform Construction Index |
| | | S.F. | Square Foot | UF | Underground Feeder |
| Perf. | Perforated | S.F.C.A. | Square Foot Contact Area | U.H.F. | Ultra High Frequency |
| Ph. | Phase | S.F.G. | Square Foot of Ground | U.L. | Underwriters Laboratory |
| P.I. | Pressure Injected | S.F. Hor. | Square Foot Horizontal | Unfin. | Unfinished |
| Pile. | Pile Driver | S.F.R. | Square Feet of Radiation | URD | Underground Residential Distribution |
| Pkg. | Package | S.F.Shlf. | Square Foot of Shelf | | |
| Pl. | Plate | S4S | Surface 4 Sides | V | Volt |
| Plah. | Plasterer Helper | Shee. | Sheet Metal Worker | VA | Volt/amp |
| Plas. | Plasterer | Sin | Sine | V.A.T. | Vinyl Asbestos Tile |
| Pluh. | Plumbers Helper | Skwk. | Skilled Worker | VAV | Variable Air Volume |
| Plum. | Plumber | SL | Saran Lined | Vent. | Ventilating |
| Ply. | Plywood | S.L. | Slimline | Vert. | Vertical |
| p.m. | Post Meridiem | Sldr. | Solder | V.G. | Vertical Grain |
| Pord. | Painter, Ordinary | S.N. | Solid Neutral | V.H.F. | Very High Frequency |
| pp | Pages | S.P. | Static Pressure; Single Pole; Self-Propelled | VHO | Very High Output |
| PP; PPL | Polypropylene | | | Vib. | Vibrating |
| P.P.M. | Parts Per Million | Spri. | Sprinkler Installer | V.L.F. | Vertical Linear Foot |
| Pr. | Pair | Sq. | Square; 100 square feet | Vol. | Volume |
| Prefab. | Prefabricated | S.P.D.T. | Single Pole, Double Throw | W | Wire; Watt; Wide; West |
| Prefin. | Prefinished | S.P.S.T. | Single Pole, Single Throw | w/ | With |
| Prop. | Propelled | SPT | Standard Pipe Thread | W.C. | Water Column; Water Closet |
| PSF; psf | Pounds per Square Foot | Sq. Hd. | Square Head | W.F. | Wide Flange |
| PSI; psi | Pounds per Square Inch | S.S. | Single Strength; Stainless Steel | W.G. | Water Gauge |
| PSIG | Pounds per Square Inch Gauge | S.S.B. | Single Strength B Grade | Wldg. | Welding |
| PSP | Plastic Sewer Pipe | Sswk. | Structural Steel Worker | Wrck. | Wrecker |
| Pspr. | Painter, Spray | Sswl. | Structural Steel Welder | W.S.P. | Water, Steam, Petroleum |
| Psst. | Painter, Structural Steel | St.; Stl. | Steel | WT, Wt. | Weight |
| P.T. | Potential Transformer | S.T.C. | Sound Transmission Coefficient | WWF | Welded Wire Fabric |
| P. & T. | Pressure & Temperature | Std. | Standard | XFMR | Transformer |
| Ptd. | Painted | STP | Standard Temperature & Pressure | XHD | Extra Heavy Duty |
| Ptns. | Partitions | Stpi. | Steamfitter; Pipefitter | Y | Wye |
| Pu | Ultimate Load | Str. | Strength; Starter; Straight | yd | Yard |
| PVC | Polyvinyl Chloride | Strd. | Stranded | yr | Year |
| Pvmt. | Pavement | Struct. | Structural | $\triangle$ | Delta |
| Pwr. | Power | Sty. | Story | % | Percent |
| Q | Quantity Heat Flow | Subj. | Subject | ~ | Approximately |
| Quan.; Qty. | Quantity | Subs. | Subcontractors | Ø | Phase |
| Q.C. | Quick Coupling | Surf. | Surface | @ | At |
| r | Radius of Gyration | Sw. | Switch | # | Pound; Number |
| R | Resistance | Swbd. | Switchboard | < | Less Than |
| R.C.P. | Reinforced Concrete Pipe | S.Y. | Square Yard | > | Greater Than |
| Rect. | Rectangle | Syn. | Synthetic | | |
| Reg. | Regular | Sys. | System | | |
| Reinf. | Reinforced | t. | Thickness | | |
| Req'd. | Required | T | Temperature; Ton | | |
| Resi | Residential | Tan | Tangent | | |
| Rgh. | Rough | T.C. | Terra Cotta | | |
| R.H.W. | Rubber, Heat & Water Resistant; Residential Hot Water | T.D. | Temperature Difference | | |
| | | TFE | Tetrafluoroethylene (Teflon) | | |
| rms | Root Mean Square | T. & G. | Tongue & Groove; Tar & Gravel | | |
| Rnd. | Round | | | | |
| Rodm. | Rodman | Th.; Thk. | Thick | | |
| Rofc. | Roofer, Composition | Thn. | Thin | | |
| Rofp. | Roofer, Precast | Thrded | Threaded | | |
| Rohe. | Roofer Helpers (Composition) | Tilf. | Tile Layer, Floor | | |
| Rots. | Roofer, Tile & Slate | Tilh. | Tile Layer Helper | | |
| R.O.W. | Right of Way | THW | Insulated Strand Wire | | |
| RPM | Revolutions per Minute | THWN; | | | |
| R.R. | Direct Burial Feeder Conduit | THHN | Nylon Jacketed Wire | | |
| R.S. | Rapid Start | T.L. | Truckload | | |
| RT | Round Trip | Tot. | Total | | |

# Index

## A

Abbreviations, 291-292, 304-306
  use in quantity takeoff, 20
Accessories
  for cable tray, 84
  wireway, 104
Addendum
  for discrepancies in drawings, 18
Air terminals
  material and labor units for
    lightning systems, 206
American Institute of Architects,
  219
Ampere values (chart), 300
Appliance circuits, residential
  description, 210
  labor units, 212
  material units, 212
Architect/Engineer
  clarifying plans with, 250
Assemblies estimate. See
  Systems estimate
Associated Contractors of
  America, 220

## B

Bare costs, 4, 25, 221
  installation, 34, 229
  material, 227-228
Bid bond, 15, 37
Bidding
  accumulating data for future
    bids, 66
  assessing company
    strengths, 59
  determining risk, 60-61
  market analysis, 59
Bids
  solicitation of, 18
Boards
  definition, 135
Bonds, 37-40
  for subcontractors, 37
Bottom line, 4
  in estimate summary, 277
Boxes
  cost analysis (sample), 254,
    270, 279
  definition, 127
  quantity sheet (sample), 278

Branch circuits. See Circuits, branch
Bulbs
  fluorescent, 182
  incandescent, 182
  See also Lamps
Burglar alarm systems, 199-200
Bus bars
  internal, 144
  for switchboards, 160
Bus duct
  100 Amp and less, 172-173
  cost modifications, 172
  definition, 167
  description, 170-171
  feeder-type, 170
  material and labor units, 171,
    172-173
  plug-in, 170
  sample unit price page, 222, 223
  takeoff, 171-172
  units for measure, 171, 172
  weight by ampere load
    (chart), 298
Bushings
  labor units, 91
  uses, 90

## C

Cabinets
  takeoff, 128
Cabinets, freestanding
  labor units, 128
Cabinets, telephone
  labor units, 128
Cable
  labor units, 195, 196
  material units, 123, 195, 196
Cable, armored
  description, 111-112
  material and labor units, 112
  takeoff, 112
  units for measure, 112
Cable, direct burial
  description and uses, 194
Cable, flat conductor, 122
  labor units, 123
Cable, lightning systems
  material and labor units, 206
Cable, mineral insulated
  description, 116-117

  material and labor units, 117
  takeoff, 117
  units for measure, 117
Cable, nonmetallic sheathed
  description, 117-118
  material and labor units, 118
  takeoff, 118
  types of, 118
  units for measure, 118
Cable, shielded power
  description, 113
  material and labor units, 114
  takeoff, 114
  units for measure, 114
Cable, undercarpet data systems
  material and labor units, 125
Cable bend
  material and labor units, 123
Cable dead end
  material and labor units, 123
Cable runs
  material and labor units, 124
Cable terminations
  description, 112
  high voltage, 115-116
  material and labor units, 113,
    115-116
  takeoff, 113, 116
  units for measure, 113, 115
Cable tray
  accessories, 84
  cost modifications, 86
  covers, 84-85
  description, 83
  fittings, 84
  job conditions, 85
  material and labor units, 85
  takeoff, 85-86
  units for measure uses, 84
Call system, nurses'
  description, 207
  material and labor units, 208
  takeoff, 208
  units for measure, 207
Capacitors
  description, 175
  material and labor units, 176
  takeoff, 176
  units for measures, 176
Cash flow, 75-76
Cash Flow Chart (sample), 77, 78

# Index

# Index

# Index

# Index

# Index

# Index

# Index